BEING HUMAN

BEING HUMAN

The Technological Extensions of the Body

*Edited by Jacques Houis, Paola Mieli
and Mark Stafford*

AGINCOURT/MARSILIO
An Après-Coup Book
NEW YORK · 1999

Colloquium transcripts copyright © 1999 Après-Coup
All articles copyright © 1999 the authors
All translations copyright © 1999 Jacques Houis, except:
Translation of "Psychoanalysis and Genetics: Clinical Considerations and
    Practical Suggestions," copyright © 1999 Donald Nicholson-Smith

Photographs and video documentary copyright © 1999 Après-Coup
Photographer: Flavia de Souza

Book design and typesetting by Guy Bennett
Cover: Being Human Installation by Christopher Kirwan
Cover Design: Christopher Kirwan

To Serge Leclaire

veillant
    doutant
        roulant
            brillant et méditant

                    avant de s'arrêter
            à quelque point dernier qui le sacre

                — STÉPHANE MALLARMÉ

ACKNOWLEDGEMENTS

For the help they have kindly extended to us in the realization of this book we would like to express our gratitude to Geneviève Leclaire, Antoinette Fouque, the European Commission, the Department of Photography and Related Media of the School of Visual Arts, and Silas Rhodes, Chairman of the Board, School of Visual Arts.

Contents

Premise	11
Colloquium Proceedings	13
Opening Remarks, *Paola Mieli*	15
The Technological Extensions of the Mind	21
Opening Remarks, *Jacques Leclaire*	48
"The Proposed Theme," *Serge Leclaire*	49
"The Technological Extensions of the Structure of the Body"	52
Opening Remarks, *Mark Stafford*	98
"Downtime," *Marcos Einis*	99
"The Technological Extensions of the Senses"	102
Afterthoughts, *Jacques Leclaire*	149
Elements for a Unifying Thread, *Serge Leclaire*	151
Comments on "Elements for a Unifying Thread," *Dany-Robert Dufour & Paola Mieli*	156
An Introduction	161
Brief Preliminary Considerations on Sameness, Otherness, Idiocy, and Transformation, *Paola Mieli*	163
"The natural interface between the symbolic and the real."	189
The Biological Truth Criterion: A Shaky Foundation, *Serge Leclaire*	191
Human Individuality in the Age of DNA Diagnosis, *Robert Pollack*	197
Psychoanalysis and Genetics: Clinical Considerations and Practical Suggestions, *Andrée Lehmann*	201

**"I don't think it matters to anyone where their eggs
and their sperm come from."**   213

   Reading, Writing and the Discourse of DNA,
      or The Mind of a Molecule, *Ona Nierenberg*   215
   Interview with Renée Fox,
      Renée Fox/Mark Stafford   242
   Some Reflections on Medically Assisted Reproduction,
      Paola Mieli   257
   Allah Mean Everything! *Amiri Baraka*   277

"A process of insidious but irreversible metamorphosis."   291

   The Reciprocal Creation of the World and the Subject,
      Dany-Robert Dufour   293
   When Science Remakes the Body, *Jean-Pierre Lebrun*   305
   Medical Discourse, Science, and the "Talking Cure,"
      Annick Galbiati   320
   Towards an Epistemology of the Unconscious,
      Antonello Sciacchitano   332

"Why do people say artificial mind and not artificial soul?"   355

   What Do Cyborgs Eat? Oral Logic
      in an Information Society, *Margaret Morse*   357
   "Your Wish Is My Command": Human Communication
      with Magical and Mechanical Agencies
      in Norbert Wiener's Cybernetics, *Salvatore Guido*   387

"Ain't science a wonderful idea?"   401

   An Advocation for Immortality, *Jim Yount*   403
   The Worlds of Bodies, *Nicole Malinconi*   435

**"You knew that sooner or later you would meet yourself
either coming or going."** 437
E-mail, *John Perry Barlow* 439
The Anger of Friendship, *Mark Stafford* 448
Some Notes on the Technological Extensions of the Senses
in the Age of Television, *Claus-Dieter Rath* 454
Because We Are Digital, *Charles Traub & Jonathan Lipkin* . . 460
Variations on the Technical Body, *Dennis Phillips* 472

**"Maybe they are everywhere, hearing all the messages
we are constantly sending, and under no circumstances
do they want to answer."** 475
Alien Abilities and Behavior, *Seth Shostak* 477

Panelists and Contributors 497

Premise

The current pace of technological creations establishes a novel vision of the world. New technologies, from virtual reality to artificially aided reproduction, are altering the scope and power of the human being. They are altering the relationship of people to themselves and the social and natural environment. How are such changes transforming the ways we relate to each other and the world? How are they affecting the structure of our minds, the character of our thoughts, our subjective experience of the body?

This book collects selected materials from a research project on these questions initiated by Après-Coup Psychoanalytic Association. In 1991 Serge Leclaire and Paola Mieli decided to undertake this research with the attempt to explore the effects produced by recent developments in science and technology. They shared the opinion that the vitality of psychoanalytic discourse is, among other things, related to its capacity for exchange and communication with other discourses – all the more so at a time when scientific and technological changes seem to open up to "another vision of the human." Serge Leclaire died suddenly in August 1994. Stunned and saddened by the loss of our most intrepid and creative interlocutor, and dearest friend, Après-Coup decided to pursue this project in the spirit of the convictions we shared with him for many years, loyal to a psychoanalytic ethics of which he offered a lifelong example.

After a colloquium in New York in 1996, the research continued with workshops and presentations. By bringing together the representatives of different fields, from artificial intelligence to space technology, from genetics to cryonics, to poetry, a forum for an unexplored interdisciplinary exchange on the recent applications of technology was created. This encounter highlights the concerns and the interests that such different fields have in common – as opposed to the exclusivity that is the consequence of technical and scholarly specialization. At a time when we are inundated with reports in the media of advances in medical, communication and information technology, this book asks whether it is possible to develop a vocabulary for the transformations we are undergoing, rather than leave their implementation to trial and error.

Colloquium Proceedings
Being Human: The Technological Extensions
of the Boundaries of the Body

Serge Leclaire and Paola Mieli – New York, 1991

Colloquium[1]

Saturday, November 9, 1996 – Morning session

Opening Remarks

PAOLA MIELI: In *Civilization and Its Discontents,* Freud remarks that men are always in search of a very simple thing: they strive after happiness. Yet this simple program doesn't work. Freud specifies three main obstacles to the achievement of happiness: the power of nature; the feebleness of our own bodies plagued by illness, physical decadence and death; the sufferings caused by relationships among human beings.

Humans have always applied their energy to the transformation of the external world. They have studied the human body in an attempt to overcome illness and postpone death. Through the introduction of political and social structures they have tried to regulate human relations.

Today's technological revolution can thus be inscribed in a long and well-documented tradition. A number of the technological discoveries that have been made in the course of human evolution have radically transformed human life: from the spoon to the compass, from the print-

---

[1] An international colloquium addressing the interface of contemporary technology, human subjectivity and the representation of the body. Sponsored by Après-Coup Psychoanalytic Association and MFA Photography and Related Media Department of the School of Visual Arts. Held on November 9th and 10th, 1996 at the Union Theological Seminary in New York City.

The Organizing Committee included: Christopher Kirwan, David Lichtenstein, Paola Mieli, Mark Stafford, Charles Traub; the Assistant Committee of the Colloquium included: Flavia de Souza, Helena Gibbs, Salvatore Guido, Edward Lucero, Vittorio Marchi, Ona Nierenberg, Ana Vicentini.

The present collection is an edited version of the proceedings. Redundancies, incoherent passages, and halting speech have been eliminated or altered with the approval and participation of all the panelists who responded to our inquiries. Nathan Felde withdrew permission to publish his comments.

ing press to gunpowder. Yet, the accelerated pace of today's technological revolution seems to point in a direction altogether different from previous experiences. The impact of the new applications in artificial intelligence and biotechnology on subjective reality, causes individual and social changes that are rife with consequences.

Researchers and scientists are inclined to discuss among themselves the purpose and the results of their research. Their debate hardly ever exceeds the boundaries of a professional group. And yet their discoveries affect every one of us, every person, every citizen: they transform the individual as well as the social structure.

The purpose of the project *Being Human* and of today's gathering is to establish an interdisciplinary forum for the discussion on the effects of contemporary technologies and their applications on the human and the social body. It is an attempt to develop a lexicon that might more adequately capture the sense of the transformations among which we live. *Being Human* is an invitation to step out of the specificity in which we tend to operate, and contribute instead to the creation of a common program, a common understanding of the individual and social effects brought about by differentiated research.

If we agree with Freud and assume that suffering is also, if not mainly, an outcome of human relationships, we might, all together, be tempted to address the question, "how do we live with the other?" in these highly accelerated technological times. As researchers and citizens we might thus be able to contribute to the work of the legislator. As scientists, legislators, and artists, we share a responsibility towards present and future generations.

Indeed, the way in which the applications of new technologies contaminate different fields of activity could favor a cross-disciplinary exchange. A contemporary version of the Renaissance Humanist *Bottega* ought to be possible: a studio and a place of practical learning where curiosity was satisfied and stimulated through the study of anatomy, mathematics, optics, cosmology, literature, art, and a chorus of humans was exposed to all kinds of technological wonders.

■ BEING HUMAN:

The fact that my colleagues and I at Après-Coup, a group composed of mainly psychoanalysts, have decided to convene the present gathering should not come as a surprise to anyone. It may surprise, however, those who are inclined to make one single heap of all disciplines beginning with PSY, those who confuse psychoanalysis with psychic healing, or therapy, or psychology, or psychiatry; those who believe that our practice amounts to listening to people's troubles, to being supportive, to dispensing medications, providing a diagnosis and resigning from any form of social responsibility – beyond becoming a little star in our little institutions, doing almost nothing and perhaps making good money.

We separate ourselves from such practices. Psychoanalysis is something else. There is a specificity in psychoanalysis: the recognition that the human being is divided by the presence of the unconscious, the presence of a knowledge of which the individual is unaware, but that doesn't stop producing effects in daily life. Rational awareness does not exhaust the complexity of our knowledge, nor of our drives and desires.

There is a basic understanding in psychoanalysis: that humans are *social and sexual beings* and are inscribed as such in a symbolic context (language, culture, ethical tradition, social expectations etc.) that affects their subjective identity, ability to produce, disposition and desires. This is the condition upon which an unavoidable dialectical interchange between individuality and collectivity is predicated.

There is a specificity in the practice of psychoanalysis as well: the establishment of an ethical position whereby human beings are called upon to assume a subjective responsibility for their lives, a position that, in the very acknowledgement of the principle of psychic causality, categorically subverts any notion of determinism, be it biological, social or psychological.

Serge Leclaire has remarked: "The present state of science and of biology in particular, bears adequate witness to the fact that biological order, if not biological 'truth,' consists, above all, in an organization of symbols (names, numbers, or letters) assigned to elements made visible by experimentation: cells, molecules, proteins, antibodies etc., which can thus be differentiated. Natural order in general and living (biological) order in particular, can today be read as a symbolic order, a system of laws that, as such, does not intrinsically differ from social order: symbolic reality, 'human nature' as, essentially, a domain of speech, language, writing, is at work there. To put it simply and briefly, I would say that the opposition between nature and culture today takes the shape of an opposition between two different types of symbolic activity."[2]

Insofar as the medium of psychoanalysis is language, and psychoanalysis addresses the symbolic systems that human beings generate, and by which human beings are determined and affected, we believe with Serge Leclaire that psychoanalysis is a discipline that can create a bridge of understanding between different symbolic domains.

It is by no means irrelevant that the subtitle of today's conference refers to the Body. The human being has a body, of course. It is a body. In light of the new developments of science and technology, it is essential to critique together any philosophical or ideological discourse that either extrapolates the mind from the body by viewing them as separate entities, or reduces human complexity to a form of materialistic determinism. The analytical understanding of the ways human beings are both the producers and the effect of a symbolic – biological and social – order, enables us to structure an idea of the body without falling into idealistic or mechanistic ideologies.

I should like to stress that in psychoanalysis the body is an erotic locus. Any intervention on the body involves an intervention on its erotic geography; any production of the body – of its mind – is charged with erotic connotations. For reasons that I hope we will discuss together, the discourse about, and the practice of new technologies, especially in the field of medicine, but also, occasionally, in that of digital communications, tend to disregard (should I say deny?) this simple matter of fact.

2 *Being Human*, 192

■ BEING HUMAN:

Yet, new technologies in their direct applications to the body involve erotic reconfigurations. This is certainly true in the case of assisted reproduction, gene prevention, organ transplantation; but it is also true in the case of virtual reality and digital communications.

In order to address the effects of contemporary technological productions on the individual and society it is essential to reintroduce in our common debate the understanding that human beings are sexual beings, that the satisfaction of their needs neither exhausts nor silences their desire, and, finally, that their lives are affected by their rational decisions as much as by their fantasies.

It is my hope that this forum – today's conference as well as the future activities that will take place under the aegis of *Being Human*, through our web-site, meetings and publications – will be able to formulate suggestions for a creative and productive application of new technologies, and stimulate industrial producers as well.

I would like to conclude these remarks by expressing my gratitude to Serge Leclaire. For those who are unfamiliar with his work, let me point out that Leclaire was always attuned to the dialectic between the collective and individual subject. He thought that by the very fact that psychoanalysis is inscribed "in the immemorial history of the vicissitudes of the ternary structure which today has been quite weakened by the growing hegemony of the binary structure," psychoanalysis occupies a fundamental place in culture, a place from which it is essential to contribute to the current debate and unmask the simulacra of responses to the uncertainty of the subject in its relation to the real.

Without the enthusiasm and dedication of Mark Stafford, we would not be here today. I am extremely grateful to Charles Traub for his generosity, his suggestions, and his overall support. A special thanks goes to Christopher Kirwan, responsible for today's installation and for the design of the web-site. His creativity and poise have energized this undertaking.

I thank from the bottom of my heart Helena Gibbs, Salvatore Guido, Ona Nierenberg, Ana Vicentini, and Flavia de Souza for their invaluable assistance in the conceptual and technical formulation of this colloquium.

Après-Coup is a nonprofit organization, completely supported by donations and membership fees. I want to thank all of you who contributed with your donations to the present achievement. In particular, I would like to thank Hans Deichman and Luisa Castiglioni from Omina Freudeshielfe Foundation, the Reigefield Foundation, the European Commission, and the friends of Le Cercle Freudien Psychoanalytic Associations, Paris, and APUI Psychoanalytic Association, Paris. Special thanks to Annick Galbiati and Danièle Lévi.

I finally want to thank the Department of Photography and Related Media of the School of Visual Art, the members of Après-Coup Psychoanalytic Association and the staff of the Union Theological Seminary for making today's event possible.

Before introducing today's panels, I would like to make a remark. For the sake of organization, we divided the colloquium into three panels and named them the Technological Extensions of the Mind, the Technological Extensions of the Structure of the Body, the Technological Extensions of the Senses. The distinction "Mind, Body, Senses" is strictly functional; we consider it fictional. In fact, for the most part, these categories tend to establish an artificial separation in human subjectivity, compromising its understanding. We expect this very distinction to be contradicted by the course of our discussion.

■ BEING HUMAN:

The Technological Extensions of the Mind

Panel Discussion

PAOLA MIELI: Let me introduce the Chair of the panel, Dany-Robert Dufour, Philosopher and Professor of Semiotics in Paris.

DANY-ROBERT DUFOUR: I'd like to introduce you to the moderators of the first panel: Mark Stafford, writer and publisher, New York, and Nathan Felde, former Director of Nynex Science and Technology Lab, Cambridge.

The panelists are: John Perry Barlow, Cognitive Dissident, Electronic Freedom Foundation; Marcos Einis, Psychoanalyst, Paris; Vyacheslav Ivanov, Professor of Linguistics, UCLA; Lewis Kirshner, Psychoanalyst, Boston; Erik Parens, Philosopher and Bioethicist, Hastings, New York; Stuart Schneiderman, Writer and Psychoanalyst, New York; and James Bailey, Technology consultant and writer.

MARK STAFFORD: I'm going to begin the questions on the technological extensions of the mind by reading you a short quote from our chair, Dany-Robert Dufour: "The subject creates the world, which in turn creates the subject." Mr. Bailey, how would you address the ways in which prosthetic technologies of the mind have been altering the structure of the mind?

JAMES BAILEY: The subject is continually recreating the world at the same time that the world is creating the subject. We may be pretty much limited to processing information one step at a time, but circuits and computers are not so limited. And we are already seeing kinds of computing that look much more like DNA than 1–2–3. They do billions of things at once; and like everything else that does billions of things at once, they are capable of adapting and evolving on their own. And they're the best hope we have for processing the very large databases that are spewing out of our check-out scanners and our satellites. So, to get a hint of how much our minds and our technologies interact, you could try banishing the words "in turn" from your thoughts. It's

22　Colloquium Proceedings

Mark Stafford and Dany-Robert Dufour – Being Human Colloquium, New York, 1996

Charles Traub and Panelists – Being Human Colloquium, New York, 1996

■　　BEING HUMAN:

not easy. Stop thinking about computers in the language of physics. Start thinking about the technology in the language of biology and psychology and sociology.

MARK STAFFORD: Dr. Kirshner, do you think that there is no longer this reciprocal relationship between the subject and the world?

LEWIS KIRSHNER: Well, I think that maybe we speak on somewhat different levels. Certainly there's an interplay between the subject and the world, because the subject is not an isolatable entity that can be defined as a thing, separate from the processes that enable and constrain it from existing. I think sometimes we fall into this fallacy of thinking of the mind as a thing, and that we can work upon it. I mean, if it is a thing, it is a very funny kind of thing. And certainly, from the Freudian point of view, it's not a linear system at all, but one that exists and can exist in different states, and at different times at once. I think one might say, as a psychoanalyst, that the mind, the subject, has always been a fluctuating entity. At all times it depends on the complex kind of ebb and flow about what we take as belonging to self and what we put out as non-self. Whether it's parts of our own body, or other people's bodies, or ideas, there's this boundary, but it's not a fixed boundary. There's traffic across it. You don't need a passport to go over it, and I think that the boundary may be changed a lot by the new technologies. But in terms of what I would view as more fundamental structural properties – like this constant problem of definition of self and non-self, of dealing with otherness, differentiation versus fantasies of oneness and sameness which get into some of the motivation behind technological extensions – I don't see that as something that is fundamentally changeable by technology.

MARK STAFFORD: The American philosopher John Searle has described the use of the computer as a metaphor for the mind, as a "desperate" metaphor. Marcos Einis, what are the consequences of substituting the computer for memory, writing or data analysis?

MARCOS EINIS: Many things can change with technology. With Leclaire we always wondered, especially lately, about the new forms of the demand brought about by the existence of the computer. I've pondered the strange paradox the computer produces among those who work with it, the programmers, the professionals. I concluded that what drives the computer is a true existential paradox, something that can be summed up in the terms of a real existential problem: in order to exist, the programmer has to disappear. This may sound like nonsense, but it means the following: when the programmer goes to work in his machine, he must, in order to meet with success, in order to make the system work, disappear. He has to merge with the functioning of the machine itself. He has to think like a machine and if he does not think like a machine, this is the strange paradox, he makes a mistake. We worked on the entire question of error, and agreed that, in fact, error was the most human thing in the encounter between man and machine.

JOHN PERRY BARLOW: Well, I disagree. I think that both the computer and the human neurology are parts of the supportive ecosystem for thought. And particularly that part of computer technology that connects various parts of human neurology. I want to go back to the invocation of Leclaire in which he says that both life and the mind and its creatures are symbolic systems. I believe that there is no useful distinction to be made between information and life, that both of them are biologically self-organizing systems, adaptive, self-reproductive, and evolutionary. And the computer is simply another environment in which those creatures of Mind, with a capital M, may dwell and mutate and evolve.

MARK STAFFORD: Stuart, can we agree that artificial intelligence is creating the notion of a disembodied mind?

STUART SCHNEIDERMAN: Let me just back up for a second, because I wanted to question what Professor Dufour was saying. You know, taking his formulation I would say that, in my view, the subject inherits the world and tries to find his way in it or find a place in it, and maybe influence it a little bit. I think we flatter subjects when we think that they create the world. I should mention that I don't know much of anything about artificial intelligence. So, you should all keep that in mind. But there is the great idealistic project in Western Civilization which starts with Plato and goes through Descartes and a lot of contemporary continental philosophers, which says that what we want to do is create a disembodied mind which is not influenced, or not corrupted, because it's always presented in terms of corruption, by human experience, human judgement, human error, as you mentioned. And there is in the idea of artificial intelligence, it seems to me, a dream or a fantasy that somehow we're going to find out how God thinks.

MARK STAFFORD: Mr. Bailey, do you think that in artificial intelligence there is that kind of fantasy at work?

JAMES BAILEY: Oh, I've seen it. And I watch it play out among the artificial intelligence scientists I used to work with. They clearly have this fantasy. They also have the fantasy of making themselves immortal, and of putting their own personalities and characteristics into a machine so that it could live on forever. I think that we have to forgive the artificial intelligence people, the crudity of the hardware they have right now. I would like people to ask themselves: "If I could run a cement mixer at the speed of light, could I make it intelligent?" And the answer is, maybe. I mean, the speed of light is a very powerful thing; you can do a lot very quickly. But you're left grasping and wondering why you started with a cement mixer, and is there not somehow a better tool....

LEWIS KIRSHNER: In the '60s, there was a science fiction story about this strange ship that lands on a beach somewhere, and scientists come out to see what's in it. When they open it up, it's some space device, and there's a thing inside. They have physicists and biologists and geologists, and they take every kind of test of it. And they're all very perplexed. It's a strange organization of atoms. And finally an analyst comes and sits down and says, "What's your name?" What really makes the nature of a human being? Something about that story seems very basic and has to do with having a name, and that means you have a history. We want to surmount the pains of our history. I think that's part of the fantasy. It is a narcissistic fantasy to be able to,

THE TECHNOLOGICAL EXTENSIONS OF THE BODY

in some way, get rid of the weaknesses, failings, pains, whatever limitations, and have a perfect mind-body that's totally under our control. This has animated human beings forever, from the beginnings of recorded time.

MARK STAFFORD: I would like to ask Erik Parens to comment on the discourse of contemporary technology and science that seems to oscillate between two types of conceptualization of the mind. One involves a division between the body and the mind and the other sees the productions of the mind as the result of a deterministic mechanism, a physiological and chemical mechanism.

ERIK PARENS: I must confess that I'm not sure I quite see what the conceptual difficulties are. For example, a lot of people feel compelled to choose up sides. Either one is a defender of mind and social institutions and the like, or one is the friend of neurobiological accounts of what it means to be human. And it seems to me that we don't have to choose. Let me be more flat footed, which is what I'm accustomed to being. Let's think, for a moment, about a child who has a great deal of difficulty in school. Well, it seems to me plausible that one reason the child is having fabulous difficulty in school is that the child comes from a dysfunctional family, has had a terrible upbringing, doesn't have food in its stomach, and its teachers are demoralized. That seems to me one powerful way of understanding this particular condition. It seems to me that, at the same time, it may be that the neurobiologists could help us understand the same condition. Whether to be a friend of mind and social institutions, or a friend of neurobiology, seems to me like an utterly false choice, one we all need to try to give up.

VYACHESLAV IVANOV: We cannot speak about causality as being necessary in all respects. The problem we have with the mind and neurosystem is whether really we can establish a cause and effect relation. The usual idea that the brain is directly connected to the mind may be too simplified. We think that there is some relation, but it is very complex and we have different parts of the brain working partly independently. Let me give you an extraordinary example. After severe damage to the left hemisphere, a composer may produce brilliant music. The first case of such kind was found with Ravel, the great French composer, but since then, several other cases have appeared. I know personally an extraordinary composer who even thought that after the damage to his left hemisphere, his musical production went much easier for him.

MARK STAFFORD: Erik, do you think that in contemporary neurobiology there has been a denigration of the symbolic function?

ERIK PARENS: I'm on thin ice here. It's my sense that the thoughtful neurobiologists are fascinated by the loop. They are thinking about, for example, how it is that a psychological trauma might leave a neurobiological trace.

LEWIS KIRSHNER: One possibly related point here concerns the causes of human distress. In *Beyond the Pleasure Principle*, Freud seemed to advocate the point of view that there's something inside the human being that causes unhappiness and pain. I mean it's good to try to remedy external problems, no doubt, but the fantasy is that all of the problems can be remedied: if we can find the proper neurobiological interventions at the right level, somehow we'll be able to erase the traces of trauma, we'll be able to erase the predisposition to depression, and so on.

MARK STAFFORD: Have the resources of the computer resulted in a transformation of our sense of time?

STUART SCHNEIDERMAN: When people write with computers, it's not the same thing as writing with a pen or a pencil, or even a typewriter. Why is it that when people write with computers, they write too much, and most of it needs to be cut away? The reason you have such big books these days is that people are filling them up with whatever comes to their mind. And I think that part of the issue is this subjective time. People who write well, as well as I can understand it, are people who write for an audience, for readers. The best writers are not the ones who just let it flow onto the paper. So that in some way, the experience of using a computer to write, because it removes so many obstacles to the transmission between your brain and the screen, makes

THE TECHNOLOGICAL EXTENSIONS OF THE BODY ■

people enormously self-involved with the products of their own minds. They lose track of the other person. They never put themselves in the other person's shoes and they produce an awful lot of bad writing, if it doesn't get edited at some point.

VYACHESLAV IVANOV: As far as computer problems are concerned, there have recently been several discussions connected to the problem that are contrary to what is usually supposed. Many computer programs lead to quite approximate decisions. So, numerically, computations very often lead to quite inexact results, if we speak about the rigor that was common to traditional mathematics. Computers can do wonderful things, and help us a lot, but they make our decisions much more approximate.

DANY-ROBERT DUFOUR: Where there is symbolization I do not see how it is possible to think as anything but a trinitary or threefold subject. I would say that even in dialogue, when there are two, there are always three, because the one speaks with the other about something else. Which is to say, as almost all languages put it, that there is an "I" which speaks to a "you" about an "it." An "it" which is usually absent. At the root of symbolization, there is one: the subject. There is the subject's other, the one I am speaking to, and there is another other, the one who is absent. I must integrate the one who is absent in order to found, to enter into, subjectivity and intersubjectivity. It is the common space we inhabit. Before we speak we are already trinitary subjects. We are subjects of trinity because we are subjects of intersubjectivity. So we ought to study this threefold model as it relates to production – the production not of intersubjectivity but of knowledge. To tell the truth, I do not think the production of knowledge necessarily occurs in a trinitary space. I believe that one of the terms must be excluded, and this goes back to the origins of philosophy, to the Platonic dialogue, where "I" doesn't have a discussion with another but with an absent who could appear at any moment. I therefore think we should clearly distinguish two fields: the field of symbolization and the field of a logic of the production of knowledge, the latter implying a "two" rather than a "three." I think this is a key question that may warrant further study.

■ BEING HUMAN:

JOHN PERRY BARLOW: I agree with Professor Dufour. If you go back to the foundation of computer science, Claude Shannon's invention of the bit, and Gregory Bateson's later discussions of Shannon's theory, in which he said that information is a difference that makes a difference, then you will see that the difference is the relationship where the real information exists. In any instance of mind, what mind arises from is the relationship, and not the poles of that relationship. It doesn't really make any difference whether it's the difference between the synapse firing into the neuron, or whether it's the difference between societies, or the difference between men and women. As to the point of using computers to create mind, a terrible error was made in the early days of artificial intelligence, where they were trying to emulate human intelligence. And the best thing we learned from that was that we didn't know very much about what intelligence was. In more recent years, and especially in thinking machines, what we've been trying to do is to grow intelligence by taking lots and lots of relationships and accumulating them into a space where mind can start to dwell.

MARK STAFFORD: Stuart Schneiderman has written about the social mind. And I would like to ask him to comment on ways in which the social mind is being considered in the applications of new technologies.

STUART SCHNEIDERMAN: Well, by the social mind I meant simply that society has institutions, ceremonies, rituals, things like that, which are the product of human minds for whom those institutions made a certain amount of sense. Now, with my background, of course, I consider language one of those institutions, and the interesting thing about language is this: you would never say for example that John's mind is thinking, and you would never say that John's body is walking. John thinks and John walks. I think it's interesting the way we all use language: "I was walking down the street and I saw a body." That means a corpse. Otherwise, you saw a person, and the person has a name. You don't say, "I'm going to bury his body," except poetically. You say, "We're going to bury John." So, there is a certain amount of

THE TECHNOLOGICAL EXTENSIONS OF THE BODY

intelligence, and there is a certain amount of knowledge, and there is a certain amount of wisdom that's contained in these social institutions. That's what I meant by a collective intelligence.

VYACHESLAV IVANOV: I would like to comment on the link between language and personality by paying special attention to some cases where a personality is split, pathological cases. You know the famous case of the German poet Hölderlin, who, in his later years, thought of himself as two or three separate persons with different names. The great linguist Roman Jacobson found, in his poems, that there were no personal pronouns like "I" and "you." So it is possible that the personality may be, so to speak, abolished. But just the pathological aspect of this, shows us that this is absolutely abnormal. Language is used so that it shapes our personality. And I think that maybe the most interesting aspect of the comparison between the computer and the human mind is just in this respect. The problem is that we did not think, until now, about the possibility of anything like personality in connection with the computer.

MARK STAFFORD: Dr. Kirshner, I wonder whether you think that an artificial mind, or artificial intelligence, is comparable to an artificial body part?

LEWIS KIRSHNER: You mean a computer program that is an artificial intelligence program that solves problems?

MARK STAFFORD: Yes.

LEWIS KIRSHNER: No, I don't think there's any difference.

ERIK PARENS: I'd like to understand. Could you say more about that?

LEWIS KIRSHNER: I think that if a person gets an artificial hearing aid for example, and is able to use it, and it enables him to hear better, this has the same meaning as if he had a computer that helped him remember all his appointments for a week. You know an artificial intelligence that wakes him up in the morning and says, "Today, you have an appointment at 9:00. Don't forget to brush your teeth."

MARK STAFFORD: Stuart, can I ask you to comment on that? I

would imagine that there is some difference between an artificial body part and an artificial mind.

STUART SCHNEIDERMAN: If you had, for example, an artificial mind, in place of the one you have at the moment, which, presumably, is not artificial, you'd be someone else. You know, if you had an artificial hand or tooth, or something like that, it wouldn't change who you were. It certainly wouldn't change your personality. I think personality is very peculiar in this debate. I don't have too many answers. But just to raise a couple of issues. In the first place, personality is something that people believe, nowadays, that children have at birth, that it's not entirely, or even primarily, something socially or developmentally determined. Whether that's true or false, I don't know. The second point is, of course, that we know that Dr. Kramer suggests that Prozac has this extraordinary capacity to make you into somebody else, to change your personality. And it does this, apparently, by fiddling with your brain chemistry. Whereas psychotherapy in the past would have said, well, we want to restore you to the wonderful state you were at before this trauma overcame you; or we want you to become who you really are. Prozac is quite different. They're saying, we are going to make you into somebody you have never been. And, you know, this kind of personality-altering drug, through biochemistry, is something that we ought to think about and address.

MARK STAFFORD: Another traditional way of transforming people is education. Professor Ivanov, would you like to comment on the impact of neurophysiology on our ideas about education?

VYACHESLAV IVANOV: I think that the main problems are critical ages. You know, we can develop some extraordinary capacities, but only before a certain critical age. Probably, the whole system of education should be reconsidered, from the point of view of the developments connected with the great Swiss psychologist Piaget, and some other people who are following him who have studied the speed and the possibilities of human development. It seems that, in general, we can acquire certain possibilities very early. I think, for instance, it can

Colloquium Proceedings

Stuart Schneiderman, Marcos Einis, Vyacheslav Ivanov – Being Human Colloquium, New York, 1996

James Bailey and Lewis Kirschner – Being Human Colloquium, New York, 1996

■ BEING HUMAN:

be proved that any person, with the exception of some genetic difficulties, can develop certain musical abilities. And if they are not developed, it depends on the lack of musical environment for a small child. You understand that a person cannot acquire a language if the language is not spoken around him, or with him. This is evident. But, in general, society is guilty of the slow development or underdevelopment of many of the possible abilities. Really, what we call a genius is possibly a normal person, and all of us might have become absolutely wonderful if the society, that is, the traditional system of education did not hinder this. So I think that what we call education may be really the system of obstacles that society traditionally builds, not to let people develop their possibilities.

MARK STAFFORD: Mr. Bailey, perhaps thinking about your own work on the development of mathematics, post-algebra, and calculus, – what you describe as after-math – maybe you could respond to some of those suggestions about the educational system.

JAMES BAILEY: I'm certainly one who believes that at least geometry should be reassigned to the history department and that we are teaching our kids the maths of the past. I also happen to believe that calculus and algebra should go over to the history department as well. Mathematics really is the way we learn how to think. It's that thousand hours we spend at home doing math problems that really inculcates in us that we should think one step at a time, and that we should perceive the world sequentially, and, not incidentally, that we should communicate our thoughts about the world in words and numbers.

MARK STAFFORD: Dany-Robert Dufour, would you like to comment on some of the structures of mathematical thinking that are implicit in this kind of educational practice? Can you conceive, along with Mr. Bailey, of consigning geometry, calculus, algebra to the history books, and conceive of a different kind of mathematics being taught?

DANY-ROBERT DUFOUR: Obviously, this is an extremely complex question. I spoke earlier about knowledge. I believe that two types of knowledge are possible, are produced. Two kinds of knowledge can be produced by natural intelligence. The first type of knowledge is of a narrative kind. This means that we spend our time telling stories, telling ourselves stories without really knowing whether these stories are true or not. In any case, they help us live, these stories we tell, alone, or with others sharing them. These stories do not need to be true or false. They either work or they don't. They allow the subject to join others and belong to a community. Thus, it is possible to belong to a community by sharing false stories. There are other types of knowledge, though. There is another kind of knowledge which does seek to answer the question of knowing whether something is true or false. Here stories are submitted to the test of truth. Obviously, here we enter a different world. I also think we enter a different structure. This is what I was trying to explain earlier on. So, quickly, let's say that the first type of knowledge involving stories is of a narrative and trinitary kind, and that the second type of knowledge, involving exposition and persuasion, where affirmations are made by saying yes or no, true or false…let's say that this second kind corresponds to another type of knowledge, another dimension of the mind, which is that of the binary, which permits the creation of technologies of the mind.

MARK STAFFORD: I would like to ask Mr. Bailey if the emphasis we have placed on the binary productions of the mind is driving the quest for a universal language.

JAMES BAILEY: Oh, I think there is an aspiration for universal language. I think that at a very, perhaps, primitive level, we can watch it happening on the Internet, where if you present your information in words, whether it's in English, French, German, whatever, nobody pays attention to you. A homepage full of words never gets hit, whereas a homepage full of images not only seems to transcend national boundaries, it seems to communicate better. Images are one possible example of a more universal mode of communication; and after images come little movie snippets and things like that.

MARK STAFFORD: Could I ask Professor Ivanov to comment on the historical connections between the image and the linguistic sign?

■   BEING HUMAN:

VYACHESLAV IVANOV: I think that, from the beginning, mankind used at least two different systems: like painting, for instance, cave paintings and natural language, and some aspects of the natural language are continued in the artificial languages of mathematical, symbolic logic and computer languages. But in natural languages, we have several possibilities of expressing emotional and imaginative kinds of apprehension of the universe, which are very difficult to code in other, artificial means. I think that what we need now is the understanding that each kind of symbolic system traditionally used by human beings, has its own field of application. So, we cannot mix up these different fields. And what we need is the environment of different languages that will not prevent each other from making what can be made with this particular language. I think that the natural language remains the widest of the possible means of communication, but of course, other languages can be added.

JOHN PERRY BARLOW: Both the text and the image have a fundamental thing in common: they are both information. They are about something. They are not experience. And I think that part of the whole process that we're engaged in here, the most interesting part, for me, is trying to figure out how to communicate experience itself. I mean, there is a vast difference in any kind of compression that I might make of this moment, and everything that's going on in my sensorium as a consequence of the phenomena in this room. To the best of my knowledge, the only universal knowledge that we've ever been able to come up with for experience is music. Which is to say, it's impossible to misinterpret a song, in a sense. I mean, you never get music wrong, because everybody understands it on some level. In the very process of hearing it you experience it and, presumably, you experience it at least to some extent in the non-symbolic, non-representational sense intended by the creator of that music.

MARCOS EINIS: I would like to add something. We asked ourselves: "How can we program a machine which doesn't simply function through differential diagnosis?" Physicians know what is meant by a differential diagnosis. If one system is affected, another one is not. This differential diagnostic is very well adapted to what you call a binary differential mode. We imagined a system which would not be limited to a differential diagnostic mode. We thought that to introduce another scene (we called it a second scene) supposed being able to imagine a system which would split-off from an original scene, a "conscious" scene. The system would then begin to act in another scene, a second scene we could call unconscious. Serge Leclaire had a very simple idea which involved a system switching from one scene to another scene. When it confronted a signifier of the body, for instance, the computer, did not search its own conscious system, it switched to a system with a double scene. The system had to be capable of analyzing phonetic difference. When it switched from a conscious scene as it joined a signifier of the body, why would it choose one signifier over another? We analyzed each word in terms of every other signifier it could join through its difference. There is a word that has always appealed to me and was dear to Leclaire as well: *hirondelle* (swallow). This word, in French, also means cop. Now, there is a great phonetic difference between swallow and cop (*hirondelle/flic*), whereas the semiotic difference is only of minus one. This approach required a lexicon or a small dictionary to run the system. We were going to introduce ten, twenty, one hundred words, which would allow us to work inside this second scene. This meant a system with a very large capacity for work and memory.

MARK STAFFORD: Professor Ivanov, could I just ask you to comment on whether this kind of research meets the criteria for your suggestion that there might be a way of modeling the computer in a non-exact way, reflecting the reasoning of the right hemisphere?

VYACHESLAV IVANOV: I think that it's maybe a first step toward it. Theoretically, I think it involves the problem of how we can trespass, go beyond binary thinking, not only in computer sciences, but also in the humanities. I mean, structuralism. Structuralism is represented for instance in Levi-Strauss' classical works on mythology where you always have binary relations between the raw and the cooked, and so on. The problem is whether we can think not only in terms of binary oppositions. Let me give an example from political life. You know it is difficult here in this country to not think in terms of two parties. The failure of Ross Perot is a good example of the difficulty of "not-binary" thinking. To give a more serious example, I think that what is really interesting in early Christianity is just the exceeding of dualistic thinking. In the Dead Sea Scrolls, the Essene sect was always thinking of the world in terms of binary poles. In Christian thinking, we have one important person, who taught and his followers who taught that there is no difference between Greeks and those who believe in Judaism, so there is no binary thinking. I think that if we go beyond black and white, we have a system of different colors, and that is what is attempted with the systems we are discussing now in connection with computers.

JOHN PERRY BARLOW: There's a parable here which is illuminating in a couple of different directions. When Liebnitz invented binary math, he was very pleased with his discovery, because in Liebnitz's mind it meant that from ones and zeros you could construct the universe. And also, more to the point, as far as Liebnitz was concerned, it proved dualism as the foundation of everything. As sort of a didactic point, he sent a letter to the Emperor of China, proving dualism through binary math. And, after a very long time, the Emperor of China sent him back a copy of the *Tao-te Ching*, translated into German. What the Emperor of China was saying was yes, but no, it's all one, but it's not dualistic, you can take the ying and the yang and create the universe that way, but there's a continuity of flow. Now, the problem with computer science to this point has been that it's been taking the Liebnitz model to a large extent, partly out of necessity and partly because of the structure of our own thinking. Reality is analog, it is not digital, and I don't care how many times you slice it up, you're always leaving something out. But an interesting possibility now arises, not so much in computers as extensions of the mind. We're talking about two different things here.

THE TECHNOLOGICAL EXTENSIONS OF THE BODY

There are computers as mirrors, and there are computers as telephones, and there are computers as ways of being an extension of what one might be pleased to call one's own mind. And then there's the environment that develops within all the world's interconnected computers which, in its aggregate, has a lot of room for error, and interesting serendipity.

JAMES BAILEY: You wanted some attention to the error, and there's this error of the unexpected pleasure, which is I think what John Perry's referring to, which is when you log indirect experience. This is the library consequence that everyone believes has gone away, but in fact, has come back. Times a godzillion, in the form of the Web and the Internet, where we can have serendipitous experiences.

MARK STAFFORD: I'd like to ask Dany Dufour to pick up on a suggestion made by Professor Ivanov with regards to binarism and democracy ... I wonder, Professor Dufour, if you would like to say a little bit more, about the connection between binary thinking and the notion of democracy?

DANY-ROBERT DUFOUR: First, I would like to suggest a solution to some of the problems which underlie the initial question; namely, how do we create a machine able to simulate the artistic processes observable in man? Why not take a man, give him some Prozac, and simply remove the left hemisphere of his brain to make him totally creative? All kidding aside, there are obviously many problems that could be tackled. There was one this morning that may be emblematic and which we should be able to address: the mind/body question. There is a simple way to distinguish minds and bodies. Bodies fall into tombs. Such is our fate. This does not prevent them from returning, from knocking on the door of symbolization to be admitted, into symbolization and historicity, into history. What can be observed is a recidivism of bodies. Bodies are recidivists...in vain, you could say, because they fall nonetheless. If, in relation to this place held by death, which seems to characterize the very life in which we are caught, sexed life which implies death...if we try to relate the problem of the mind to the place of

■ BEING HUMAN:

death in life, we can say that the technologies of the mind remain. They remain, even when the bodies disappear, when the bodies fall. There is a memorization of the technologies of the mind. It would be possible to write a history of these technologies of the mind which remain. First came oral tradition with archival techniques storytellers shared with one another. Then there were cries which remained when the bodies fell away or disappeared. Places like the one we are now in were created, as were many places in the world called libraries or universities, where bodies show up as the passers-by of time, as passers-by in time, who come seeking technologies of the mind deposited there, which present the characteristic of being cumulative. The technologies of the mind are cumulative, meaning they engender each other and combine just like a living organism. There is, you could say, an organicism of the technologies of the mind. They combine, they grow and they escape the mastery of the bodies present. So I think the question we might now ask is at what point do or don't these technologies of the mind, which are cumulative, return to the bodies to, let's say, remove error from the body. Because there is error, not in the mind but in the body – the error which results in bodies dying, the error of sexuation, the error of sexuality, or the horror, if you prefer. Thus the question, I believe, is to know (and this is related to the discussion we had this morning about the discontents of civilisation) how these cumulative technologies of the mind will be able to revisit the error which condemns bodies to disappear? And I think this is a big question facing the democracy we live in.

STUART SCHNEIDERMAN: There were a couple of other things that I wanted to mention. There's the issue of whether a computer can simulate artistic creativity. And again, that's a question that's been around for a little bit of a while. Along with, you know, if you have a monkey with a typewriter, and the monkey can type at random, for eternity, will he ever produce *Hamlet*? I think that's kind of interesting. I don't know what we're going to make of that, but I think it allows us to bring into focus a little bit this question of what computers can do

Colloquium Proceedings

Erik Parens and Nathan Felde – Being Human Colloquium, New York, 1996

Robert Pollack, William Richardson and John Perry Barlow – Being Human Colloquium, New York, 1996

■ BEING HUMAN:

and what they can't do. And the third point that I wanted to make about sexuality, some people were mentioning it over the break, was that all these artificial disembodied minds seem to be male. And there's a sort of curiosity about that, that the world of computers is a male world. You remove the body which is supposedly what sexuates people, and you get these artificial minds and they're invariably masculine.

JOHN PERRY BARLOW: I want to respond to both Professor Dufour and Stuart here. I have a little parable relating to what you just said. A few years ago, I was talking to a film school class at NYU about Virtual Reality, which I did in those days, and I always got asked an inevitable question about virtual sex. And this was such an inevitable question that I had a canned answer which was, basically, well you know, Virtual Reality is a little like having had one's body amputated, and I think of sex as relating fundamentally to bodies, and I can't imagine what would be less sexy than a place where you didn't have one. And there was a young woman in the front row, clearly embodied, who said "But don't you think that when it comes to sex, the body is just a prosthesis anyway?" and I said "A prosthesis for what?," and she said, "Well, that's an interesting question, isn't it?" And that is, to me, the interesting question, and you know, it's part of what Professor Dufour is talking about when he says there is this other part that lives on as the body rises and falls. I think that is what I was trying to get at this morning when I was talking about the ecology of mind. There is the ecosystem in which the creatures that I think really inhabit the universe live, and it's not particularly embodied though they live there too. And, there is also the matter of the soul. I think that what Freud as a scientist was unable to grasp about human suffering was that there is a soul, that it does come into the world, and it comes here to learn about love in the presence of fear and faith in the presence of doubt, and as a consequence, the world has to be a place where suffering takes place. Otherwise, it wouldn't be very much good at its purpose.

JAMES BAILEY: Why do people say artificial mind, and not artificial soul?

THE TECHNOLOGICAL EXTENSIONS OF THE BODY

JOHN PERRY BARLOW: Well, I don't hold with the notion of artificial intelligence, either, I mean, that was a problem I was having all morning.

JAMES BAILEY: It's interesting, there's something about the soul which would not allow us to say that there are artificial feelings, or artificial soul, whereas we do say, or we do think that there is artificial intelligence. We talk about it as though we know what we're talking about.

MARK STAFFORD: If we're going to use "soul" as a term, let's connect it with the transformation of identity and notions of personality that are problematized by the uses of pharmacology.

JAMES BAILEY: Well, the problem of identity gets raised by Prozac and it gets raised by depression, by the clinical phenomenon of depression. Because the clinical phenomenon of depression concerns what they call self-esteem, self-confidence, and self-respect, and morale, and everything that goes into "who one is." This is what gets attacked in depression, this is what gets diminished in depression, so that one ends up thinking "I'm worthless, I'm a failure." There are these repetetive, depressive thoughts. And we know, I believe, or at least most people would accept, that there are certain medications that can affect this. Prozac and the Prozac family affects this. We also know that there are forms of psychotherapy that can have an effect on this by modifying the way one thinks. Those are cognitively based therapies, which don't pretend to give one extra insight or understanding, or a sense of the meaning of this or that. Cognitive therapy is based on exercises, the exercises which are based on the idea that depression is a bad mental habit, and we're going to replace the bad mental habit with a good mental habit. And the only way you replace a bad habit with a good one is by doing it a lot, over and over again, so that it becomes a form of training – a kind of rote learning, very different than the concerns of psychoanalytic training.

LEWIS KIRSHNER: I think the distinctions are valid. The question is so complicated about medication, because we know that about 80%

of the benefit of a drug like Prozac is a placebo effect, I mean, is equalled by the effect of placebos. But placebo work isn't studied anymore because there's no money in it. The drug companies want to sell drugs.

STUART SCHNEIDERMAN: There have been, for some time, a lot of mind-altering drugs around. You know, this is prior to Prozac. There are drugs that are supposed to modify your perceptions and your sensations. Perhaps they do. There are drugs that are anesthetics, like cocaine. Are we talking about a wish to gain mastery of the mind, or some sort of understanding that when you undergo a psychotherapeutic experience, even if it's only about words, spoken and written, that in fact, at some level, that is going to modify the chemistry of the brain? I mean, there are different alternatives. I know John wanted to say something.

JOHN PERRY BARLOW: You know, one of the good things about having written Grateful Dead songs for 25 years is that I'm perfectly free to admit that I'm an acidhead, without fear of reprisal, unlike, I'm sure, many of you. And I think that, in the psychedelics, there is a nexus of several interesting points. What you have there is a purely pharmacological intrusion, or so it would seem, which for many people has a profoundly spiritual effect. I would not believe in the soul as I do, quite, were it not for the fact that I discovered it as a result of taking psychedelic drugs.

STUART SCHNEIDERMAN: There is also no money in computers that work, as there is no money in the development of placebos. That's why we invented software, so they don't work, and you can make a lot of money off them.

MARK STAFFORD: I think we should ask Mr. Bailey about that, since he was involved in marketing and designing programs for a computer, which could not find a market, am I correct?

JAMES BAILEY: Somebody should put in a good word for error here, and error is wonderful. Error is the seed corn of evolution, and if there were no error, there would be no biological progress. And, unfortunately, if you build a computer that does only one thing at a time, as soon as it starts to go wrong, it's damn tough to bring it back on course. And so, you go through a lot to make sure you never make an error because recovery is very hard. If you have a computer system, or a biology, or a culture that does zillions of things at a time, then you can not only be tolerant of error, you can be celebratory about it. For artists, I think this is second nature. They embrace error. They know about error. Physicists abhor error. And I think that until we get computers that are not designed by physicists, until we get computers that embrace error, and turn it into a positive note, we'll never have computers that can do the things that artists do. That doesn't mean, however, that we can't design computers around error, and someday have computers that do what artists do.

ERIK PARENS: When you first said error, I thought that you said eros, which is something about which we've spoken nearly not at all. I think pleasure was mentioned once. I am still trying to wrap my mind around this assertion that there is no useful distinction between information and life. I'm trying to grasp the meaning of that claim. To a lay person like myself, who knows absolutely nothing about artificial intelligence, this seems like an odd claim, in so far as we think of living things as desiring things, and we think of information as another sort of thing. So, could both of you, for the sake of the lay among us, say more about that?

JOHN PERRY BARLOW: Just at the rude level of things, ask yourself what it is that weaves the nucleotypes into the DNA molecule over and over again. There is information, which is the substrate of carbon-based life. I tend to think of carbon-based life as being a thin film that floats on the surface of the real thing. And within that, there is all the variation of relationship, and the interaction of one thing and the other in an infinite web of energetic exchange. There are now algorithms that grow better algorithms, and can grow around all kinds of things.

MARK STAFFORD: I'm just going to close with one final question. We have a fantasy that machines will provide for us in some way, that they will replace the bodies of other human beings, that they are able to perform in any circumstances. Can you comment?

STUART SCHNEIDERMAN: If we could develop a machine that could think, could we thereby eliminate intellectuals? Is that the great hope that sustains this enterprise? I don't know the gender of robots, I assume they're masculine, because to say anything else would probably be offensive.

From the Audience: Mr. Dufour was talking about the technological mind and the repositories for it in the ongoing transmission, in libraries, in universities, in archives. One of the things that concerns me is that this implies a very well-structured civil society. What seems to be happening is that much of that – fundamental, basic materials – much of it is being bought by Microsoft. Can the ongoing technological mind and the democratic process be sustained?

From the Audience: I think it was Mr. Felde who said something about the power of the computer being its power to erase and to delete. And it just started me thinking about whether the erased/deleted material could ever return. I'm writing a book on the phantom limb, and the question of absent material or absent matter returning is one that interests me, and I was thinking, from a psychoanalytic point of view, if the memory of the computer that's been erased could be like the return of the repressed?

DANY-ROBERT DUFOUR: I would like to reply by bringing up aspects of this morning's discussion, specifically the question of error. Because some errors are worth more than others and are therefore saleable. Let me explain. Stuart asked earlier if we had ever seen a monkey sitting at a machine who could write *Hamlet*. Maybe not a monkey, but an evolved monkey like us, yes. There was at least one. The difference between a monkey-monkey and an evolved monkey like us, is that the evolved monkey like us uses an artificial system which we call a signifying system. You can write *Hamlet* if you possess a signifying system called the twenty-six letters of the alphabet, with spaces, periods and commas. I don't need to demonstate. This has already been done by the great writer Borges when he wrote the short story "The Library of Babel." All the books that were ever written, that are being written, and that will ever be written are already known. They are already held by the library of Babel. The library need only be infinite and contain all the possible combinations of the twenty-six letters of the alphabet plus two or three symbols. I mean that when you have a signifying system, everything is already written. Then we need only execute, intelligent monkeys that we are. Which is another way of saying that we are capable of manipulating this system. We manipulate it in every possible way, so that, at a given time, something called *Hamlet* was written. This has many implications. To write *Hamlet,* one has to enter a process of desubjectivization, to become the subject of the signifier, the subject of the signifying system. And the strange thing is that, in desubjectivization, the subject can find himself; which seems like a key paradox. The multiple errors which lead not to *Hamlet* but to a whole series of unfinished works are not marketable. But, among these errors, there are a few called *Hamlet, Remembrance of Things Past,* etc. These are marketable and they become part of the system you were talking about.

JOHN PERRY BARLOW: I just wanted to say something about your concern about the collapse of civilization owing to Microsoft. I worry about this some too, but I actually spend most of my time going around scaring the hell out of lawyers with thoughts on the future of copyright in the digital environment. And I can tell you that Microsoft is fighting a rear-guard battle because we are now entering into a world where anything that human beings can do with their minds can be reproduced infinitely at zero cost. You know, Bill may own the Louvre at this moment, but the second that anybody buys anything from it, they can distribute it as freely as they wish.

JAMES BAILEY: I'm not an expert on the copyright law, but my suggestion would be to change the copyright law so the only thing protected by it is the irreproducible. Then it's a free-for-all, so to speak. If we look at both the anabolic and catabolic process simultaneously, if

■ BEING HUMAN:

you will, with regard to what is synthesized, and what is oxidized or combusting in these computers, and if you think about it in a very simplistic way, it's bits and electrons, and it's very easy to use a kind of model of combustion and synthesis. Then you can see you're describing a reactor of sorts, that allows these reactions to be looked at in parallel, which turns the flaws or errors, or discards of one individual into the fodder and fruit of another person's desire. So, a more complete view of computing systems which would include both sides of that process, I think would be appropriate. And that keeps me looking at the metabolic process, and I think maybe a metabolic computer is a reference to a craze, looking at biology as a model for how computers might be developed in the future. So, I think it is possible to keep that in mind, as we think about how we should evolve these things that we've taken for granted and don't seem to be concerned about evolving.

THE TECHNOLOGICAL EXTENSIONS OF THE BODY ■

Saturday, November 9, 1996 – Afternoon session

Opening Remarks

PAOLA MIELI: It is now my great pleasure to introduce Jacques Leclaire, Director of Research at L'Oréal in Paris.

JACQUES LECLAIRE: Without taking too much of your time, I would like to say a few words about my father and then read a brief text he wrote in July 1992, on the topic of this meeting. The fact that, along with Paola Mieli, he was responsible for such a gathering, bringing together scientists and psychoanalysts, mixing objectivity with the subjectivity of action and thought, perfectly expresses certain features of his personality. These were tolerance, receptivity and openness to others and to difference. A very important part of his work, even more so in later years, involved crossing the boundaries of classical psychoanalysis in order to rethink its foundations in light of the changes taking place in our society, in light of society's evolution. Aiming to shift the focus of psychoanalysis toward civilization and the citizen, his actions, although sometimes controversial, were always constructive and cosmopolitan in outlook. His travels abroad were never opportunities to export dogma or colonize the analytical milieu, but to listen, to understand, to be enriched by the new and by another view of the human. He knew how to transmit this wealth to us, his children, his wife, his family, over the years in his sometimes muddled and hesitant way, but always with a great deal of finesse "as if butter would not melt in his mouth," as he liked to say. I regret not being able to pursue the conversations we had at the beginning of this project, about biology and the life sciences. Indeed, my current profession is at the center of this scientific development and seeks to create the cosmetics of the future, of which the basic role is to contribute to the well being, appearance and pleasure of citizens while respecting their genetic and cultural identity. We would have had so much to say to one another on this topic.

I will now read the text he wrote in July 1992 on the general philosophy of the work that led to this meeting:

THE PROPOSED THEME

Serge Leclaire

The proposed theme is a contemporary version of the original question of the relationship between action and speech.

Technology has put means at our disposal in every field of the reality discovered or invented by scientific processes. Now, this reality conditions and determines to a great extent the place and function of each individual in the world. We propose investigating the interaction of the various orders of reality including the reality of the psyche ordered by the agency of the subject.

To this end it will first be necessary to lift the double impasse to which we are condemned, by: on the one hand, the belief that the word partakes of a transcendent status that dominates all human activity; on the other, the practice that excludes subjectivity from the fields of scientific inquiry.

The working hypothesis we have chosen consists of allowing that the common thread running through the different fields of research and consequently the various orders of reality, derives from the fact that the operations of thought – the processes of abstraction, symbolization and conceptualization – are necessarily and universally at work there.

The "creation" of the world in its reality continues apace, according to the modalities recounted in the myth of genesis, as a succession of separations and nominations ordering the immediate profusion of an ever-original chaos' real.

The operator of this uninterrupted succession of nominations, symbolizations and separations, is the subject. This, at least for now, is the theory yielded by the Freudian fiction of a "psychic apparatus," a remarkably pertinent theory within its scope. Through the practice of a method, psychoanalysis, this apparatus tends to reveal the functioning of a mental order of reality which includes the use of thought (abstraction, symbolization and conceptualization), the creative (poetic) unleashing of the operations of language, the treatment of memory traces and memory lapses, the analysis and management of affects, of fantasy formations and the desire derived from them.

But in no field of knowledge can one have the last word, no more than in any human relationship.

The current state of technology, be it nuclear, genetic, immunological or, especially, informational, confirms the commonplace intuition that no order of reality is immutable and that any phenomenon called natural can be included in a new system and, therefore, technically processed to modify its spontaneous course.

We will also examine the modalities according to which the order of mental reality is implicated, affected by the "return" effect of the productions of artificial intelligence and communication techniques, as well as the modalities according to which the subjective agency contributes to the genesis and invention of new technologies. In a general way we will consider the networks of interaction among the various orders of reality, the involvement of decisive factors of transformation in each of the fields examined and their reciprocal effects in changing the whole of the different systems.

During each opportunity for discussion, we will make every effort to bring out the elements pertinent to the elaboration of an ethics suitable to the current mutation of our relationship to life and death. Whether it concerns the mastery over certain lethal factors or the possibilities of intervention in determinations involving life, it is without prejudice, other than having to give each other the means to think through the ongoing changes, that each is invited to enter into this meeting.

Jacques Leclaire and Jacques Hassoun – Being Human Colloquium, New York, 1996

THE TECHNOLOGICAL EXTENSIONS OF THE BODY ■

The Technological Extensions of the Structure of the Body

Panel Discussion

PAOLA MIELI: Several members of the panel of the body got sick, I don't know if this has to do with the body or the mind. Roland Gutmann, Director of the Center for Clinical Immunology and Transplant at McGill University in Montreal, and Paolo Casali, professor of Immunology and Pathology at Cornell University Medical Center, both send their regards and regret not being present. The third one, Jacques Cohen, Director of the Institute for Reproductive Medicine and Science, St. Barnabas Medical Center in New Jersey, faxed me only last night, to say that, after reviewing the last draft of the questions we proposed to debate this afternoon, he felt unprepared to answer them, and preferred not to come at all.

Let me introduce the participants. Dr. Dorothy Warburton, professor of clinical genetics and pediatrics at the College of Physicians and Surgeons at Columbia University in New York; Jim Yount, Director of American Cryonics Society in Cupertino, California; Perry Hoberman, who is an artist in New York; Bill Richardson, psychoanalyst and Professor of Philosophy at Boston College; Bob Pollack, Professor of Biological Sciences at Columbia University in New York; Jacques Hassoun, psychoanalyst practicing in Paris.

I also would like to invite Erik Parens, who was present at our morning discussion on the extensions of the Mind, to join our discussion on the Body. He is a philosopher and a bioethicist, at Hastings Institute in New York.

I want to thank all the panelists who contributed to the preparation of the questions we are going to address in the afternoon, and also thank Salvatore Guido and Ona Nierenberg for their irreplaceable help in conceiving today's and tomorrow's debate.

Robert Pollack writes, "In human beings, as in every other living thing, DNA tells each cell exactly how to produce the thousands of other molecules that maintain the cell's shape in its place in the body." With the transfer of information theory to biology, DNA became a code, an alphabet, in short, a readable text written in the language of life. But, the DNA molecule is a very special kind of text, because it accounts not only for the message itself, but for the writing and the reading of the message as well. Through this metaphor of being a readable text, DNA is considered to have an intelligence. What kind of intelligence is possessed by the DNA molecule, which is a reader, a writer, and a message, all at the same time?

DOROTHY WARBURTON: I asked Seth Shostak, who is searching for extraterrestrial life, what was the meaning of intelligence to him and he said it's the ability to send radio waves. So, I guess if we use that definition, a DNA molecule, at least all by itself, is not an intelligent creation. A DNA molecule is a molecule, it is an inorganic structure, and I don't think an inorganic structure has the ability to have intelligence. I think that DNA surrounded by all its accoutrements, which we all have – our body structure, our mind, our history and so on – creates something that we call intelligence, but I don't think the intelligence derives from the DNA itself. People have said that DNA has only one thing on its mind, and that's to make more DNA, and I think as human beings, we have a little bit more on our minds than just making more human beings. There's another part of this question which hasn't been articulated yet, but since I have the floor, I'll mention it. And that's something about DNA giving us a constitutive idealism. I think what's implied by that question is that somehow, there's the platonic ideal of the DNA molecule that tells us what kind of human beings we should be. I'd like to just point out that as we begin to understand DNA more, we realize of course, that it's very, very variable from human being to human being. There are three billion pairs of DNA in a human being, and about one in every three hundred of them are different among each one of us. So that means that we each differ in a million base pairs of DNA. There is no ideal human being. There is no real human sequence. There's only a representative sample.

54 Colloquium Proceedings

BOB POLLACK: I think intelligence is the outcome of 3.5, 3.8 billion years of evolution of natural selection, operating on sequences of DNA to select those sequences which have the capacity to survive in competition, and by cooperation. In the last couple of hundred thousand, or perhaps couple of million years, among the sequences that have survived are those that encode the social line of which we are one of the many current descendents, the line of mammals, primates, hominids, and finally, us with the superb capacity to depend on each other. I think the notion of intelligence is predicated on the notion of a social structure of interdependence. One is not intrinsically intelligent; one is measured as being intelligent by one's peers, more or less. In that sense, we are here with our intelligence through DNA, but the reverse is not true.

ERIK PARENS: I can't stop thinking about what John Perry Barlow said this morning, with respect to what Dr. Warburton said. Dr. Warburton's assumption, like, I confess, my own, is that we can distinguish between inorganic and organic things. Now, my understanding is that Mr. Barlow was suggesting that, in fact, this is a mistake, because what we call inorganic DNA is, in fact, made up of atoms, which indeed, if they don't have intelligence, they have desire. Now, I find this a provocative but very difficult notion to wrap my mind around. So this is where I am at the moment.

PAOLA MIELI: Bill, do you have something to say about the notion of desire and life in the DNA molecule?

BILL RICHARDSON: I'm troubled by the word "intelligence." I'm afraid my own preparation is from the schoolroom, so please forgive a certain pedantry: I don't think the proper word is "intelligence" but "intelligibility." Both words come from the Latin translation of the Greek *logos* which originally meant a "gathering process" into an expression of what was accessible to human beings. So, in the interchange between this gathering process and the human center that functions as our point of accessibility, the word *logos*, by the time Greek culture had evolved into what we know chiefly through Plato and Aristotle,

BEING HUMAN:

had come to mean both language and reason on the one hand and pattern of knowability on the other. Thus the word *intelligibility*, as I understand it, means *understandability*, namely a pattern that can be read by some human being which has what the Greeks call *logos*, which we, after them, call "reason" or "intelligence." What we are talking about here, I suggest, is not so much *intelligence* in the DNA, so much as the *intelligibility* of the DNA, which is readable by the intelligence of the human researcher.

BOB POLLACK: I want to place a frame around this discussion, on the matter of natural and unnatural, organic and inorganic. However life begins, it's clear that it has always, since it began, used only natural materials. The elements that make up all living things, the elements that make up DNA: carbons, hydrogen, phosphorous, nitrogen, oxygen, are all elements one finds created in stars throughout the universe. The ability of DNA to replicate itself, or be replicated, its descendant ability from presumably an earlier molecule called RNA, which has the capacity both to replicate itself and gather energy for the bond breaking and making necessary to replicate itself, those abilities are intrinsic in the chemistry. And I think that it is a poisoned gift of the Greeks that we should separate the natural from the spiritual, or the body from the soul. It seems to me the hard, necessary fact to deal with, if we want to understand extensions of the body, is that included within the body is a brain which has these brain states which we call spiritual. But there is, in terms of evidence, no reason to call upon extra-natural, supernatural, unnatural entities at any point to understand any aspect, I think, of human behavior. The only unnatural thing in our philosophy, I find, is the notion of perfection, perfectibility, and the platonic ideal, which, as I understand Darwin, cannot exist in the living world for natural selection to work. For natural selection to generate us from chemistry requires there to be no platonic ideal of us, but only variations from among our ancestors and us.

JOHN PERRY BARLOW: I wanted to offer a clarification, and a somewhat terrified defense. I had a feeling that, sooner or later, if I started thinking about life, I would find myself amongst biologists who actually knew something. Just to clarify, I'm not talking about the desire of the atoms. And also, I'm not talking about something spiritual. I may have broached the subject of the soul, but it was with a different intent, and a different point of focus. I'm talking about the desire that is inherent in the chemical reaction. I'm talking about the information that is stored in, I mean you could say desire in that sense, in the sense that you could say that the pendulum has the desire to reciprocate, that there is a desire inherent in physics that causes the pendulum to go back and forth. And I'm talking about the space of information that is between the mass of the pendulum and gravity and the fact that it's in motion, and the space of information that is there between, say carbon and hydrogen.

PAOLA MIELI: This leads us to another question I wanted to address to you, that has to do with the relation between subject and object. Can we posit an object outside of a technological mediated vision? Scientists may claim they are reading the book of life but aren't they in fact writing it?

JIM YOUNT: This has always been a problem that investigators have had: how do you separate the observer from the observed? And there's never been any good answer to that. A problem in doing investigations is: do we contaminate by our observing? For example, it was a good thing to visit the moon, to walk on the moon, but we certainly left our footprints behind. And as we become able to manipulate genetic material, as we're able to go in and read DNA, to understand the cell, we finally trespass, and are able to become more than just observers. First of all, I don't think there's any immediate danger from anything I've heard, in terms of the genome project, for example. It's just exploring, understanding, and determining what we are in a very basic way. It's the next step that we have to be careful of, and I hope that discussions like this may even help to enlighten us. First of all, we do not want to put the brakes on technology too much. There are diseases that we need to be able to cure. There are problems, genetic problems

BEING HUMAN:

we need to be able to deal with, soon. On the other hand, caution is certainly called for. As we go from observers to manipulators, it is important that we do this carefully, and it won't hurt to observe a little longer and make sure we understand before we get in and start messing with things.

PERRY HOBERMAN: I find it hard to make a real distinction between just the observation of a system and interfering with it. It's not just about the question of either correcting some genetic mistake or some kind of accident. It's also the systems. We're creating the systems that we use to understand these things. A sort of off-topic example would be fractals or artificial life; where they work, they seem to do the same things that natural systems do. But we really don't have any evidence that that is what natural systems do, and yet, we're using them to understand natural systems. So I think we are always kind of inside of what we're doing.

PAOLA MIELI: Speaking of DNA and gene therapies, I would like to dwell for a moment on the fact that gene therapies are still in their experimental stages. Only mutations screening is widely practiced, a technique which can identify, in individuals who are not presently sick, genetic traits which predict possible diseases in the future or in the offspring. How do we respond when we may know in advance the diseases we will get, and the probable cause of our deaths?

BILL RICHARDSON: The question of interference is an ethical question, it seems to me, and therefore must be answered in ethical terms. What do I do for someone else? Who is affected by defective DNA? Etc. – All those issues will be dependent upon the circumstances of the individual cases which are at stake. Its hard to articulate a general principle without knowing more about the details of the individual cases to which it is expected to pertain.

DOROTHY WARBURTON: Well, as a possibly encouraging note, I would point out that an organization like the American Society of Human Genetics, which considers these questions a great deal, and did at its recent meeting, is extraordinarily cautious about advising gener-

Colloquium Proceedings

Dorothy Warburton and Vyacheslav Ivanov – Being Human Colloquium, New York, 1996

Jim Yount – Being Human Colloquium, New York, 1996

■ BEING HUMAN:

alized screening for any condition before a lot of preliminary work has been done to try to decide what the impact would be on individuals and on society. That is easy to say and it's not easy to do. But there are some general principles that I think you can think of. One of them is that such tests should be strictly voluntary, and should be entered into with the knowledge of what it is you're going to find out. And there are people who don't want to know whether they're going to get cancer or Huntington's disease, or whatever it may be. That is their choice, and I think it should be left at that. The other is to find out what actual impact such knowledge will have on individuals. It's often difficult for even one individual to project the impact it would have on themselves. I can think of a recent study, for example, which looked at the effect of presymptomatic testing for Huntington's Disease. Huntington's Disease is a devastating neurological condition which usually has its onset in middle age, and leads to psychiatric symptoms, neurological symptoms, and eventually death. And you know that you have a 50% chance of getting that disease if one of your parents has it. So there's the ability to be screened at this point and find out whether you're carrying the mutated gene and will get sick, or you're carrying the normal gene and will not get sick. A recent study looked at what they called devastating outcomes in people who had been screened, devastating outcomes being defined as things like suicide, serious mental breakdowns, and disruption of family, and found a very small proportion of cases in which this had actually occurred. Interestingly enough, it happened just as often in those who were found to be normal as in those who were found to carry the mutated gene, because knowing that you're not going to get the disease can cause all kinds of difficult feelings, such as guilt as well. One can do such studies and get information.

JACQUES HASSOUN: I am struck by something. The human is first discovered as different, and then as mortal. I would simply like to recall something that may have escaped the attention of some of you, even if you are in the habit of reading the Bible very closely. Abel, who was very nice, did not leave much of a legacy. Cain, the assassin, is the father of blacksmiths, of singers and poets. Civilization starts with Babel. Maybe the question we could ask ourselves as analysts is the following: Could it be that at the root of our existence as subjects is the fact that there isn't a single language, that we are plurilingual? Isn't this what allows us to apprehend our death as other than a sentence? I felt like answering the question in a provocative way, the way it was asked. You might believe you can escape death. But what is this? Who can escape death?

JOHN PERRY BARLOW: I actually wish there were somebody here from the insurance business, because the problem that I think we're looking at is as much actuarial as biological. If we have a good way of finding out whether or not somebody has an increased risk to be exposed to some disease, we have to move from a society where we base our insurance on risk, to one where we base our insurance on responsibility which is collectively shared. Right now we have a system of collective risk, and there's a gamble that you won't get sick, that the insurance companies are willing to take. If the odds are fundamentally changed, they will not take that gamble, and we may have to become responsible as a society for the people that insurance will not insure. There's another question which has to do with privacy and information, once information about your genetic makeup is in some database somewhere, the chances of it leaking are very high. We don't have good policies in place for that.

BOB POLLACK: A couple of things have gone by since I've had the opportunity to speak, so let me summarize my own opinion of this class of issues. I think there is now the creation by the technology of DNA analysis by the human species what I would call a body of useless knowledge, information about individuals which cannot help them in their fate, but which can condemn them to the loss of insurance, the loss of job, loss of family, certainly, the loss of remaining time free of knowledge. So, in analytical terms, I think I'm interested in hearing from psychoanalysts, what are the consequences of the foreclosure of

■   BEING HUMAN:

the narrative of an individual life which begins with suppressed memories, but which in the ordinary turn of events, leaves open time for the discovery of those memories. Once the knowledge of what one will die of is given to one in one's adult life without symptoms, it seems to me that precludes the opportunity to undertake a relearning of one's early days. The second thing I'd like to say is that this conversation has shifted, without notice, from a general discussion of DNA in the natural world to a specific discussion of human DNA. And that shift speaks to an obsessive behavior by the individuals of our species which is, while not unnatural, certainly pathological. Our species is in fact not so much the reader or writer of our own DNA, as the killer of the DNA of many species. We are now obliterating about 1 in 200 species a year, if Wilson is correct, and certainly obliterating on the order of 1 in 1,000 species a year. This is, on the scale of evolution, a catastrophe equal to any asteroid that ever struck the planet, and at the same time, we are, in numbers, about a thousandfold in excess of any other animal our size. So we succeed at the expense of the natural world, and that is an issue for us as a species, far larger in scale from the question of how we keep any one of ourselves alive through DNA.

PAOLA MIELI: Your remarks are extremely important; can we hear what the other panelists think? Let me add a further issue: the question of gene manipulation applied for non-therapeutic purposes, gene alterations intending to enhance particular desired qualities, for instance height and intelligence. Some supporters of biotechnologies and plastic surgery argue that to eliminate the elements of subjective difference can be used to end social discriminations.

ERIK PARENS: To return to examples we used this morning, we are already using psychopharmacological agents to produce certain behaviors that are highly valued in this particular culture. Ritalin and Prozac are pretty good examples. If we can produce these blunt desired affects in ways that people seem to really want with psychopharmacology, it would seem to stand to reason that we could do, sometime in the future, the same with genetics. There are all sorts of reasons why we ought not to do it, but the question is, can we, in principle? And it seems that there's evidence that we have already begun. We have already taken a step or two down that road with psychopharmacological agents. Whether we can, in principle, give a child a growth hormone so that she will grow taller, whether we can in principle do that, is different from the question "Ought we to?" My own view is that A: yes, probably we can, not very terribly well right now, and B: we ought not to be using technological means to respond to a social problem. Discrimination is a social problem; it's not a medical disease. So, I want to try to distinguish those two....

BOB POLLACK: I'm amazed that at the 50th anniversary of the end of WWII, these things still need saying, but I guess they still need saying. It isn't simply that one shouldn't deal with a social problem with a medical fix, but that one creates a brand new and particularly evil social problem by doing so, and that evil social problem is the acceptance, the tacit acceptance that there is an ideal human being that other humans can be measured by their distance from. But it is, as I said earlier, the very essence of how we got here today, through natural selection, that we have no ideal to aspire to, neither the ideal of a good person, nor the ideal of a perfect body shape, nor the idea of a perfect spirit or mind. It is in the variation between us that we have a hope of our species surviving, not in the attainment of an ideal. So it is not even medicine, it isn't bad medicine. It isn't medicine at all, to say, "Smooth out the differences among people for their sake, for society's sake, or for anybody's sake." It isn't the alleviation of suffering; it's the attempt to attain the unattainable ideal. I should say also that we're talking about DNA here, and that the administration of growth hormone to a short kid is an issue of medicine of a sort, but the administration to a fetus of the DNA that encodes growth hormone in a way that gets it into that fetus' germ cells, that is sperm or egg cells, is a completely different event, an attempt to change the human species in some small way by the manipulation of species DNA in order to steer the species toward an ideal. And that, named "eugenics" about a hundred years ago, is a profoundly stupid, biologically wrong, and politically explosive thing to do under all conditions.

PAOLA MIELI: Is biological determinism used as a shortcut to relieve ourselves of certain social and ethical responsibilities?

PERRY HOBERMAN: I think that it's obvious that it is used that way. The story about the human growth hormone. Drug companies have…there's a lot of money to be made in that, for instance, shortness is a disease. So, I think it is going on right now. I think there's also this danger that it's not happening quite the same way as a eugenics based on race, or ethnic group, and yet if it plays out far enough, we do start to depend on that as a sort of discrimination. And at that point we forget how to deal with it any other way.

JIM YOUNT: Well certainly, there's a tendency, as you point out, to look at the biological causes of social problems and to throw up our hands and say, because people are disposed to be alcoholics, the family's been alcoholics for many generations, and we've found the gene that makes them alcoholics, we can't do anything about it. This is a danger. If we start buying into this, then we will cease to look at other causes, cease to look at other ways to deal with the situation, to deal with it through therapy, for example, or through support groups and the other more common, ordinary ways of dealing with problems. On the other hand, we need to be careful not to ignore the tools that have been placed at our disposal. If indeed there is a gene which causes people to become alcoholics, we might have a tool there, we might be able to change that gene. Or we may be able to determine that there's a tendency for an individual to be an alcoholic and make the person aware of it so that he does not place himself in a position where he will be tempted as much. For example, don't get a job as a bartender if you're carrying those genes.

DOROTHY WARBURTON: Contrary to what gets into the popular press, I don't think that those people who are studying the biology of behavior are really biological determinists. I don't study behavior personally, but I speak to people who do. They talk constantly about the fact that it's multifactorial, with multiple interacting causes, some of them are genetic, some of them are environmental. The hope of discovering something about the genetics of behavior, and particularly psychotic behavior – which few people would consider just normal biologic variation – is that if we find out that the genes are genes that are involved in alcoholism, we may be able to offer a treatment which is much more effective than that currently available. We've known about mental illness and alcoholism for thousands of years. And our ability to fix them by conventional therapies is really not very good. So I think we should be hopeful that perhaps the biological information will lead us to more effective treatments, and not be afraid of that. I'd also like to say that it isn't as if there were no genetic determinants to behavior. There is absolutely no question that there are genetic determinants to behavior, they just don't tell us the whole story.

We're sitting here and at least half the people in this room understand that they carry a single gene difference which is associated with violence and early death and that's of course the Y chromosome gene. I don't know that that recommends a direct treatment. But I think that is a reasonable baseline against which to measure the discovery of other single gene differences between people.

PAOLA MIELI: What is repressed does not disappear, but returns in another form. The history of genetics as such is inextricably bound with the history of eugenics. How does the repressed eugenic dream of perfectibility of the human species function in the development of today's genetic technologies?

ERIK PARENS: This desire for the perfect that goes awry has always been with us and it is very much with us at the moment. We've already talked a little about some of the technologies that we're using to those purposes. My concern about the term eugenics is that sometimes it lets some of us off the hook too quickly today because we can say, well there isn't a government that's enforcing this practice. And in fact, that's important, there isn't any government that's promoting any eugenics. It is we, all of us, we yuppies, that is, there aren't that many in the room

I'm sure, but those of us who are, and go and get amnio, we are already in the midst of a practice where we are trying to produce children of a particular sort. Now, this seems to me both different from Nazi eugenics, and troubling and problematical. So we clearly need a language to speak about the respects in which this practice is problematical, but I'm concerned that eugenics is not the best term, unless we make it very clear how what we're talking about is and isn't like the old-fashioned eugenics.

PERRY HOBERMAN: One really problematical area that we're not talking about is that as the general project proceeds, and as we come up with more and more genes to test for, it's obvious that not everyone is going to have access to either the testing, or cures, through gene therapy or anything. Basically, no matter how good, even if we can perfect...I can't imagine what that would mean, but even if we were to be able to rid people of disease, it would not be everyone, in fact, it would probably be a minority. That in itself is very problematical.

BOB POLLACK: Just a little background fact. Most people on this planet die of starvation or infectious disease, not inherited conditions. About a third or a fourth of people are infected with malaria, and about that number are infected with TB planetwide. And in this country, about 40–50 million people have no access to medical care, so none of these DNA tests are of any relevance to them. If you take cystic fibrosis, the most common inherited condition of people of European origin, there are a couple of tens, or perhaps hundreds of thousands of people at risk at any given time of having a baby with cystic fibrosis, who have absolutely zero chance of being tested in advance to have the freedom of knowledge to make the decision not to have that kid, and that's an irrational structure in which to insert further tests for those who can afford them.

PAOLA MIELI: Jacques, what do you think about the return of eugenic dreams?

JACQUES HASSOUN: What worries me most is the image of perfection it aimed for. It seems to be that for many, unless I've misunderstood, for many you can imagine a moment, a time in the history of society, when the human (I don't mean the subject. I mean the human.) could have complete power over disease and dysfunction. Ultimately, the divine biologist would replace the plain divinity. So, while I was listening to you, I thought of something Freud wrote about a child, little Hans, I think. "Long before you were born, I knew." What did he know? Maybe he knew something having to do with culture. He knew something about what has been called the love of a child. He may also have known that a child dreams of perfection, of the omnipotence of that famous magical child. Freud also knew that even the child is mortal. Well, it seems to me that there is a totally worrisome and terrifying tendency to overlook the death that is present in all of us. Can we strive for immortality? For perfection? What is perfection? Every time I think of perfection, I think of that highest level where the perfect one was the one who was castrated. So he could not make children. He could have no descendants. He could not create any mortals. This is a way to eliminate death, mortality, the death drive. My reaction may sound provocative, but sometimes on a Saturday afternoon one should be provocative.

PAOLA MIELI: Human relation to death is in fact the thread of our entire discussion. Let me insist on the notion of the perfectibility of the human species. Does the dream of elimination of the phenotypic differences raise the question of an esthetic ideal of the human body? Does the dream of intervention on certain genes associated with determined behaviors raise the question of the pursuit of an ethical ideal?

BILL RICHARDSON: I think Jacques has just said it. We must approach these ideals as part of an imaginary world that yields to the fact of death, and all that death implies. I'd like to address the question of death by returning to a question that was posed earlier concerning the determinism of genetic origin, e.g. the case of a genetically transmitted illness which cannot be prevented unless an intervention is made. The question was asked: what do psychoanalysts do about that? It seems to me that it's the same issue as facing the imminence of death. What's pertinent is that the necessity imposed by genetic structure and destiny toward a certain kind of death are both beyond our power to control. There is a moment when the subject in analysis must say "yes" or "no" to the inevitability of death that comes in any form of castration, particularly if it comes in the form of one's genetic determinism. In other words, it is the task of the psychoanalyst to help the analysand reach that point of freedom that enables him/her to say "yes" or "no" to his/her ineluctable limitations. That places a final period to every genetic dream. One more word: It seems to me that what's missing in using the word "intelligence" for genetic structure is its relation to a symbolic order. When one speaks about the researcher, or the philosopher of social discourse, in this context, I think one is talking about a subject who discovers and articulates that symbolic law, in this case with regard to DNA. The role of the symbolic order as it pertains to language as such remains to be discussed. Given that much, it seems to me, one should talk differently about the difference between subject and object. There is a way of conceiving the relationship between the researching subject and the object of research as encompassed by something broader than either. In the object of research that is the DNA phenomenon, there is already something that makes it accessible to the researcher. Just as there is in the researcher something that enables him/her to gain access to it. But it is this third dimension of mutual accessibility that is prior to both, in a way the foundation of both, which makes the research possible. Here, it seems to me, is the encompassing law of symbolic order, governing at least two modalities; both indelibly marked by the fact of human finitude.

PAOLA MIELI: The common discourse of medicine defines two factors influencing the particularity of human subjectivity, the genetic, and the environmental. Psychoanalysis introduces a third element, also essential for the constitution of human identity: the symbolic context, that set of language, culture, traditions, ethical and religious values, etc., in which a person is inscribed and which offers the ground for subjective identifications. Transmission is not only genetic but also symbolic, as the transmission of symptoms and traumas among generations shows in analytical practice. If we take the example of assisted reproduction, new technologies create situations where up to five different people can be directly involved in the birth of a child. What are the consequences on a human being of this kind of manipulation of the origins?

DOROTHY WARBURTON: Well, I've thought about this, and it seems to me that these five people you're talking about only matter in the context of actual human contact with the child that we're talking about. That is, I don't think it matters to anyone where their egg and their sperm came from, if they have no knowledge that those people exist, and no way of contacting them. I think that what matters is who you are physically in contact with, who you think are your parents. There are many people who have been adopted. This is basically no different than adoption. It's just a slightly different technology that gets you from one set of parents to another. But the biological situation is identical to adoption, so I think that we could think of it in that way.

JACQUES HASSOUN: It seems to me that this question can be asked in terms of ethics. From that point of view, can you really say that it doesn't matter whether this child who is born into such and such a family comes from such and such an egg and such and such a sperm? Put thusly, the question seems problematical to me. It brings up the problem of origins, specifically in terms of the body's real. What seems illuminating in this context is how someone becomes wedded to their name. And I do mean, "wedded." To your child you transmit a name that comes from who knows where. But how does the child in question take up this name? How does he appropriate it? I wonder whether each of us isn't faced with similar questions. It happened to me once that someone rang my doorbell and asked, "May I speak to you? I know you only see people by appointment but I wanted to speak to you. I was a child you knew when I was three, in daycare, and you spoke to me about the fact that I was adopted, even though I was only three years old." She said, "Listen, I just wanted to say thanks." And I

■ BEING HUMAN:

never saw this person again. She was twenty-one. Eighteen years had passed. Eighteen years later, she came, she left, and I never saw her again. What was she looking for? I certainly don't think she wanted to thank me. Maybe she had come to see, to tie the imaginary to someone who, by speaking of her father, had allowed her to accept her name.

PAOLA MIELI: The people who take care of the child are the ones with whom the child establishes a special relationship, and they become determinant for the child's subjective identifications; yet the real of one's own origins counts and has a certain place in the psychic economy. The quest for knowledge about one's own origins is a quest that concerns all human beings. A secret creates effects by the very fact it has been kept secret. If silence becomes part of a social practice, for instance where new technologies applied on the body instigate it, this is not going to be without social and individual effects.

DOROTHY WARBURTON: The knowledge that this has been the way one was conceived would be a very difficult thing to deal with, and I can imagine all kinds of problems would come up later, particularly if, unlike adoptive parents, there is basically no way to find out who these people were. I just meant that in a biological sense, you don't know where your sperm and your egg came from. You can be told that this happened in your past, but you have no inherent way of knowing this.

PAOLA MIELI: Assisted reproduction disconnects the act of procreation from erotic lovemaking, at least in appearance. What do you believe are the consequences of such a separation?

JOHN PERRY BARLOW: I was sitting here thinking about death, and now you want me to talk about its alter ego, sex. I think that sexual desire is only one of the forms of desire that we have to take into account, it's probably the one most keenly felt, and it's probably the one that relates clearly to all other desires, but it is a desire. And when I look at everything we've been saying, both this morning and this afternoon, we are talking about a dialogue between acceptance, inevitability, desire, and control, those things. I remembered a line from

Eschylus, when one of the Titans is taunting Prometheus, he was having his heart eaten out, or his liver, I can't remember which, he says "You, Prometheus, stole powers from the Gods and for whom? Things that live and die." What he was saying in that was that Prometheus had given us technology which was the ability to control in spite of the inevitability of our fates. And we have this desire which is based in part on the body, but in larger part on the fact that we are alone among species in not being satisfied with the universe that God gave us.

BOB POLLACK: That's right, punish the Y chromosome. Paola said something previously, which I think, is an answer. She said, "Every secret creates an effect because it is a secret." But it seems to me that's only so if the secret is known. So the question that that statement of Paola's raises is, "What is the value of a secret versus the value of a lie?" To prevent a secret from being known. I think this is a serious question in the matter of what you say to your kids, and what people think about their origins. I don't know how it is possible to know the biological truth about one's origins, or the passionate truth about one's origins, because there's such a great incentive for parents to say what it is they think they want their kids to know, and how else will a kid know what the real passions were between the parents? So, I think your question answers itself. These are data that are not recoverable, except perhaps through analysis of a family, not just the kid, but the parents. What a kid is told about his/her parents' passionate reasons for conceiving the kid has probably very little to do with what they were thinking at the time.

PERRY HOBERMAN: The thing about the secret or the lie is that the kid doesn't know, but of course the parents know. So, it is hidden knowledge, and it seems inevitable that that carries some consequences, because obviously, if you're not going to tell the kid how it was conceived, you're keeping something from the child. As far as the parents are concerned – separating lovemaking from conception – it's just a more extreme version of what we already have with contraceptives. It just separates it further. I think it seems that the consequences would really be for the child, if the child knew, because there's a big difference in knowing that your parents conceived you and knowing that you came from a sperm bank that some college kid got paid to contribute to.

DOROTHY WARBURTON: Most children aren't conceived in some sort of beautiful, erotic lovemaking anyway, most children are conceived, probably, in some sort of violent act, which has nothing to do with love. We don't talk about it (or maybe psychoanalysts do) but most of us don't know, in fact, how we were conceived at all.

JACQUES HASSOUN: I am really surprised by this statement. I don't really understand the violence you refer to. But the question I wanted to ask you may be this: There is the child's shattering discovery that his parents are mortal. There is that moment, or when the child has his own children and discovers that his parents are mortal. At that moment the question arises of what desires he is the product of, not necessarily because the father jumped on the mother after drinking half a bottle of whiskey. Maybe this isn't the way the question is posed. It is posed in other terms – in terms of "the two who preceded me are mortal. What do I make of this, faced with my own children?"

Break

PAOLA MIELI: I would like to start this second part of our panel discussion by raising questions concerning transplantation and implantation. Transplantation and implantation of prosthetic devices have provided an extended life and hope for millions of seriously ill people. How do such practices transform our notion of the body, its physical reality, its image, its fantasies? Is there a subjective difference between a substitution of body parts and the implantation of artificial prosthesis?

PERRY HOBERMAN: Well, it's hard to consider a question like this outside of how those kinds of interventions are treated in the culture at large. It has a lot to do with the representations of those kinds of prosthesis and transplantations. A lot of what's happening in terms of our interventions on our own bodies are somehow literalizations of mytho-

Colloquium Proceedings

Paola Mieli and Jacques Hassoun – Being Human Colloquium, New York, 1996

Perry Hoberman – Being Human Colloquium, New York, 1996

■ BEING HUMAN:

logical creatures like centaurs and.... There've been ideas of transforming humans in various ways for all kinds of reasons, for millennia, and here we are, in certain cases, doing some of these things. In terms of the questions of one's own fantasies about the body, about eroticizing the body, I think anything can be eroticized, under the right circumstances. It's not a question of it necessarily being seen as monstrous. I'll leave it there.

PAOLA MIELI: Dorothy, what do you think about the way in which organ transplantations impact on the fantasies of patients?

DOROTHY WARBURTON: I find this very hard to think about. I can't believe it doesn't have some effect on one's feeling of self to know you have someone else's heart in your body. I think it must have, especially since we have such mythological connotations about what hearts are and what they do, that they're related to our emotions and so forth. Even though we may know that this is untrue in the biological sense, I think that we still probably believe it somehow. And I think it must be a problem that people who have transplants must face, but I don't have any particular insight into this question.

PAOLA MIELI: Animal parts can be engineered and harvested for transplantation into human beings. What once was relegated to the realm of the monstrous is now being worked out in the laboratory. Could interspecies intermingling be considered monstrous?

BOB POLLACK: The baseline for interspecies mingling is, of course, eating. And we know from the panic in England about mad cow disease that the ingestion of the tissue of a foreign species can carry with it a serious risk of infection, unless the infective agent is killed by cooking. In fact, the mad cow disease is possibly an agent of a novel sort that isn't killed by cooking, and it raises in any event my real answer to your question. I'm not particularly sensitive to the monstrosity of carrying a baboon heart, say, but I am extremely sensitive to the idea that by moving tissues around from one species to another, we may possibly, inadvertently expand the host's species range of micro-organisms that we don't know about that live in the donating species' organs, and thereby create new infectious agents that would plague us as a species.

PAOLA MIELI: Is there an esthetic barrier that prevents us from transplanting the visible structure of an animal into a human? For instance, let's say the hand of a monkey on a man?

JIM YOUNT: Of course, of course there is. On the other hand, we have to deal with tools that are available to us. If it was a question of someone who had no thumb, if they could have the thumb of a baboon, would that be acceptable to them? Is it monstrous? And if it's acceptable to them, should it be monstrous to us in society? I don't think so. I think also the word "monstrous" is kind of loaded here. It suggests, it almost suggests the answer, "Yes, it is." It's certainly not always. So, if the transplant didn't show, there certainly would be no esthetic barrier. I guess the question is, "Does it bother the person himself?" And if it's apt to, it seems like some questioning should be done of the person who is to receive (for example) the thumb of a baboon, before he receives it. There may be some candidates who are not very good for this, even though they say, "Sure, I wouldn't mind having the thumb of a baboon," but maybe they have a psychological makeup such that they really would. So I think that there is certainly a place where people in the mental sciences, psychoanalysts and clinical psychologists, could work to determine if there is going to be a problem if something like that is attempted. I wouldn't reject it out of hand, even something that is as obvious as, say, a thumb. On the other hand, if someone wanted a tail, just for the fun of it, they've always wanted a tail, then we'd have to think about that one.

PERRY HOBERMAN: At this point it sort of does seem science fictional, but once we go down that road, there's no turning back. It's not just that we're inviting the animal into us, certainly not with the kind of research they're doing with pigs, they're injecting them with human genes, to make the organs more transplantable. We're making pigs more human so that part of the pig can become part of us. Obviously, I can't talk about this from a scientific point of view. From a more cultural point of view, it seems that it's in popular representations already, and has been for a long time, so it's clearly not culturally out of the question.

■ BEING HUMAN:

The Technological Extensions of the Structure of the Body

PAOLA MIELI: Biology stresses the continuity between animals and humans. For instance, we can share organs, and share large portions of DNA. However, we decide the fate of higher animals when it comes to appropriating their bodies. For instance, to specially engineer them and utilize their organs. Is there then a continuity or a discontinuity between humans and animals?

BOB POLLACK: It's a tricky question. There's a continuity at the level of DNA among all mammals. There's certainly a dramatic continuity between us and our nearest neighbors. We differ in DNA content by a percent or less between us and chimps as a species to a species, but there's obviously a trick in that because we have consciousness and a perception of our own mortality and language. None of these are available to other species, and they cumulatively make a discontinuity, but the real continuity between us and animals is not between us and other animals that look like us, but between us and invisible single-celled animals in us. In that sense, it's not simply us dominating other animals, but other animals patiently waiting for us to die so they can live off us. This is the real lesson of continuity. There are as many legions of bacterial cells in a person's gut as there are human cells in a person's body. It's not clear to me which organisms have the best chance of surviving in the long run, considering that those bacteria look like 3 billion year old bacteria, and we've been around only a short time. So, the dominance of humans over other species is far less apparent than the real story, I think.

PAOLA MIELI: Jacques, what do you think about continuity and discontinuity between humans and animals?

JACQUES HASSOUN: It seems to me that this question has already been raised on several occasions in terms of ethics. Twenty years ago, Leon Polyakoff held a conference at Cerisy on this question. You can call someone a pig. You can call him a camel. You can say he's an ass. There is still a discontinuity in the effects of discourse. There is a radical difference, a radical discontinuity between humans and animals, even though, in their identification games, children have always used tales that involve talking animals. Which raises the question: Why must we have a talking animal before we have a talking human? What kind of identification is involved? Is there identification on the child's part? There certainly is discontinuity where there is characterization of the human. In my culture, to call someone a pig is the worst insult. In my culture, calling someone a son of a bitch supposes warfare among the clans. So what does it mean to wear a piece of animal body as a part of one's own body? It seems to me that when the piece of animal body becomes a part of the human, the human humanizes it by writing its name on it.

LEWIS KIRSHNER: I agree that in all these prosthetic metaphors, whether it's mind or body, the subject is somewhere else, and the subject always has the possibility of incorporating the new arrival into his subjectivity, or externalizing it, treating it as foreign. And, you know, that's exactly the problem that certain patients have after heart and kidney transplants, who become psychotic. They have a lot of trouble for their own reasons, subjectivizing the foreign organ. That's the possibility because the subject is not affected directly, in my view, by the adjuvant organ. Now that may be different when we speak about the extreme case of a sex change operation. How is the subject changed in that sense, because is the sexuation of the subject changed? I wonder how people would feel about that question.

PAOLA MIELI: Since we are speaking of displacement of body parts, let's address the issue of body parts as a commodity. Obviously the development of biological technology has led to a worldwide industry that draws billions of dollars and profits. Dorothy, how is this market influencing the choices and decisions made by the researchers in your field?

DOROTHY WARBURTON: I think it's influenced a great deal. I think there's more and more biotechnology that's being financed not by the traditional methods of government and foundation grants, but by industry. And I think that does influence the choices of what will be investigated, because companies want to perceive that there's a big mar-

■ BEING HUMAN:

ket for their product, and therefore they want to work on common diseases, and not on rare ones which will have very few takers. We have the problem of things like orphan drugs, which are drugs which some people need, but not very many, and therefore drug companies don't want to manufacture them. And I really deplore this. I think it's a very bad precedent. I think it also influences the way research activities are carried out. It used to be that when you went to a meeting, your fellow researchers would be bubbling over to tell you what interesting things they'd just found. Now you can't pry it out of them, because there are questions of patents, and priorities, and all sorts of commercial considerations, and I think this is a very significant problem.

PAOLA MIELI: Jim, what would you say about this?

JIM YOUNT: Well, I think I have a bit of a unique perspective here, in that I am the Director of the American Cryonics Society, in which we freeze people. And we've had a number of researchers who have started with an interest in suspended animation, in being able to cool down a subject, keep that subject some time, and then ultimately revive the subject, which is, of course, a goal of cryonics, to be able to lead to the technology where a healthy person, or a person who perhaps has cancer, and will someday die, will be placed in suspended animation and could, if necessary, be revived tomorrow, but can be kept for some period of time until a possible cure is found. It's a goal that may well be decades off, but one that we're interested in. There have been two different commercial companies that have been started with technology that was originally developed by grants of the American Cryonics Society, where they've worked on blood substitutes, or the solutions that are used to cool and keep organs during the time that they are being transferred from one location to the other. Now, from one standpoint, that has hurt us in that it has taken researchers and put them into a field that isn't really cryonic suspension. But on the other hand, these same researchers, during their lunch breaks and after hours, continue to do cryonic suspension research. For example, one of our researchers recently told me that he tries to put a hamster a day into suspended animation, and he does. He kind of squeezes it in with the other kind of research that he's doing. Fortunately, he's good enough in his field and has enough influence in the company that they allow that, they kind of look the other way when this happens. So in some ways, this has helped what we're doing, the concept that there is value to transplanted parts.

BOB POLLACK: I'm sorry, I just have a vision of a freezer-full of hamsters. My mother is alive at 83 because she has an aortic valve from a pig which was implanted in her 20 years ago, and the half life of these implants is 10 years, so she's 10 years past where she's expected to be alive, and as far as I can tell, she's very happy, bearing this newly signified extension of her body which came from a pig. I think this is, of all the questions we've had so far, the least troubling to me. I think that the body, as long as the mind is sound, the body can subsume any number of instruments, any number of pieces of plastic, metal, wood, or tissue and still be a human body with a human mind. The boundary conditions of where one comes from and where one is going are, to me, much more troubling than the materials that the body is made of. And I think that, therefore, the notion that we can beat death by freezing is an extremely troubling, and, to my mind, wrong-headed notion, far from any question about moving pieces around so that one can extend one's single, mortal life.

PAOLA MIELI: Bob, in the paper you presented for the collection of papers that we gathered from panelists for the preparation of this colloquium, you say that scientists cannot accept mortality. And you tell us that while biomedical science claims, as its intention, I quote, "To cure us of nature's defects, it is, in fact, animated by another purpose: to acquire a mythical immortality at whatever cost to its players and anyone else." Can you say more about this?

BOB POLLACK: It is extremely easy to fall into the scientifically plausible, but logically implausible trap that since everything that one dies of is separately treated as a problem in nature, the cumulative effect of the success of all parts of molecular biology and molecular medicine would be the prevention of death *per se*. I think that that just doesn't hold up when you examine the place of death in nature. People often ask, When does life begin? and the obvious fact is, life never begins, it only ends. Life began once and it persists through the fertilization of living cells to make a new living organism, and then, individual organisms die, but life persists. I think my colleagues by and large have not absorbed the lessons of evolution in their agendas of medicine.

JIM YOUNT: I am an immortalist. I believe in the desirability of extending my life, at least. I don't impose that on anybody else, that desirability of extending my own life without end, but this doesn't mean that I think I'm an immortal. This does not mean that I do not recognize that something really bad could happen. There are always things that could happen that would bring an end to my physical existence. There are a number of people who do not believe in any kind of a spiritual continuation. They do not believe in the soul. They do not believe that we're going to heaven. They do not believe that we will be reincarnated in some way. And I'm among that number. Because of that, my only chance of having the same thing that the religious people have had for years, and that fuels the religious movement, and the cathedrals, and all of the things that have allowed other people to seek those dreams in that theme, is a physical immortality. So, no question in terms of myself. That's what I am. That's the goal I seek, and anything, any technological advance that goes in that direction, I'm for it. Does this mean that I want to impose this on other people? No, absolutely not. I think that each person should make those choices. Is it a terrible thing that whenever there are advances in science which extend the human life span, that this may inevitably lead to people saying: "We've conquered all the diseases, we're all living to a hundred twenty, let's see if we can make it a hundred and fifty, and if we make it a hundred and fifty, can we make it a hundred and sixty or two hundred and maybe live forever." Yes, I think we're going in that direction. I don't necessarily think it's a bad thing. If it's a question of you asking me to die because you think it's a good idea that I should die and become worm food, or because you think it's a good idea because we need the room, or some people want to have kids. Well, that's not a solution. A solution is not an ending of my physical reality so you can do what you want to do. I am willing to work with you on questions of population, and where do we put the people, and how do we have a better lifestyle. I'm not saying that you shouldn't have your kid, but I'm also saying that you should not tell me that I have no right to be an immortalist and to work towards physical immortality. So I take a very different view on that.

PAOLA MIELI: In the material you sent us in preparation for this symposium, which is extremely interesting material, you say that immortality, I quote, "Is three to four decades from becoming a reality." Technically, do you think that this is possible?

JIM YOUNT: Will we be able to live forever, I guess, is the question, and how soon? And of course you have to know we can't live forever, the heat death of the universe is going to wipe us out, lots of other things will wipe us out. Could we live a good long time, a thousand years? Could we be free from the usual lifespan limit? Could we consider aging a disease, and treat it, and find that, indeed, it's possible to take a pill and live forever? Yes, I think so. If a pill was available, and you could take that pill, the effect of the pill would be that you don't age and you go back to, let's say, the age you were healthiest, happiest, age 25, 30, something like this. And you could stop taking the pill anytime you wanted. Would you take the pill? I would. I think that there would be a significant number of people who would take this pill. Will this pill be available, and how soon? I quite frankly don't remember the quote of three to four decades, but that might be the case. When it's available, it's going to be a strange new world that we live in, and these kinds of choices that we're talking about now will have to be made by all of us. We'll have to decide whether we're going to take that pill or not. And, of course, it's going to make some very new and interesting and different problems than we have right now.

■ BEING HUMAN:

The Technological Extensions of the Structure of the Body

DOROTHY WARBURTON: If there's any kind of biological determinism that I believe in, it's the inevitability of death. As we understand more and more about how cells grow and age, we see that there are a tremendous number of changes that take place that you can't fix with a single pill. While the dream of immortality may be nice, it's not restricted to scientists. I would like to say to Bob Pollack that I don't think that's what drives my research, and I don't think he should really put this on all of us. If this is his problem, then it's his problem, but it's not my problem. Scientists as human beings are no different from other people, and I think we all have this dream of immortality. I'm sure I have it too, though, in my heart, I don't believe it.

PAOLA MIELI: Many people have commented recently on the physically and morally intrusive quality of various procedures for life extension. In the United States, people are compelled now to write living wills in order to protect themselves from life extension technology. And, as the *New York Times* wrote last March, a study finds that doctors refuse patients' request for death. Is there a relationship between such a refusal of death and the dream of immortality that we are talking about?

BOB POLLACK: My faith, such as it is, has the presumption of a redemptive end of days when, perhaps, bodies are reconstituted with their minds inside them. But that's not for historical time, and it seems to me that medicine and science work in historical time, neither at the moment of creation, nor at the redemptive end of days. In that historical time, death is an inevitability, and prevention of death at the cost of pain seems to me unethical, immoral, biologically unnecessary, and wrong. When people wish to die, it's not because they're suicidal at those moments; it's because they see no content to their life but pain and suffering. And it's my understanding that that's the usual content of a living will. It's not to say kill me, it's to say do not keep me alive in pain with no hope of coming out of that pain. That state is not, in my mind, comparable to the state of being alive as we talk about it, and I don't see a difficulty with acceding to such a living will.

PAOLA MIELI: So why do doctors deny people the right to die? What do you think, Bill, about this?

BILL RICHARDSON: If it's true, it seems to me that they are suffering from the same fantasy of immortality that pushes the idealistic fantasies of youth. What's at stake here is not so much the fantasy of immortality, I think, as the fantasy of omnipotence. Expanding knowledge breeds the sense of expanding power – as Foucault has helped us to understand – and this is very seductive. If I understand what was said this morning, the omnipotence of knowledge leads to the dream of surpassing death. The purpose of genetic research, as I understand it, is first of all to find ways of curing disease, as we know it, and of avoiding disease, as we know it. Its first purpose, then, is not to make us immortal. Death is a fact. We begin with that, and we deal with that as best we can. We deal with it either with a God, or without a God, whether we be scientists, or analysts, or sociologists, or philosophers. In any case, death must be addressed. Science can do nothing to change this ultimate fact.

ERIK PARENS: Well, I'm sitting here sort of stunned by Jim Yount's confession. On the one hand, I think, well, we could have a thousand year analysis. That might be interesting. When I think, though, about all those infinite number of prophylactic dental visits, not to speak about having to bear my own company for the rest of time, it doesn't really appeal to me. What does interest me more, and it goes back maybe to Bob's earlier comments, I would like to live a normal life span. If some genetic intervention can come along that can help me rid my kid of cystic fibrosis or PKU or whatever, I mean, what does that have to do with the Nazis and eugenics?

BOB POLLACK: It doesn't have anything to do with the Nazis and eugenics. The Nazis and eugenics are about killing people, not keeping them alive. The best guess I know of for the life expectancy of our species is something between 85 and 95 years, and I think it would be a marvelous agenda of medicine and molecular biology to bring everybody to that life expectancy, and at the same time learn how to value the last of those years as well as the middle of those years. That isn't the agenda of modern molecular biology or the agenda of modern medicine.

JACQUES HASSOUN: We should not forget that there are still people in the world whose life expectancy is thirty years. There is some indecency in thinking exclusively in terms of a ninety, one hundred or one hundred and twenty-year life span. Why not two thousand years? Like the Thousand-Year Reich? What ideology is behind the wish to live so long? I don't claim to have an answer, but what ideology is there behind the idea that you can play hide and seek with death? Freud claimed that we always die from internal causes. What does this mean in light of "the war against accidents" etc., etc. Ultimately, how does this internal death take us? I can't speak of this without thinking of Primo Levi's work, since Primo Levi was addressing the problem, not of survival, but of life, which is very different. Well, does living a thousand years involve life, or survival? The second question which occurs to me in this context is, what does it mean to transmit a name? Since, in the end, transmitting a name is all we really end up doing, what does it mean to transmit a name within a fantasy of immortality? Isn't transmitting a name always caught up in the fact of knowing that we are mortal? I'll conclude with a third question. There has been much discussion of omnipotence. This immediately made me think of what I believe Saint Bernard said when he was talking about the Virgin Mary and omnipotence. It also brought to mind Suplex, the omnipotent supplicant. Ultimately, aren't we in this fantasy of omnipotence? Aren't we still caught up in the supplication of wanting to go beyond? Lacan says that the pleasure principle is dumb. It just says "yes" endlessly. The question of death is raised in a very precise way, in the beyond of the pleasure principle, the only beyond known to psychoanalysts as psychoanalysts. The unconscious is ignorant of negation, death, time, and sexual difference. The subject may be the one who tries, his whole life, to extricate himself from this ignorance, and become able to say no, to know his mortality and that he is caught up in sexual difference. I might be wrong but, to me, these questions seem to be imbedded in the whole of this afternoon's propositions.

PAOLA MIELI: Before inviting the public to intervene, I would like to know if any of the panelists would like to make comments. Jim, do you want to comment on what has been said?

JIM YOUNT: Well, a little bit, not too much. Comments have been made about the quality of life, and sure, we're all interested in that. Nobody wants to live a long time if the quality isn't very good, and nobody wants to continue as a person with infirmities and problems getting around, problems of being isolated, problems of memory, not having a good memory, not being able to remember your friends' names, not being able to go places and do things. Sure, quality of life is very important, but, we have to realize that these are choices. In other words, if indeed I wanted to live for a long time, I always have the choice of ending it, and that should be anytime that I want, anytime that I'm bored. That right should be preserved. If we're able to use extraordinary means to have lots of interesting animals available for a long, long time, is that good? We really don't have to answer these questions in an absolute sense. It's happening. Rather, we need to be prepared to deal with the reality as it is unfolding around us. For example, if, as psychoanalysts, someone presents themselves to you who indeed believes in living forever, wants to live forever, it is important to understand why they're doing this, where they are coming from. This knowledge may be important for their treatment, not necessarily to cure them of this desire, but to understand the other problems that they have, and the way that they are presented. So if indeed this, one of our goals, is to try to look at the knowledge that's coming from other disciplines, then, we need to do this in a non-judgmental way. Being able to understand that we have different life-views, and that we may have to deal with patients, and friends and colleagues that have these other life-views. That's it.

PERRY HOBERMAN: The idea of some kind of actual immortality is completely based on medical practices that don't exist now. It's hard

■ BEING HUMAN:

not to see it as a kind of fantasy. In the same way that the fantasies of immortality that might be associated with doing something great that people will remember you by don't seem like a good thing because it takes you out of what's going on here and now. In other words, you've invested most of yourself in something that's a fantasy, and obviously the idea that there's a hereafter and that that's where it really counts has led to all kinds of problems.

PAOLA MIELI: Would anybody like to make any comments?

DOUGLAS TRUMBULL: Well, I think I've addressed this to Mr. Pollack, my thought being that our planet is plagued with famine, and ill health, and wars, and a lot of stupidity, however you want to characterize it, as well as the demise of a lot of species, etc., etc. And one of the issues that hasn't been discussed here is the fundamental issue of the transformation of ourselves into some kind of improved form, some kind of more enlightened or more wise form, which will behave in a broadly different way. If Seth Shostak, for instance, contacted through SETI some advanced civilization that could send us a new genetic code, would we actually be willing to transform ourselves in a fundamental way, you know, to become more enlightened, or in any way so that we would not continue to behave as we do?

BOB POLLACK: I would understand the question better if I understood whom you mean by "we," all of humanity? Well, all of humanity divides itself into about 5,000 languages, not one of which is comprehensible to people speaking the others. So we have chosen a strategy of survival as a species which is intrinsically self-isolating. Now whether that evolves from primate behaviors in general to protect boundaries, or whether it's a coincidence and a consequence of the uses of language and identifying family structures, I don't know. But in any event, we're stuck in what we heard described before as the Tower of Babel. The Tower of Babel is a tower of arrogance, and a risk to ourselves, and we have not been dispersed across the planet with one language, we've been dispersed across the planet making new languages constantly. In fact, just to say how serious the problem of language is to your question of how we could transform ourselves, routinely, different populations across the planet have slightly different frequencies of various versions of various different genes, in that sense they have different genomes. Populations sharing a language have very similar genomes compared to populations with different languages and at the boundary of language, one finds the boundary of DNA sequences, suggesting in fact that language has served as a vehicle for the self-reinforcement of genetic differences. In other words, the DNA that makes the ability to have language now converts the DNA itself in turn. So there is a case *a propos* an earlier discussion of the two forms of language, the two symbolic languages of DNA and human language, each transforming the other. But in that process we cease to become functionally able to affect ourselves as a single species because we can't understand each other.

DOUGLAS TRUMBULL: But is it possible that this information is going to be a part of our own genetic transformation?

BOB POLLACK: We're sitting here and we have to put a machine in our ears to understand the difference between English and French. With 5,000 languages, I don't think it's possible.

CLAUS-DIETER RATH: Rereading the title of today's two sessions, the Technological Extensions of the Mind and the Technological Extensions of the Structure of the Body, I became aware that the notion of technological is something which can be considered as already a mental answer to technique. There's a difference between technique and technology. Technology is already a logic of technique or a *logos*. It might be an ideology of technique, of the possibilities of techniques. It can be narration about. Then it can also be scientific research about techniques. I think it is very important to come back to what some representatives of the Frankfurt School, like Herbert Marcuse pointed out about this difference between technology and technique. They said we easily tend to identify technology with reason, to identify the reason of technique with the technique as reasonable. If the fascination of so-called virtual reality is not exactly this, there is a kind of ideology of feasibility, of something without any boundaries. It's not a question of extension of boundaries of the body, but an abolition of boundaries, maybe a kind of virtual reality where everything should be possible, and the danger is that some take it for a real possibility, not only a fantasy or a game, because we should see that virtual is the contrary of real or actual, so not to recognize the end, the death, the castration, as we say in psychoanalysis, is something very dangerous.

DANY-ROBERT DUFOUR: I would like to share my surprise at the nearly unanimous condemnation by the panel of the position Jim expressed. It also surprises me from the perspective of psychoanalysis. Funny that psychoanalysis should abandon its neutral position, so to speak, in order to formulate a prescription. I am also surprised by the unanimity of this condemnation. Be it moral, ethical, political, ideological, everything was expressed in the condemnation. It seems to me that we are wrong to act shocked by this position. The adventure of immortality did not begin today or even yesterday. It has been around for at least 2,500 years. At least since the birth of philosophy. The soul was mentioned this morning. Well, since the birth of philosophy, the soul has been reputed to be immortal. It was on the basis of this immortality that a certain number of languages, artificial languages, were created. Geometry was born, as were arithmetic, cosmology, and music, all of them exact sciences. These sciences were created once it was thought that stable, repeatable utterances existed, that there is a being about which truths can be expressed which stand the test of time or eternity. I therefore think we have been on this journey for a long time, and we hardly have the right to act as if we did not know it. When I say that we are on this journey, I mean everybody, not only Jim. He says so. We do not. I think that the turning point we have reached (I am trying to link what was said this morning to what is being said this afternoon) is that we are in the process of applying the mental technologies we have perfected for 2,500 years to the term excluded 2,500 years ago, namely the body. Nietzsche is the one who proclaims the end of the division, the one who says, listen, the time has come to stop our fear, our squeamishness, to express our pretense, and to finally apply what we know, the mental technology we have created now, to the term that has been excluded, the body. I believe that our century, in a sense, follows this Nietzschean prescription or prophecy. The program is expressed quite clearly: "Never have I met a woman I would want to bear my child, for I love you O Eternity. I love you O Eternity." The old story has to be done away with, Nietzsche says, in order to do without sexual difference. He committed to a program that our century applies, and that we apply. I therefore find…maybe the word is too strong but I don't know another…I find a certain hypocrisy in condemning Jim when we are all on board. To conclude with this topic, you know that there is also in Nietzsche a beautiful image that happens to relate to what we discussed this afternoon about animals, man and superman, because, of course, the superman is what is involved in every discussion. He says that man is not an end point. Man is none other than a rope stretched out between the animal and the superman. He can be thought of as the tightrope walker. Obviously the tightrope walker always runs risks. Not least the risk of falling. It is possible that this will happen to us. It is quite possible. But I don't think this gives us the right to act like hypocrites when the entire division of thought of the work in which we participate, assigns to each of us a place in the execution of this program.

JOHN PERRY BARLOW: I am also a little surprised at our condemnation of Jim, since I found implicit in everything that went on here this afternoon, the denial of death which is as strong as his, if subtler. I think that to spend this amount of time talking about death without talking about the extent to which our refusal to accept it as a society has caused what I would consider to be our dominant pathology, is amazing and is evidence of our own denial. We are a technological society, and our religion is control. And for this reason, I think death has become shameful, and in the words of Philippe Cabry, "wild among us." And the reason that people don't die at home, the reason that the very sight of a dead body is horrible, is because it is an act of weakness to die in the modern world, and it is considered a shame. I just spent three days in Oxaca, Mexico, helping celebrate the day of the dead. I spent three nights in graveyards with people who take a very different attitude towards death than we do. And I must tell you that it is a much less neurotic society than the one we live in. I think that in addition to the social and spiritual pathology of this, there are other practical issues to discuss, like the fact that we're spending somewhere around 80% of our total health care dollar on the last six months of life. We would do better to learn more about acceptance and less about how we can deny it better with technology.

From the audience: We talk about extending the life of the human being, but at the same time there's a death of the social space, it seems. I was wondering if someone could say something on the relation of the body politic to any of these ideas of extending the body.

BOB POLLACK: The natural model of the body as a model for the state carries with it implicitly the idea that dissent is equivalent to a tumor or an infection, that one state, one body, carries within it the idea that the viability of the state as a whole is put at risk by the behavior of individuals who break out of the constraints of state law. So I think that, as with many arguments of Plato, that is an argument which fundamentally leads to fascism and not to democratic societies. I'm happy that the democratic society should turn out to be unnatural and not like the body, that's fine with me. Just as medicine is unnatural in that it prevents death when naturally people would die. I do not want to be boxed into the argument that because I find mechanical immortality a waste of time, therefore I welcome death. I don't. I think it is a noble enterprise to extend life as long as it is biologically possible. I think the biological limitations are not an excuse for not trying. And so, finally, let me say that on the matter of the body as metaphor, that great platonic vision is backed by the other great platonic vision, that there is an ideal society as there is an ideal body. I think it needs saying again. I think both fail the test of physical reality as first argued by Darwin. That is that the living world extends itself through time by variation and the survival of variance, not by progress toward any ideal, neither in the person, nor in the state.

DANIÈLE LÉVY: It seems to me that this afternoon's question has been "what is a body?" We don't know very well what a body is. As we just heard, even a political body is not very clear. Perhaps that is why the entire discussion has been centered on immortality. Of course, a subject cannot help wanting not to die. So, what do we know about what a body is? For scientific discourse it is an organism, something that must be fine-tuned to function well, with an emphasis on function. For the psychoanalyst, as Paola reminded us just this morning, it is an erotic body. This means that the body seeks pleasure, which is not the satisfaction of need but something else, structured by its immersion in language. This is what psychoanalysts have discovered and are trying to make audible. So what does an erotic body mean? It means that it seeks pleasure at any cost, and tries to avoid displeasure at any cost. This involves a mechanism of radical avoidance that we may be seeing at work this afternoon. So I find very interesting the question of the technological extensions of the body, or rather, the effects on the subject of the technological extensions of the body. It seems to me a very difficult question to deal with in the absence of a somewhat solid idea of the body. This means at least a double body: the biological body and the body for the symbolic subject insofar as he invests it with pleasure and the refusal of displeasure. I had a thought about prostheses and organ replacement. Organ replacement satisfies the biological body. It restores a failing function. It is extremely difficult (I know of no examples) to eroticize a prosthesis. Which raises the question of whether a prosthesis or a new organ belongs to the subject's body, the subjectified body, the non-scientific, body symbolic. Where internal organs are concerned, this does not seem surprising, since internal organs do not seem to be represented in the psyche. Everything happens, or nearly everything, as if the internal organs were purely biological. It's something that we cannot see, unless we've taken a look, which means opening and therefore changing the image the subject can have of his body. It is

■ BEING HUMAN:

clear, though, that people have to work through a whole process in order to adapt to this new thing, after, say, a mastectomy. Or an internal artificial organ...the psyche has to work through it. I'll end by telling an anecdote about a man who had lost his sight. He had gotten used to living with four senses, without his eyesight. He spoke about the intensification of the other senses in a person who is missing one. What happened when, through a miracle, one more miracle of science, he recovered his eyesight? He became deeply depressed because he could no longer recognize himself in the world. He no longer knew the world he had made for himself.

PERRY HOBERMAN: It's hard to see how prosthetics or transplantations couldn't be eroticized in the sense that clothing is a kind of prosthetic, and that certainly gets eroticized. The absence of a limb will become eroticized in certain cases. I'm not a psychoanalyst, so I don't know. I'd like to actually hear psychoanalytically whether there's any obstacle.

PAOLA MIELI: It is a very interesting question. I wouldn't be so radical in taking the position that the prosthetic part of the body cannot be eroticized. Lets think of the way in which lost body parts can be experienced as present: for instance, in the case of a missing limb, when it is felt, perceived, in spite of its absence. It is absent, yet it is present in the subjective experience of the body. This indicates an aspect of our sense of the boundaries of our body, how, in fact, we can include and exclude, according to a mapping which is radically subjective and which involves imaginary, symbolic as well as real elements. The possible eroticization of what is missing redefines the idea of boundaries. Think also of what glasses are for a myopic voyeur. It seems to me that the substitution of an organ or the acquisition of a prosthesis, including a synthetic prosthesis, may require a certain amount of time to be integrated, but it will not necessarily escape being eroticized.

DANIÈLE LÉVY: I thank you for this remark that allows me to clear up a misunderstanding. To answer you, there is Steinbeck's one-legged whore in *Of Mice and Men*. When I said, "eroticized," I meant for the subject himself, a zone of his body able to bring him pleasure, or displeasure, for that matter. My idea was that a prosthesis was a zone that could not be inhabited the way the subject inhabits his body.

BILL RICHARDSON: Isn't there a distinction that's being made between the body as erotic and erotic parts of the body? After all, if I understand the drift of Merleau-Ponty's analysis, namely where the body is a conscience "incarnée," the incarnation is the incarnation of an entire subject. And the erotic that I understand Paola to have been referring to is the entire drift, the entire drive of the subject as a whole, and not simply the functioning of erotic parts, which I thought I heard you implying, perhaps wrongly.

PAOLA MIELI: I would like to return to your comments, Dany, on the history of the defeat of death. A contradiction emerged this afternoon: on the one hand there is the declaration that the project of some researchers is the defeat of death; it has been said, for instance, that the dream of immortality animates to a certain extent biotechnologies. On the other hand, there is the tendency of denying the idea that the defeat of mortality is something that concerns us all, at least in terms of a certain aspect of our psychic life.

JACQUES HASSOUN: Yes. Certainly. To conquer death is a fantasy. I would even say that you can't do better as fantasies go. Death cannot be imagined! Well, if it can't be imagined, how do we give it a symbolic dimension? If we are only dealing with the real, if we say that death cannot be imagined, then we might say that we are only dealing with a corpse. I think that this is what Leclaire told us during the famous Cerisy conference on this question. Now, it seems to me that death does not escape the clouds, the real, imaginary, or symbolic clouds. So, what about death on the symbolic level? Call my point of view archaic or barbarous, but it seems to me that death, *vis-à-vis* our relationship to the symbolic...death seems to involve the total defeat of language. This reminds me of what François Peraldi says about the fantasies brought about by funeral homes, with the infinite smile of the embalmed corpse who never stops smiling infinitely. So how can you be

The Technological Extensions of the Structure of the Body

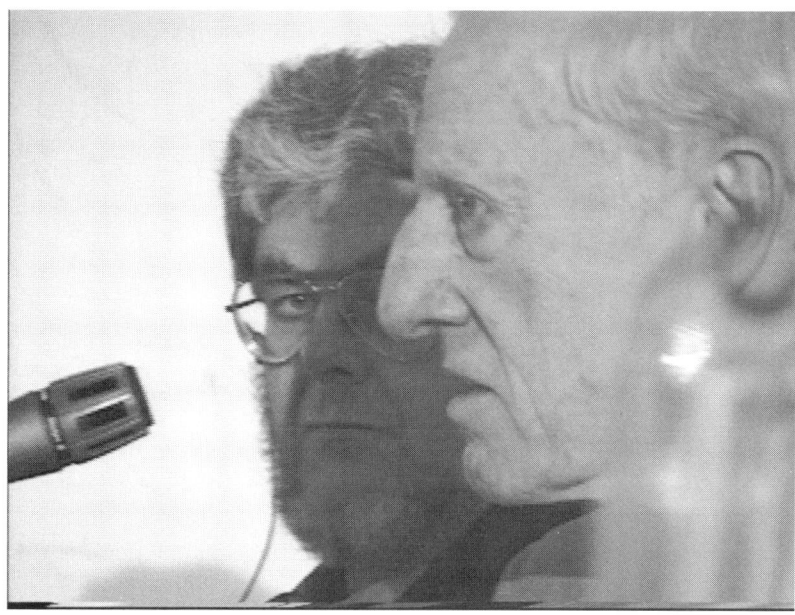

Robert Pollack and William Richardson – Being Human Colloquium, New York, 1996

Jacques Hassoun – Being Human Colloquium, New York, 1996

THE TECHNOLOGICAL EXTENSIONS OF THE BODY ■

an analyst and deny death or say it can be conquered or pushed back to infinity? I would like to know. It seems to me that this is Freud's central point. He really cut into psychology with *Beyond the Pleasure Principle*. How can there be psychoanalytical thought that excludes death?

PAOLA MIELI: The very presupposition that the body is an eroticized locus entails not only pleasure, pleasure at any cost, but also, as Freud says, a beyond of the pleasure principle, which inscribes death in the repetition of *jouissance*,[1] in the very core of the human being. Dany underlined an issue: the impact of a political or ideological program on the defeat of death. We heard about it this morning *a propos* the extension of the mind. Its social and ethical implications are worth addressing.

DOROTHY WARBURTON: I just feel like we need a reality check here. There's death and there's death. We're all sitting here talking about death at a ripe old age and how we can't avoid that, and I think we would all agree. Right, Jim, we wouldn't *all* agree. But there is the death of children from diseases which can be cured. There's the death of children from malnutrition. There's the death of women at the age of 40 from breast cancer, when they're at the peak of their productivity. And to say that to go on with medical research to try to ameliorate that condition is because we are denying death, I think is just simply ridiculous. Medicine has always tried to avoid the avoidable death, and that's what we're trying to do. We're not trying to avoid death in the end, but only the avoidable death particularly in the prime of life, or even before the prime of life.

BOB POLLACK: I want to agree with Dorothy first. That is to say, I tried to say this before, but I don't think I made myself clear. There is a finite lifetime to individuals of our species as there is to individuals of any species. Whether that lifetime is 85 or 115, I think it is a reasonable

---

1 *Translator's Note:* Throughout this book, we maintain the word *jouissance* in the original French, instead of the terms commonly used to translate it, "pleasure" or "enjoyment," because *jouissance* accounts not only for pleasure but *also for its beyond,* according to fulfillment of erotic and death drives.

■ BEING HUMAN:

agenda of medicine and molecular biology to push the survival curve so that the maximum number of people get to the maximum age in good health, and then die. An agenda that includes that as its primary purpose should include an understanding of all phases of life, including the last moment before death, so that the last moment is not seen as a failure or an embarrassment. With regard to DNA's contribution to this discourse on the end of life, I think it is fair to say that, while I have no belief in chemical immortality or suspension of life, I have confidence that in about 40 years, to pick a number, a biblical number, 40 years, we will know whether and to what extent death is the consequence of a set of inherited versions of different genes, and to what extent death is a consequence of an inevitable and tropic accumulation of random mistakes. To the extent it's the latter, we will have to admit we can't do anything about it, to the extent it's the former, we will be tempted to manipulate it, by gene technology. As a society, we are not prepared for either eventuality.

JOHN PERRY BARLOW: I want to try to get back to your original question about the body and death and psychoanalysis, and the way in which we should be treating them. Through an odd set of circumstances, the constant in my life has been death. And I appreciate very much what you're saying about preventing the death of the young because all of the people I grew up with and was close to in my little town in Wyoming, died by one means or another. The woman whom I had intended to spend my life with dropped dead two days before her 30th birthday, and I have had to study death closely in another sense besides the obviously medical or technological. I want to get back to the question of the prosthesis. The woman who asked me if it didn't seem that when it came to sex the body itself was a prosthesis, and I said, a prosthesis for what? I think the answer to the question lies in the study of this "what." And the reason that the study of this "what" is not being couched in the terms I would prefer, in psychoanalytic theory, is because psychoanalytic theory regards itself as a science and therefore cannot admit the spirit. Now, I don't know if there's a backdoor at this point where we can start thinking about those things, but I recently watched somebody get past his own sense of limitation on mortality and also the pleasure principle in a very direct way, when I spent the last days of Timothy Leary with him and finally argued him out of having his head frozen. The reason I wanted to do this was because I thought he was doing a very heroic thing in dying publicly without turning himself into a medical experiment, or closeting himself away as doing something that was weak and shameful in dying; by doing it joyfully, having a good time on the way down, going to parties and clubs in L.A. till the very last minute, and throughout maintaining this terror of the other side that was based on his conviction that the only immortality would be derived from the reanimation of his head. And there was an interesting moment that took place, there was a switch that toggled when he suddenly started to feel the other side in some palpable way, and a different light came into his eyes. That night, I convinced him that if he allowed them to cut off his head and freeze it, it would take the heroic act of his death and allow the media to turn it into something ghoulish and horrible again, which is the way we treat it in this country. And the only way I was able to convince him was because suddenly he was able to imagine another form of immortality besides the purely technological, he was able to feel that. Now, I'm not talking about the soul necessarily in any religious sense, but I believe that there is another side, and I believe I saw him see it. And I don't think it would necessarily do harm to the science of psychoanalytic theory if we admitted that there might be.

DOUGLAS TRUMBULL: This may seem stupidly naïve, but I just wanted to say one thing. It seems to me that the issue is not just the denial of death, but death means the total cessation of our individual consciousness as a human being, of our awareness and our sense of self. What we're seeing in fact is a tremendous compensation in play, a quest for just the opposite, which is an increased consciousness. This increased consciousness is being sought via digital technologies, or optical technologies, or whatever the hell technologies we've got. It's not just about the prolongation of life. It's about denial of death via increased awareness. If you look in this book of Leonardo right here which is a good example of a lot of the art forms that are in play today everywhere, you'll see that the art is very psychedelic. It's very dreamlike. It's very out of body. It's like a drug experience. And I've seen this everywhere in the world of Virtual Reality. Everything seems to be a quest for some increased state of awareness about the nature of the universe and life, in an age where drugs are really a bad thing, and people really aren't supposed to do that. I think people are really seeking a kind of drug-free high, or drug-free alternate experience of their own consciousness and life, via these new technologies.

PERRY HOBERMAN: I just want to mention, because it hasn't been mentioned, that the other great sort of idea about immortality has to do with uploading or downloading one's brain into a machine. In terms of the idea of a body as prosthesis, there is this idea. It doesn't seem like a good idea, but there is this idea of becoming the machine. If biology is one symbolic system, and language is another symbolic system, the reason to me that we reach what seems to be almost a sort of impasse is that when we talk about actual physical immortality, we're leaving the biological symbolic system in some way, at least as far as what we know as humans. So, it may be possible that we would become something that's not human.

Sunday, November 10, 1996

Opening Remarks

MARK STAFFORD: I would like to offer a brief review of some of yesterday's discussions. For those of you who were not present yesterday, let me explain that we consider the division of our panels into mind, body, and senses to rest on purely fictive distinctions. But just as we want to avoid working within imaginary divisions of mind, body, and senses, we also want to note that it is illusory to work with a notion of a "complete" body. Jacques Hassoun's observation that technology plays a part in our struggle and reckoning with death, offers an excellent introduction to this paradox. The notion of a "complete" body is as fantastical and problematic as the idea of the disembodied mind. As we heard yesterday, we cannot be in our bodies without the presence of language, the most ubiquitous of our tools. But as many of our panelists have noted our dependence on tools, our sense that we have become their instruments leads us towards a rejection of this dependency. The ethical questions raised yesterday revolved around this contradiction in which we simultaneously acknowledge both our intimacy with technology as well as our responsibility to resist certain technological interventions. What we are resisting is an ideological position which certain technological programs support. Cognitive science is searching to understand the relationship between the biological and the psychic, but in doing so it supports an ideology that encourages the substitution of machines and chemicals for the struggles and limitations of the subject. The Internet has fostered new forms of collaboration and yet many applications of information technology produce the illusion that we can close the gap between subject and object. Our technologies are new because they offer an image of undreamt-of possibilities but are they not also the familiar dreams of

Prometheus and Faust? If we accept the "laws" of the marketplace and refuse to engage the ethical questions that their appearance poses, are we not abandoning them to the same fate that Mary Shelley's Frankenstein anticipated?

DOWNTIME

Marcos Einis

> The potential of the telephone, of radio, of conventional television itself, now seems dwarfed by the promise of distance information, communication satellites and network access.
>
> The distance separating objects and the time needed to visit them in order to bring them closer together, required, until quite recently, a representation of "real" space made up of directions and obstacles to avoid a physical linearity that the transportation revolution would soon transform.
>
> The revolution in information services, its interactive technology, the growing importance of information and the existence of new mediation artifacts intensifies this metamorphosis.
>
> Barely needing to leave his home, the Filipino reservations clerk has already received an updated database from the Lufthansa desk in Frankfurt, by means of a modem which allows him to be operational half a world away before office hours begin.
>
> By transporting and processing information without transforming matter, simply by managing signs, the new communication techniques modify the scale of time and space.
>
> Abstracted from its historical and physical context, the event becomes a chain of "compressed data" circulating within integrated circuits. Things are stripped of their reality just as the order of the mental reality of time and space is substantially altered.

Until a few years ago, the act of communicating involved either traveling, or an effort or a wait, in every case an action mediated by space and time. Today their suppression has become the goal of the artifact. The outcome is general promiscuity in a complete motor inertia.

Video, telephone or on-line shopping, telemarketing, video surveillance, E-mail, tele- or video-conferencing, etc., etc. The proximity and the distance of objects can no longer be considered in terms of a distance traveled.

They can now no longer even be considered in terms of the time spent traveling over the segment of territory separating them.

Space-time has been taken over by relay artifacts which allow the circulation of units of information moving within a long-distance network in real time, and crystallize objects on the flat surfaces of screens which have become props of their presence by the mere fact of being linked.

The object no longer travels. It circulates. It appears. It is no longer in repose. It poses, simulated, allowing its materiality to be supposed, then disappears. Nor is there any need to pay it a visit. The very nature of its access ensures its movement in a network surface where the depth of the field to be traveled is superfluous. Grasped, it is already there, everywhere, at the same time. Time saved, shortened, reduced, suppressed, boxed in, denied. Immediacy concedes to the machine the right to keep time. It is now up to relay artifacts to scramble distances by multiplying paths, exploding the center, recentering eccentricity.

Along with the abolition of distances and the downing of downtime, the soft revolution signals the death of a certain way of conceiving of time and the birth of a new proximity: the digital. It inhabits the message and lives off the sign. It renders them insignificant by virtue of the unexpected omnipresence it imposes. It happened before it was even thought. Preceded by its simulacrum, trapped in its closed itinerary, it was annulled at its

■ BEING HUMAN:

beginning. As for thought, it can only pursue the digital by following its path, by entering its circuit and becoming one with it. The object is in and through the circular process of input-output only, where the meaning of supply is none other than to instantly reactivate a demand and maintain the cycle in a formidable signifying shortcut. The distinctive function of language is replaced by a binary signalization where supply and demand reciprocally and simultaneously test one another without a moment's reflection, in a network.

Conditions of exchange dominate the object of communication; even more than the content of the message transmitted, it is the how it is communicated which matters, to the extent that this eventually institutes a utilitarian logic of "transmission feasibility" which supposes the very exclusion of its object.

In any "thinking network" strategy, effort seems exiled from the future and distance becomes simulacrum. But the network is first and foremost a structured architecture of means and conditions making possible, after a process of analysis and coding, using machines able to manage sign and symbol, the linking only of identical bodies. In "instrumental" communication, all communicative identity must not allow itself, in its action, the least contradictory, polymorphous, or variable aspect.

Breaking this rule runs the risk of complicating modeling and blocking standardization and permanence, the *sine qua non* conditions of "communication."

Henceforth, the word, in its function of equivocal mediation, is no more than a natural obstacle to be removed from the path of the code's operational functionality and an eventual additional source of noise in the specular reception of the object of demand.

Beyond the relationship to space and time, through the nature of the artifacts and the conditions of their use, through the overproduction of signs and their dizzying circulation, it is the very essence of human communication which is placed between paradoxes.

THE TECHNOLOGICAL EXTENSIONS OF THE BODY ■

Colloquium Proceedings

The Technological Extensions of the Senses

Panel Discussion

MARK STAFFORD: Thank you, Marcos. First of all, I'd like to introduce our two moderators. Timothy Binkley is the Chairman of the MFA Computer Art Program at the School of Visual Arts. Paola Mieli is President of Après-Coup and psychoanalyst in New York. Our panelists consist of Dany-Robert Dufour, philosopher, Paris; Michael Groden, Professor of English, University of Western Ontario; Richard Teitelbaum, composer and Professor of Music, Bard College, New York; and Claus-Dieter Rath, psychoanalyst in Berlin. On the panel on our left hand side, we have Seth Shostak who is an astronomer and Director of the Public Program at SETI Institute in California; we have Claude Rabant, psychoanalyst in Paris and Editor of *IO International Journal*, and, on his left, Douglas Trumbull, Vice-Chairman of IMAX and CEO of Ridefilm Corporation. Thank you.

TIMOTHY BINKLEY: The human body is limited. We are all confined to being in one place at one time and, I think, most importantly, at one scale. We are here now. We occupy a certain size space and a certain size time. Yet, new technologies have extended and redefined this limitation. In what way does the elimination of distance, which requires time to traverse, transform our subjective experience of the body, and our way of relating to other bodies? I'd like to ask this question first of Claude Rabant.

CLAUDE RABANT: You are throwing the ball right away into my little court, but I'm going to send it back very quickly, because, in my opinion, the body has never been something that could be kept in one place. In my opinion the body is something which, by definition, is always disseminated in time and space. Thus, it is in some sense the body's definition to be elsewhere. For me, if we have a body, it is because we are somewhere other than where we are. For example, when we dream, we travel in all sorts of ways. On the one hand, the body

■ BEING HUMAN:

The Technological Extensions of the Senses

Timothy Binkley – Being Human Colloquium, New York, 1996

Mark Stafford, Paola Mieli, and Dany-Robert Dufour – Being Human Colloquium, New York, 1996

THE TECHNOLOGICAL EXTENSIONS OF THE BODY

can't be separated from its senses. On the other, the fact of having senses means to me that we are always projected into another space. We are always projected into another body, inhabited by other bodies. The fact that we now have technical means at our disposal simply gives us, in my opinion, stronger perceptions. It gives us more refined ways of perceiving what is inside our own bodies, for instance. Listening to what Marcos Einis was saying earlier, I thought about the fact that transforming time and space was the very definition of the sign. By definition, the sign is what transforms time and space. Therefore, we now have different signs, different modes of writing. I don't even know if they are more efficient. They are different. We seem to go faster, but it isn't the same speed. We used to go very fast when we imagined ghosts, when we imagined ghosts visiting one end of the universe or another. That may have been even faster, so, to a certain extent, technology limits some speeds. Technology limits certain capacities that we had to imagine. So I think that the very definition of the sign, the very definition of writing, of the trace, is to transform, to change time and space. I think that the only thing we have to do today is to try to discover how we can not be afraid of what we are driving, of the new signs and traces we produce.

TIMOTHY BINKLEY: Seth Shostak, what do you think about this idea, of limited space and time?

SETH SHOSTAK: Well, I'm going to play devil's advocate and suggest that your premise that we are limited in space and time will change in the near future. If we define what Tim is here, imagine a machine that sort of took him apart, piece by piece. It would start at the top left, and well, the machine says, here's a phosphorous atom, and it's located here, here's a nitrogen atom, that's the next one over, and here's an oxygen atom, and so forth. If you made this long list of everything that was in Tim, this is obviously a mechanistic approach, if you did that, just worked out, the number here, about 10 to the 28th components, 10 to the 29th, that's a 1 followed by 28 or 29 zeros. It's a very long list. In fact, if you wrote those numbers onto CDs, it would take 10

■   BEING HUMAN:

million, million, million CDs to write all those numbers. That's a larger number than even is in my brother's Country&Western collection. In fact, that would cover Manhattan to a height higher than the Empire State building. So that's a lot of CDs, and that's a less efficient way of representing Tim than just bringing Tim, but we don't need all that. Yesterday, Dorothy Warburton suggested that there were only a billion, no only a couple of million base pairs in your DNA. Each one consists of four different types of compounds, so that means we could write your DNA onto one CD. That's true of all of you. The sperm and egg combination that made each of you contained less information than you can find on one CD in your collection at home. I don't know how much of his college education Tim remembers, but let's say a million CDs worth. That's still a small number. That's giving him the benefit of the doubt, actually. If we do that, then all we need is another machine over here that reads the CDs and puts the atoms back into space, and now we can play him back every hour or two hours, and we can make, in the course of the next week, another thousand Tims, each with the college education, the personality, and, not to mention, the body of Tim. So I think that that kind of technology would really change our perception of space and ourselves.

TIMOTHY BINKLEY: Next, I'd like to ask Douglas Trumbull, who's been involved with the changing of sensory experiences of space and time, what he thinks about this idea of limited space and time.

DOUGLAS TRUMBULL: Well, I'd like to talk a little bit later about some of my thoughts on the compression of space and time via film and media. I feel that as a result of this technological revolution that we're in the midst of, that still has a long way to go, we are seeing an imbalance between body, and mind, and spirit, and that refining the physicality of ourselves is no longer nearly as important for food, defense, transport, commerce, and personal success. More and more of these functions are directly related to intellectual rather than physical activity. This has created an imbalance. I know for myself that I have to forcibly withdraw from this revolution in order to retain my sense of self and indulge in body work and physical exercise and other simple human things that have nothing to do with my intellect, to keep myself in a sense of balance.

TIMOTHY BINKLEY: I always tell my students that in addition to insisting that they consult documentation when they encounter problems with the computer, they should meditate as well. I have the same feeling. Michael Groden, what do you think about space and time limitations?

MICHAEL GRODEN: As a literary scholar, my connection with this particular conference and panel is that I'm working on a project to try to put Joyce's *Ulysses* in a hypermedia presentation. And in doing this, I've encountered issues of space and time both in terms of the way Joyce has been approached and in terms of the way computers have been approached. And the specific parallel that I've noticed is that early on, one of the ways that people attempted to gain control of *Ulysses*, to make some sense of it, was to suppress all aspects of time in it and to spatialize it, to turn it into a fixed object that removed questions of passage of time, which is central to all narrative. What I've noticed in following the ways in which fiction written on a computer, hypertext fiction, and other aspects of computers, are talked about, seems to me a repetition, a return of this process. People talk about spatial elements of hypertext; the instantaneous movement from screen to screen fosters this. But we talk about navigation through a story, we talk about getting lost in cyberspace if the navigation breaks down. One of the main programs is called "Story Space." One of the things that might be happening is that the elements of traditional narratives get lost in spatializing them, things like history, politics, and all of the questions that are crucial for time, might be happening again. And in terms of the question of placing a traditional work like *Ulysses* in a computer, hypermedia format, it seems to me absolutely crucial to try to avoid repeating that process. I don't exactly know how to do this yet. The temptation has already been mentioned of that sort of instantaneous possibility of moving from a detail in a literary work to an explanation of what

■ BEING HUMAN:

it is, eliminating all the time it takes to find your dictionary, to find your reference book. It is extremely tempting. What seems to me vitally important is to not lose the element of time and everything that goes along with that.

RICHARD TEITELBAUM: My work has mostly been involved with computers in terms of real time, interactive systems, not so much involved with Internet kinds of communications, but in terms of using a computer as a self-reflexive system whereby I can externalize my internal prostheses and flash them back to me, and deal with them as a kind of feedback system. I started doing stuff like this a long time ago using alpha-waves, and one of the things that was quite interesting to me in the biofeedback process was that if I was using alpha-waves as a kind of meditative tool, that if I wanted to enhance them, it was necessary to get into a sense of absolute unison with my own physiological processes. So in a certain way I was collapsing time into a certain situation where it became outside of time.

TIMOTHY BINKLEY: Paul Virilio remarks that, "Thanks to cybertechnologies, history has just struck the wall of worldwide time. With live transmission, local time no longer creates history, worldwide time does." What, then, is the relation between worldwide time and the subjective time of individual daily life?

DANY-ROBERT DUFOUR: To try and answer this question, I would like to pick up on Claude Rabant's remarks, which called into question this idea that there was no elsewhere for the body, by saying that the body was already elsewhere, that it was inhabited by a past that never stopped returning in dreams, in what are called the summons of the unconscious. So I am, of course, in agreement with this hypothesis, on condition that we add something. This would be that the elsewhere exists only as long as a "here" is constructed for the subject. It means that the elsewhere, which is absent, can be re-presented (I insist on the "re," dash, "presented"). This is the scene of re-presentation. An "elsewhere" can be re-presented only when a "here" has been effectively constructed. I myself call this the symbolic function. The symbolic function is the capacity a subject has to re-present, to render present what is elsewhere, to turn there into here. This reminds us that when we say "new technologies" it implies that there is an old technology. I would even say that there is an archaic technology. It is within an archaic technology that we are shaped, or not, into subjects with a "here." This archaic technology is, of course, language. But it is more than that. It is the situation of face to face bonding entered into by, in principle, every newborn baby. During this face to face bonding a certain number of exchange mechanisms – of exchange with the other, of his presence in a situation of co-presence – are constructed. Thus the symbolic function is formed in the archaic technology of face to face bonding, during which a here and now can be built for the subject, from which he can summon what he wants of the world, any exterior, any before or after, any exteriority. It really is an extraordinary, even magical thing, the symbolic function. Right here, I can instantly summon any object in the whole world. I can show you anything. I need only evoke a unicorn, say "pink elephants," for them to be represented in your mind's eye, for you to see them. This is the function of representation. So, now, how do the new technologies relate to this? Well, I think that the new technologies – to the extent that they abolish distance – as Marcos noted, allow us to go faster, to exit the here and now more quickly, and more quickly summon an elsewhere and a past. It is a very interesting function indeed, which might be called *jouissance*. I don't know how to translate that into English. It means to go out of yourself, to go see if you are elsewhere; and if you are, to find yourself there. It is an interesting function because we meet the boundary of the here and now, and play with symbolic categories, which procures a lot of pleasure, the vertigo of the here and now, *jouissance,* in other words. This is the point where the subject is ravished from himself, ravished in every sense of the word, enraptured. Rapture involves going outside of yourself, leaving the here and now. We enjoy leaving the here and now. It really is the only enjoyable thing in life, I think. Why? Because we meet, we cross the boundaries of the prison of the present, the prison

BEING HUMAN:

of the here and now. We play with something that is related to vertigo, and vertigo, as you know, can be very pleasant or totally disagreeable. When is it and isn't it pleasant? I'll end with this. It seems to me that it is more or less pleasant when the category of the here and now, the symbolic function in other words, is more or less well-established. When the symbolic function is basically established, if the new technologies allow us to bring the elsewhere and the past into the here and now faster, or allow us through telepresence to leave or go elsewhere, to leave one's self to go elsewhere, then, in such a case, the new technologies promise an increase in *jouissance*. But if the symbolic function is not sufficiently established, if the ability to summon the elsewhere and the past to the here and now, if the capacity for representation does not function, then there is every reason to fear that the new technologies will lead only to new suffering. There is reason to fear that the subject will now become completely lost, without a here and now, completely dispersed and disseminated, in multiple places. In this case, instead of procuring new enjoyment, the new technologies risk procuring great suffering. I would call this a risk of psychotization of society.

TIMOTHY BINKLEY: I'd like to ask Claus-Dieter Rath about this relationship between subjective and worldwide time.

CLAUS-DIETER RATH: One feature of the new technology is the ability to easily conquer distance. Maybe the notion of worldwide time refers to an idea of a kind of global unity. Marshall McLuhan has called the global village an idea, a fiction, an image, a dream of everybody being together, and of things happening in a united world which are decisive for my subjectivity. I doubt we are living in this kind of world because I think we still live in very local spaces and contexts. It may be that it refers also to events, that something can really mark us in the sense of what Monsieur Dufour has argued here.

TIMOTHY BINKLEY: Distance communication among us always involved the privileging of certain senses over others, and for many people, digital technology realizes the ideal of total communication.

Even if cyberspace communications do not engage the entirety of the body, only parts of it. Two senses are privileged over the others, sight and hearing. Does digital communication promise the overcoming of sensory limitation? Are there any real or imaginary consequences of the central role required by sight and hearing in the domain of communication?

CLAUS-DIETER RATH: Normally, we assume five sensory faculties: sight, hearing, smell, taste, and touch. Electronic media mostly involve sight and hearing, and to a small degree touch, because we touch the keyboard, for instance. But in terms of extensions of senses, smell or taste are still very limited. I don't know if it would be desirable to develop them further. We can see that the eye has become the most important. We have to take into account that the normal looking space is the space between my eye and the screen, which is either a computer screen or a cinema screen, or television set. And then, from there, we can enter other dimensions, but at the same time, only through a limitation that is linked to this kind of extension.

TIMOTHY BINKLEY: Dany-Robert Dufour, is total communication possible?

DANY-ROBERT DUFOUR: I was talking earlier about the archaic technology in which we are formed into subjects: orality, the fact of speaking to someone who answers you. Let's see how a certain number of exchanges take place, involving, notably, feelings about the other's presence, which is not, after all, such a small thing. So, as for communication, in this communication that is, after all, a mode of communication, what stands out, and literature is very informative on the subject, (I am thinking of writers such as Bataille, Blanchot, Beckett and, surely, many others) is that we are never able to say what we want to say. We say too much or too little. We are never tuned to our communication. There is always some kind of surplus. We miss the mark or we can't get it out, and something is always left over. So the fantasy of a total communication would be the promise of a fine tuned message where we say what we want to say. It would be fine if we could say

■   BEING HUMAN:

what we wanted to say, but the problem is, as Marcos pointed out yesterday, that if we cannot say everything, if we say too much or too little, it is because there is another text. There is a text beneath the stated text. There is a subtext, if you like. Or there is an archtext that never stops acting like a parasite to the text of our statements. It edits, it adds something. There is, then, a kind of double text made up of this will-o'-the-wisp pre-text and the text of the edited or expanded message. So the fantasy, the dream of total communication is surely possible, on condition there be no more subject. It would be a message without a subject. I think that there are, indeed, messages without a subject. Mathematics produce propositions without a subject.

TIMOTHY BINKLEY: Let's go back to this notion of the body as a potential obstacle to total communication. Is it an obstacle to total communication, and, if so, is total communication any way to overcome that obstacle?

DOUGLAS TRUMBULL: I think it's fair to say that audio communication, particularly digital audio communication has enough bandwidth to closely replicate what human beings can hear throughout the audio spectrum. So that's quite accurate, and typed or written information is pretty accurate. Whatever you write down is pretty much what shows up at the other end. It may be open to interpretation. Visually, we're into an era of tremendously increasing bandwidth, which, via bandwidth, is allowing us to transmit images of greater and greater technical clarity, and the resolution of an image does in fact convey more information. We're also, both in movies and the world of computer graphics, entering a world of three-dimensional representation, where we use both of our eyes, and therefore we see solid space rather than two-dimensional space. This also adds a tremendous amount of information to the exchange of the content. One of the most interesting things I've come across was when I worked on a movie called *Brainstorm* in the early '80s which had to do with the idea of increased sensory exchange, via actual neuronal content. I had no vocabulary for it, but I recently met a Dr. Michael Persinger in Sudbury, Canada, where he has a laboratory. There he's, in fact, doing what I was talking about in the movie *Brainstorm,* and he's in the process of decoding and encoding certain brain functions to where he can, with the flip of a switch, make you see God, or the white light, or have an orgasm, or feel good or bad, or be depressed or euphoric, or sense a presence behind you. And his work is really very interesting to me because it's easily encoded. It's easily transmitted. It doesn't take a lot of bandwidth. It's just the manipulation of pulsed magnetic fields via magnets around your head, just like an earphone. You could do it with something. The technology's quite simple. It's just a matter of coding. I think it indicates to me that there is a very near future possibility that you could send emotional information over the Internet, direct emotional experience of some kind, as an augmentation to the visual and acoustical kind.

TIMOTHY BINKLEY: Claude Rabant, is the body an obstacle to communication, and is total communication possible?

CLAUDE RABANT: First of all, I think that if we did not have any obstacles, we would have no motive to communicate. The theory of communication is the overcoming of obstacles. From this point of view, I don't think we have to worry. There will always be enough obstacles so that we will always want to communicate. Personally, I don't understand the term "total communication." I don't understand it because I don't see how a communication could be total, except, eventually, in two ways that have always appealed to my imagination, I must admit. I mean the possibility of teletransporting yourself physically from one place to another, of being able to phone yourself. Now, that would be true total communication, and, after all, one can imagine it happening some day. I don't see why it couldn't, in theory. What would be interesting about the telecommunication of the body itself is that it would make transparency and opacity completely coincide. That would be worthwhile. It would be totally opaque and totally transparent at the same time. That would be a true total communication. I would like it to exist. Otherwise, we are always caught between transparency and

■ BEING HUMAN:

opacity. We always have some transparency and some opacity. They mix, and this is why we are never happy. The other type of total communication that could exist, is total autocommunication with our own brain. If we could auto teletransport ourselves into our own neurons, that too, I think, would be worthwhile. By auto teletransporting ourselves into our neurons, we could achieve immediacy regarding our own processes. Ever since I was a kid, I have dreamed of being able to continue reading with my eyes closed, to continue listening while falling asleep. For me, entire dramas can be played out from the point of view of communication, of not being able to read while I fall asleep. I am forced to start dreaming in order to continue reading. That is an extreme form of opacity *vis-à-vis* oneself. Anything that can bring us a little closer to this autocommunication, this way of autotransporting oneself in space and time, is worthwhile. In terms of this, I don't understand why distinctions need to be made, radical distinctions between new and old technologies. As I see it, there is a change of scale involving important changes. But for me, these are not qualitative changes. In this context, let me invoke an often forgotten name in the history of thought, one importantly related to our topic, Spinoza, who said that mathematics is the body of God. After all, what we are seeking is the body of God.

TIMOTHY BINKLEY: Well, on that note, I'd like to ask Seth Shostak, as someone who's trying to communicate with we're not sure quite what, what your thoughts are on this.

SETH SHOSTAK: I don't know what total communication means, either. It sounds like pure energy. I don't know what that is, but it often happens in science fiction movies that you have discussions of pure energy. For myself, I like the impure kind. Total communication, I think, there are some problems. One is, and this was brought up in the beginning, why is it that sight and sound dominate? It's only because that's where most of the information is. Doug Trumbull has already pointed out that we can reproduce sound very accurately to the point where you may not be able to tell whether you are listening to a recording or to someone speaking in the room. Is it Memorex, or is it Cousin Marvin, right? But that's not true with television, even though television brings 200 times as many bits per second into your living room, as your hi-fi system does, it's still a very imperfect representation of reality. Doug works on Imax films, they bring 2,000 times as many bits per second as your hi-fi system does to your senses, and it's still imperfect. You would need something like a few hundred thousand times so that the screen would be indistinguishable from what you see with your eye. Now, if you could make this kind of a system, and I'm sure you can within the next twenty years, then you would have reality there, you would go to a theatre, if that's where you did it, or you would put on your goggles at home, and what you saw in front of you would be so realistic that it would be indistinguishable from being wherever it is, with your eyes. So that might be interesting. The other senses have been tried, there was an attempt to make smello-vision about thirty or forty years ago. It's very nice to have smells but they don't convey very much information in the sense that Shannon would describe information. And there was a problem because you would have one smell in the theater and you would have to get it out to get the next smell, and so you have this scene in a garlic pressing factory, and then you move to a love scene, and you can't get the garlic out fast enough. So there are technical problems there. If you had tubes in your nostrils, perhaps they could be solved. That hasn't been done, but those are just very practical reasons why we have that. But total communication in the very unsublime sense that I use it here, that you can represent reality with perfect fidelity, that's just a technical problem. We'll solve that in twenty years.

TIMOTHY BINKLEY: Oh, I'm happy to hear that, as someone who's very much involved in trying to solve some of those technical problems. I'd like to shift our focus a little bit toward art, and take a look at how the new technologies are transforming the relationship between the artist and his or her product, as well as the relationship between the audience and the art. As an example, new technologies allow students

■ BEING HUMAN:

The Technological Extensions of the Senses   115

Seth Shostak – Being Human Colloquium, New York, 1996

Douglas Trumbull – Being Human Colloquium, New York, 1996

THE TECHNOLOGICAL EXTENSIONS OF THE BODY ■

to produce computer-generated music without ever practicing an instrument. And this has led to the proliferation of novel musical productions by people who don't know how to play anything. Since the machine makes the music, anybody can be an artist, composer, or performer. Richard Teitelbaum has remarked that nobody practices anymore. Could you maybe expand on that and give us some more thoughts about the idea of practice and digital technology?

RICHARD TEITELBAUM: Just a couple of things that were said that I would like to refer to. There have been a couple of references to archaic technology and not separating the technologies, and it made me think about the question of the possible uses or the influences of what we could learn from really archaic technologies which in some ways are being lost and forgotten. And a lot of things that were said about getting outside the body, and certain kinds of communication, are things that, really ancient technologies like magic, you know, so-called magic, trance, from these so-called primitive societies have to offer us, and I think that maybe one shouldn't just disregard the possibilities of incorporating those kinds of ways of thinking… maybe they're not so different from the technological, the electronic and digital realms. Now, what was your question?

TIMOTHY BINKLEY: Has the way in which the body has been involved in the production of art changed, do we have a fundamentally new kind of creativity with digital technologies, which don't require necessarily the kind of practice that has been necessary in the past?

RICHARD TEITELBAUM: Well, that's a very big question, I must say. I think it's quite conceivable that the way things seem to be going, the kind of highly skilled, let's for instance say musical practice, a violinist practices eight hours a day for twenty years, maybe that's going to become less and less of a common phenomenon, and maybe it's going to be relegated to a smaller group of specialists. That seems to be what's happening in terms of orchestral musicians. Orchestras themselves seem to be dying. I have mixed feelings about that. I think that partly the orchestras have been so reactionary in their acceptance and

■ BEING HUMAN:

promotion of contemporary music that they've sort of deserved to die. That's one way I feel about it. On the other hand, it's a shame to have the great traditions of the past disappear and be replaced by computer games. So, trying to find some way to use these new technologies in a way that transcends the very limited applications that we see at the moment is the real problem ahead of us. People like Douglas Trumbull and others are trying to deal with that. I guess my plea would be that the industry should try to hire, or certainly make use of work of artists in this regard, and as much as possible not be totally restricted by commercial concerns.

TIMOTHY BINKLEY: Yes. Michael Groden, you deal with a rather different kind of medium which doesn't involve practice, at least not in the same sense. Could you comment perhaps just a little about how these new technologies are changing literature?

MICHAEL GRODEN: Yes, in a couple of different ways. The talk about total technology seems important. Elvis Costello in one of his songs says "I talk to myself, but I don't listen." And it seems to me that communication cannot, of course, only be in terms of the message that's sent out, and literature won't exist if it ever stops being miscommunication. I mean, it has to be misinterpreted. When hypertext fiction first came out, one of the ways in which people tried to advocate its possibilities was in terms of the fact that the control the author exerts over the text is greatly reduced or even disappears, and the reader becomes the producer, the creator of the text, which was one way of getting around the dilemma of the possibility of total communication. My experience of the fiction, and that of almost everybody else I've ever talked to who's tried to encounter hypertext fiction, is that it just doesn't exist this way. The kind of fiction where you choose what screen you'll move to next, and, in a sense, always create the sequence of material that you'll read, in some ways heightens even more the sense of the other person, that other body trying to say something to you. What's being said changes quite a bit, not just the content but the structure of it, because you're trying to gather how these things relate together, as well as the material. Even this new technology which promises to change dynamically and drastically the relationship between sender and receiver, I don't think quite does that. It changes it. But it doesn't eliminate the problem. In some ways it restates the problem in different terms. I don't think that's going to change.

TIMOTHY BINKLEY: Some people feel that the very notion of art has been deeply transformed by its relationship to technical innovation. Charles Traub has proposed a redefinition of the notion of the artist as a creative interlocutor for the users and he has said that today originality resides with the receiver rather than with the creator. In what ways does technology fundamentally change art?

DOUG TRUMBULL: I can only give you my own limited angle on part of that. Many years ago, I read a really terrific book called *Four Arguments for the Elimination of Television* by Jerry Mander. In one of the sections of the book he talked about the narcotic aspects of images on an electronic screen, and I think that is a profoundly important aspect of the fact that we are seduced into looking at CRT screens endlessly, whether it's television or hypertext or the computer graphics or games or whatever, and that, in fact, a very large part of that seduction is the flying spot, electron beam phosphor illumination that sends these huge bolts of electricity back into your cerebral cortex, your visual cortex, and has nothing to do with the content. And so when we talk about this electronic communication, I think, we should never lose sight of the narcotic aspects of technology itself, completely separate from content. It has a very deep effect, because you may be spending lots of hours doing something, but you may be not getting lots out of it.

TIMOTHY BINKLEY: Claude Rabant, do you think that the very notion of art has been transformed by the new technologies?

CLAUDE RABANT: Yes, of course, I think things have been greatly transformed, but to measure it, you also have to measure to what extent the Western artistic domain was incredibly limited in its processes for several centuries, and also in what way this domain was coded, even its instruments, in terms of a certain representation of the body.

■ BEING HUMAN:

We don't have time to go into detail, but I think it would be very interesting to relate these things to each other. For instance, in what way has the Western body produced and been produced by a certain type of artistic practice, a certain type of limitation to the notion of the instrument. In the case of classical dance, for instance, we might think of the way classical dance was put into place at a certain time, linked to a certain type of power, linked to a certain type of political organization, e.g., in France, Louis XIV, etc. How did this determine a certain type of discourse, a certain type of body, a certain way of inhabiting space? Now, what would seem particularly interesting in the spirit of what was brought up earlier, about music, would be to examine… Well, it doesn't start with us. It starts at the end of the nineteenth century with lots of people, with Kandinsky, a little later with Picasso. It wasn't yet what you would call new technologies, but they were already new artistic practices, new modes of writing, of producing body, space, with modern and contemporary dance. We might examine and see to what extent, as the expressions of one or another technological manifestation open, multiply and deconstruct artistic space, to what extent these breaches, these deconstructions give access anew to very archaic things, so-called primitive music. Does this mean that an entire body of very ancient, ancestral, forgotten knowledge is, in part, made accessible and comprehensible once again, for us and by us, through the so-called new technologies? The impossibilities of comprehension that we have created, the very forms we have produced, at the same time the abstract forms and the concrete forms that the instruments are, all these forms have limited our possibilities of understanding. What we reopen is the understanding of practices and bodies of knowledge that have existed for a long time. For instance, the knowledge embodied in Hindu philosophy or in Hindu practices. A reproach that might be made to the West is to have needed such heavy machinery, as these technologies sometimes are, to be able to understand anew things that other populations have known, in a very simple way, for a long time.

THE TECHNOLOGICAL EXTENSIONS OF THE BODY ■

PAOLA MIELI: I would like to address this question to our moderator, because he's involved in production himself. What do you think about Charles Traub's comments on the fact that originality resides with the receiver? What is your idea about the possibility of a redefinition of the notion of training in the domain of art these days?

TIMOTHY BINKLEY: Certainly, some originality devolves upon the receiver in interactive art. I think that the artist still has to be very creative in interactive forms, but rather than in traditional analog forms, where artists were focused on putting their visions into physical objects as a way of preserving their ideas for posterity, and I think now the focus is on creating environments and processes, so I think the focus is different. I think the artist still has to be creative, but rather than putting something out there which passively sits awaiting a viewer or user, the artist is engaged in something that's very interactive. And I think that one of the big changes that has taken place relates to this notion of scale, and because of the change in scale in art, I think we finally have come to an art that is truly, or can truly be sublime. Cott defined the sublime as that which is, or, at least, which appears to be infinitely great, and I think that is the kind of experience we have in front of these computation machines, which are capable of doing things so rapidly that it seems as if they're doing them in no time at all. A task that would take a human being hundreds of thousands of years is accomplished by a computer in a few seconds.

Break

Afternoon session

PAOLA MIELI: Michael Groden pointed out an issue concerning communication, the relation between sender and receiver, and the way messages are received. Computer and digital technologies promise that the message that is sent will be received as such. Human communication, on the other hand, is characterized by equivocation, ambiguities, misunderstandings. On what ground can we postulate a communication where there is no space for misunderstandings?

CLAUS-DIETER RATH: We can distinguish two types of language operations now in this context. One would be a poetic language that permits you to create metaphor, and to invent names for things, to get very close to your unconscious thought, to invent names for beloved and hated persons and nicknames, to describe things in order to cope with the discontent that is the basis of each and every speech act. It's an address to the other, there is something lacking, that you would like, and the other would be there to understand, to love, to help or whatever. So, we could distinguish from this kind of speech act another one which is in a very narrow sense communication. Communication would be an ideal of total understanding. Perhaps you remember what in the late '40s was proposed by Lasswell as the so-called Lasswell formula, where he talked about the division of these different positions in the process of communication, like the communicator, the message, the medium or channel, and the communicatee. And at this time, he established what has to be understood as communication. He argued that an act of communication between two persons is complete when they understand the same sign in the same way. Here, we have a tendency towards a closure, what we can call in a narrow sense a "code," a code where you have no more poetic dimension, but you have only one meaning of one sign. It is like when you call, as it happened to me, for a flight reservation, and they say if you want the seat reservation, you push the button 2, or say "2." So it's very clear you have one thing to say, and if you say "2 and ½" or "3," or if you say "U2," it would not work. This is in a very straight sense the problem of the code which leads us also to a social problem, a problem of inclusion or exclusion of a regulation of social codes. We know the traditional etiquette which was created at court, in a kingdom, as a kind of system of behavior, of taste and so on. "The politically correct" is a code of speech. It's a limitation of things that have to be done, or not done, or said, and how something has to be correctly coded. Say "2" if you want this. Say "3"

THE TECHNOLOGICAL EXTENSIONS OF THE BODY ■

if you want this. If you have not said "2" or "3" before doing this and this, you risk becoming a transgressor. There is a strong tendency towards talking about universalization and world time, but at the same time the tendency to create a very small and closed group, a kind of auto racism. In the name of personal traits, maybe natural features, groups declare themselves a sort of people, a society they consider far superior to the other ones – a society with higher values, innocent, of course, because the others are guilty, and oneself is the victim. And these social systems of identifications are accompanied by a very coded speech, and coded language. I can add one point where this could be related to the electronic media, because we have the idea of an extension of the senses, at the same time as we have a restriction of the senses. While we have many technical tools made of plastic, glass, and metal material and so on, we have very few occasions to touch and very few to smell and to taste. On the one hand there is this extension of the senses, and on the other a reduction, a separation between bodies. There is an anti-racist slogan in France: "Touche pas mon pôte," "Don't touch my buddy," but we could say "Touche pas ton pôte," "Don't touch your buddy, but touch the key on the keyboard." Cybersex permits a very clean kind of physical contact without blood, sweat, and tears.

PAOLA MIELI: Along these lines, I would like to hear Michael Groden again. Since you are working in fact with words, what do you have to say about the possibility of misunderstanding in digital technology?

MICHAEL GRODEN: I'm out of business if we ever eliminate the inevitability of misunderstanding. If literature can't still exist in the realm of metaphor, and in the realm of the utter impossibility of total communication, it ceases to exist, and then I have to find something new to do. What's fascinating for me in terms of trying to put a literary work into a hypermedia format is that I'm sort of playing both ends against the middle here, because the work itself remains in this realm that I've just talked about. On the other hand, I want to place it into a

■ BEING HUMAN:

context that makes it easier to understand, to take advantage of all the technological possibilities that the computer opens up to us, even though I, as I said earlier, believe that the kind of instantaneous action is dangerous in the sense that it tends to suppress the possibilities, the constant awareness of time. It also of course opens up incredible opportunities in terms of making material available much more easily, and that can make the work easier to deal with. It is crucially important that the computer, besides being a repository for material, a place to store an incredible amount of material and access it very easily, is also a technology that is going to change the way we think, the way we hear and see things. It's going to make differences in the way that our minds and senses work, and that relates to the kind of project I'm involved in, because a work like *Ulysses* or any kind of written material placed into a computer context will end up being different, because it's in a different context and because of the ways in which it's written about and talked about on the computer will differ from the ways it's written about and thought about and talked about in print and on paper.

PAOLA MIELI: Can you already say something about these differences as you see them?

MICHAEL GRODEN: It has to do with the difference between thinking and writing hypertextually and thinking and writing linearly, at least, that's what I assume the differences will be. As we start to have to think and respond to these works in terms of associationally related thoughts rather than linearly related thoughts, the content of those thoughts will change. I can't be more specific yet because in my attempts to do this so far all I've encountered is this sort of incredible resistance in my own mind. I mean, I'm trained to think linearly and I find it so far very difficult to get around that, and to trying to think and write hypertextually. Maybe in a year.

PAOLA MIELI: I would like to focus a little bit on the issue of fantasy here. Is what we call "virtual" reality a way to realize fantasies? In what way did the boundaries between reality and fantasy shift with the application of new technology?

124   Colloquium Proceedings

Michael Groden and Dany-Robert Dufour – Being Human Colloquium, New York, 1996

Mark Stafford, Paola Mieli, Seth Shostak and Claude Rabant – Being Human Colloquium, New York, 1996

■   BEING HUMAN:

DOUGLAS TRUMBULL: Well, I could give a very tangible example which was very interesting to me. I'll start briefly with talking about *2001: A Space Odyssey,* which was a movie I worked on many many years ago, and it profoundly affected me because it was the first time, and actually one of the only times, that I've seen a film I felt I actually entered into, and the fantasy of traveling into space overtook me. I was no longer empathizing or connecting with characters on the screen. I actually entered the screen on this kind of trip. So, that's driven my career and it lead to this rather oddball project called the Back to the Future Ride which is at Universal Studios in Florida and Los Angeles, and this is basically an adaptation of flight simulation in a public venue for entertainment purposes, where we've taken what would be a normal third person experience of watching a movie (*Back to the Future* being a crazy story about a Delorean car turned into a time machine, and you watch the actors on the screen go travel through time). For the purpose of this amusement attraction, the roles are shifted to where you become a character in that movie, and you actually physically enter into a Delorean car, and you're transported via a huge motion picture screen that's virtually all around you. It fills your whole peripheral vision into this motion picture illusion of a trip, which is very sensory and very physical because the car is actually being activated by hydraulic rams which are controlled by very subtle and complex physical motions so that you're not just being stimulated by sight, you're not just being stimulated by sound, you're being stimulated by physical stimuli, and the totality is incredibly overwhelming. The result is that this attraction has been unbelievably successful. There are only two of them in the world, seen by over 60 million people already. It's been very profitable for Universal. My take on it is that it's not about the *Back to the Future* movie at all. It's not about those characters. It's not about that story, but in fact, this ride is successful because it's a huge physical manifestation of an out of body experience. The ride is a fantasy, it is dream come to life in your conscious life, and it has a very powerful effect. Everybody gets completely euphoric when they have this experience. It's only four minutes long, but it's very intense. And so, I just wanted to give that as an example, and go from there.

PAOLA MIELI: Actually, I took one of your rides. At the entrance to the movie theater, people were asking if there were pregnant women present, because they wouldn't let pregnant women take the trip. The ride has, in fact, a very strong physical impact. What do you think about it, about the realization of such an experience?

DOUGLAS TRUMBULL: My personal philosophy is that human beings, by their nature, seek experiences which are way beyond the literal frozen reality we're in. I don't know if anybody else could speak for this, but I'm daydreaming all the time. I'm not physically here half the time. I dream all night long. I take trips in my mind. I think about things that are not physically real yet, and so I think that human beings actually crave experiences that are beyond the limitations of physical reality, of just walking around, or driving a car, and so altered states, whatever they may be, whether they're hallucinogenic states, drug-induced states, senses of euphoria, orgasmic experiences, or whatever, are very desirable. And I think part of this whole revolution is trying to find ways to achieve some desirable and satisfying result in a socially responsible and non-reprehensible manner, which drugs are.

SETH SHOSTAK: I want to know how progress on the Orgasmatron is going, because…. I don't know. In what ways has the boundary between reality and fantasy shifted with the advent of new technology? In some sense I'm sure they have not shifted. I mean, is the fantasy of a good book inferior to the fantasy of computer animation? And I'm sure most of you would say no, fantasy is not in the hardware, fantasy is in the software, and I mean in the software of our brains, that's where fantasy arises. So, of course, our brains change very slowly. Well, mine's changed a lot in the last few years. It seems to have ossified. Anyhow, there is this shift, but I don't think that at some fundamental level, I don't think that technology has much to do with a shift in fantasy. We have new ways of representing fantasy, that's all, but there is this difference, that the advance of technology does make some things real that were only fantastic in the sense of being not real.

PAOLA MIELI: Dany, what do you think about this?

DANY-ROBERT DUFOUR: I don't know. I don't know much. In the encounter between two people in the archaic technologies I was talking about earlier, it seems to me that there is always an interposition between each party of something called fantasy. This means that we never truly meet the other because we impose a screen that is sometimes opaque enough to allow no good to pass through. This is how we can think that beautiful women are not beautiful, which is an error, or that women who are not beautiful are beautiful, which is also an error, because we interpose another dimension between the two, the dimension of fantasy, which distorts the other person. So, it seems to me that with these new technologies, we are no longer hampered by the other. Since only the fantasy is left, which can be projected or modeled to our liking without any interference from the potentially bothersome other, what we have is masturbation.

RICHARD TEITELBAUM: My notion of fantasy has more to do with subjective fantasy, so in a way I have trouble relating to the idea of even calling the kind of ride that Douglas was describing a fantasy. I've never done it, so maybe I'm missing something and I should, but I always got sick on roller coasters and things, so I've avoided that sort of thing. To me, what's more interesting is to use the technology to uncover my own fantasies or the fantasies of someone I'm communicating directly with. I don't mean to be negative about this. I'm interested in using technology to know what I'm doing when I improvise or I'm communicating an E-mail, or in chat rooms, like cybersex, for instance. If we're going to talk about cybersex, the fantasy aspect of that seems to be one of the more interesting things. You never know whom you're talking to, you never know if they're acting out their own fantasies, you're free to act out your own fantasies, so these kinds of uses of the technology I think can enhance personal fantasies, and that's the kind of thing I find more interesting, perhaps.

THE TECHNOLOGICAL EXTENSIONS OF THE BODY ■

Colloquium Proceedings

PAOLA MIELI: You speak of a self-reflective process of creation in the production of your own music, could you say something about this?

RICHARD TEITELBAUM: What I do is live interactive computer music. I usually start from a completely improvising situation. I have no script, no score, and I play something which I assume is generated by my unconscious since it's not coming from any objective directive. And then, I use the computer to capture that and flash it back to me in various guises, transformed in various ways, musically delayed and processed, transposed, inverted. And I kind of look at that as a way to look into my own mind and establish a kind of dialogue with my own unconscious. There are other people who use the computer, doing computer music, who have a paradigm and rather than thinking of it as an extension of their own internal processes, they try to create an artificial player. Yesterday, there was talk about artificial intelligence. It's akin to artificial intelligence, but that seemed to be a term in ill repute yesterday, so you might call this artificial creativity, although that's probably going to open another can of worms. Anyway, the idea there is, that they create a musical personality which is generated obviously from their fantasies about what such a personality would do, and that personality listens to the way they perform and analyzes what they do and then responds with a certain degree of unpredictability, a certain degree of roundness, a certain degree of unexpected behavior. But, to me, what's interesting is to establish a kind of a threshold area where you're just on the edge of being able to understand what the responses are and how they relate to what you're doing.

DANY-ROBERT DUFOUR: About music, since music has been mentioned several times this morning and yesterday. Shortly before leaving France, I was shown something that really impressed me, and I really don't know what to make of it. When something like this happens, I generally share it with others and see how they react. So let me try this with you. I saw two musicians, obviously crazy in the positive sense of the term, "gone" into a totally extraordinary experience. They had

■ BEING HUMAN:

invented an instrument called, quite simply, "the meta-instrument." So, what is this meta-instrument? I don't really know. But it is a kind of armor covering the forearms, and having a number of keys sensitive to the touch. There were two of them who had this instrument. On each arm this kind of armor, and then, I think, there were pedals on their feet. In front, I think there are thirty two parameters that can be parametered by the different movements – up, down, right, left, rotations, etc. These musicians hold advanced degrees from the Conservatoire and are also programmers who have invented this meta-instrument. They create a sort of sound object by waving their arms, raising and turning. They build a phonic frame for the listeners who are in the middle. It is very impressive visually because you have the impression that the movements of the body are a simultaneous translation, and create this sound object that becomes very tangible. I am not saying that it is visually perceptible, but it is felt. It takes up space. It advances. It retreats. It swells. It turns right. It stretches. It splits. It breaks into four parts, ten parts, five parts. It regroups. It returns. So I was truly intrigued by this, because it obviously worked, and it reminded me of a number of major philosophical ideas, notably about passions. As you know, there are philosophical treatises on the passions from Aristotle to Descartes. Something that could express an intelligence of the body, the intelligence at work when the body is functioning in the closest proximity to the expression of its passions, which we could call affects, if you prefer…so I was struck dumb by this experience. I don't quite know what to think of it. In any case, I think there is some kind of immediacy there. A last point. It relates to what Claude Rabant was saying earlier. I think that this is the invention of a new musical instrument, because until now the instruments that have been invented are true instruments of torture. To play the violin well, you have to twist your hand into unbelievable positions. Piano, cello, you always end up having one leg longer than the other, one foot longer than the other, callused, deformed fingers, etc. It is true that you can end up sounding good, but not always. I've tried without success.

THE TECHNOLOGICAL EXTENSIONS OF THE BODY

Whereas with this new type of instrument, it isn't the body adapting to the form of the instrument, but the form of the instrument that weds the body in a certain sense, allowing the body to generate.

PAOLA MIELI: There is a tradition in the history of music of this sort of invention. The Tethramin synthesizer for instance. It's absolutely fascinating. But I would like to go back to a couple of remarks made here. Douglas raised the question of subjective responsibility in the production of entertainments, Richard pointed out how cybersex and the erotization through the machine brings about the question of the relation between self and other. In his latest book *Folie et Démocratie*, Dany-Robert Dufour remarks how, according to his judicial rights, the subject of American democracy is structurally supposed to know nothing. This is the way he puts it. For instance, he can smoke two packs of cigarettes a day for twenty years, sue the tobacco company for having created the cancer, and win. She can drink a bottle of whiskey while pregnant, abort the next day, and subsequently win a lawsuit against the liquor company. He can put his little wet dog in the microwave to dry him and win a suit against the microwave manufacturer after his dog explodes. Is the subject of American democracy supposed not to know anything?

SETH SHOSTAK: I'm not sure why this is a question of democracy. I'm not sure how democracy figures into it. I think any society that has a critical mass of lawyers will have the same problem. I don't see this as a black and white issue. This is a matter of where is the equilibrium between collective and individual responsibility. The fact that in the United States it's shifted more toward the collective responsibility, in other words, if you decide to make yourself a hotdog by putting your pet Pekinese in the microwave, yes there are lawyers who will defend you against the manufacturers. But this is sort of an oscillatory situation, even in American society, and I think that the pendulum is in fact, swinging the other way. There have been numerous initiatives in the state of California to limit the ability to sue companies for this. The recent suit involving the woman who spilled a hot cup of coffee from

McDonald's into her lap and then claimed that she had these terrible burns. What was the consequence of that? Well, there was enormous uproar from people who liked hot coffee, because now McDonald's was going to serve their coffee somewhat cooler. There was tremendous joy among certain members of the legal profession. But the result of this is just finding an equilibrium, the price that's paid for allowing the individual freedom, of course, is to moderate what you might call the march of progress. Probably, the march becomes a little slower, and are you willing to pay that price for a certain amount of collective responsibility? I give you an example, a concrete example here, not involving a faster than light concrete mixer. A concrete example is breast implants made by Dow. Those of you who live in America know there's been a long history of legal action against Dow chemical because of these implants that made for more erotic encounters, if you had them done. Some women claimed that they made them sick. There is actually no clinical evidence that these things did make them sick. Well, someone may contest that, but that's my understanding. But the undeniable consequence is that Dow Chemical and other manufacturers are not making the materials that are used in other areas of medicine, making it very difficult for medical device manufacturers to make leads and tubes and so forth, that would be useful to you in some cases in medical treatment. So, that's the price that you pay for this lawsuit. There's a slowing down in progress, and it's only a matter of society deciding how much they're willing to pay for this.

MICHAEL GRODEN: In terms of some of the places I know these questions are leading, the computer has this obviously double direction. It can go in all different kinds of things. One is to increase the level of ignorance, drastically. The other is to potentially change the way in which we gain knowledge, and the way we process knowledge, and all those kinds of things. I sort of agree with what Seth has just said. A lot of the conditions that have created the subject supposed not to know have simply to do with material, and the way in which knowledge is gained, and in which knowledge is processed. The possibility of being ignorant, of claiming ignorance, those possibilities will be greatly increased through the technology.

DOUGLAS TRUMBULL: This may seem naïve and an oversimplification, but it seems to me that democracy comes hand in hand with a free market economy and capitalism. And capitalism, in my mind, always leads to a tremendous increase in the power of corporations, and a desire among corporations to keep people stupid consumers, and we are constantly barraged and inundated with unbelievable levels of sophisticated advertisement to sell products which we don't need, and it's that economic factor that's really driving everything. I personally experience a huge degree of frustration, disappointment, disillusionment, lack of morale, and self-esteem because I don't feel like my work is really seriously appreciated. I come to colloquiums like this because I'm truly interested in the scientific and humanistic aspects of the work I'm into. I'm not into this because I'm into entertainment *per se,* but the entertainment business is the only place I can get the dollars to do my work. I can't find a university that has a hydraulic motion base that's moving people around, trying to study the effects of physical motion on human subjectivity. It's all about money, it's all about commercialism, so I would say you can't just isolate democracy as an item. It comes with a lot of baggage.

PAOLA MIELI: You raise an important issue that relates to the question of how to influence the industry in order to develop technologies that could be fruitful for both consumers and producers.

SETH SHOSTAK: Well, the problem is that the only thing they seem to be interested in is the bottom line and I guess if you could convince them they could make money on it, they might be willing to do something they might not otherwise, but that seems kind of a long shot. I mean, the ideal would be to convince them to be socially responsible, and since the current trend is to eliminate government subsidies for these kind of developments, everybody says. Well now, industry and private individuals are going to have to take this up. But there doesn't seem to be very much evidence of this. There were attempts to put in laws like one percent for art kinds of things, and they don't seem to be happening, and I'm kind of pessimistic, as Doug seems to be, about what can be done, but this is a really fundamental question, which we haven't addressed at all.

TIMOTHY BINKLEY: I'd like to add just one other footnote to this, because I think that one thing we haven't really talked about much is the way in which the economy relates to the technology. We're not talking today simply about technological innovation, we're talking about the economic potential of bringing that technological innovation into the hands of the general populace, and in my own field it seems to me that that's really a major factor. The reason artists are using computers in such great numbers today is not simply because graphics computers have been invented and software has been invented that can do certain things. It's also because the economic structure, and the corporate eco-structure that exists has in fact found a way to deliver the technology to us at a price that we can afford. So, I think that that's an element that needs to be put into the equation, and I think that of course there are certain responsibilities that we want to place on corporations, but I think that we also need to realize that it is the ability of the technology to come together with economics that makes all of these things possible.

DANY-ROBERT DUFOUR: Previous responses to the question of the relationship between the subject and democracy have focused on political and economic issues. I think that there are also questions relating to the kind of subject produced in these societies. What I mean is that there are questions having to do with, you could say, the law. What shape does the law take in democratic societies? This question, as you know, is not unrelated to the symbolic or symbolization, because that is where the law is expressed. Now, I think that before the mass democracies that came into existence after the Second World War, the subject was always defined by another, by something else upon which it was based. I am, of course, being very cavalier, but to survey history from the Greeks to the present day, I would say that the subject, in former times, was grounded in, founded on, in Greece, in the Greek world, a relationship to nature, for lack of a better word. The forces of nature spoke, expressed themselves through oceans, volcanoes, earthquakes, geothermal activity, etc. Oracles, who could tell you the meaning of these events, were located in propitious places. Later, the subject was defined (I'm moving very quickly) through a reference to God. We were all subjects of God, of the God of Monotheism, with, obviously, many variants. Then, subjects of the King, subjects of the Republic. That one is an invention of the French Revolution. "Subjects of the Republic" means the creation of a common space. This subject is responsible to collective space. What happens with the democracies is that this legal question shifts to the level of the subject himself. You could express this by saying that the subject we are dealing with in the democracies is the subject who has the historically absolutely novel characteristic of being legally autonomous. He has only to answer to himself for himself. This obviously means a very important gain in freedom. But it also means addressing responsibility in an entirely new way. If the subject is self-defined, with a loss of reference to another level, collective or divine, then new questions arise. The subject doesn't know anymore, because when you ask yourself what you should do, you're in the situation Claude Rabant welcomed, where you can see your own neurons working. So there's an immediate loss there, I think, as far as knowing what to do is concerned. You no longer know what you're doing. You no longer know what you know. And this is the subject I think we're dealing with in the democracies. So how does this (I'm obviously going very fast) relate to the new technologies? Because the new technologies create a space, a network, which is a space of immanence. It is a space where the subjects are interrelated in an eternal present. The subject has exited temporal space, which was both prehistoric and historical space. We are dealing with a kind of post-historical subject. We are post-historical subjects outside of temporality, living in an eternal present, because every place in the world is immediately accessible through telepresence, and also able to visit us immediately at home. So questions of responsibility need to be formu-

■ BEING HUMAN:

lated in terms of this space of eternal presence. Responsibility implies a distance, supposes that I am accountable to someone else. Of course, the stronger the immanence, the less I am held accountable by others. Therefore, I think that responsibility can only decrease globally. I'm making a prediction. Responsibility can only weaken globally because of the growth of the legal autonomy of the subject, pulling us along into that strange post-historical space.

CLAUS-DIETER RATH: I would like to add something. Liberty is one of the democratic values, and we have also, of course, fraternity and equality. And this is one of the central points of the question when we deal with new techniques. New techniques become new technologies and become technocracies. We have the relation between democracy and the technocracy. Technocracy is often exercised in the name of equality. It's the question of standardization. We have to have the same chances, and we have to be treated in the same way, and everything is standardized. This is one of the very big challenges of the relation between the subject and equality, one of the democratic ideals and, on the other hand, the relation between this and the question of desire. When we talk about extension of senses relating to knowledge, I think we can talk about the extension of knowledge, but also about the extension of defenses against knowledge. We should acknowledge that sense and senses have at least three different meanings that we should take into account – the sense as a "door of perception," as doorkeeper, like eyes and ears; the sense as feeling, as in "no sense of place," or "a sense of loneliness," or of the uncanny. And the third meaning is sense making, and its lack in the case of senselessness. In these last two dimensions, we have something which can be a promise and a belief concerning media techniques, taking them as a perfect being, as an absolute reference, as a God (God was mentioned before) where all the knowledge is incorporated, a kind of obese archive which I can refer to, whenever I feel discontent. But, the big question is, and I end with this, by accepting this offer, do I avoid my own search for my own desire and my own feeling?

CLAUDE RABANT: I might digress a little in terms of what has just been discussed by paying homage to a work by Michel de Certeau called *l'Invention du Quotidien*. In my opinion, this book deserves to be better known. It warrants rereading because, at the time it was written, in the early eighties if I'm not mistaken, right before his death, Michel de Certeau had started to work on questions that happen to bear a great deal on the problems that arise anytime the theme of democracy is brought up. In particular, whenever codes, dominations, powers, ways of directing individual behavior are put into practice, all kinds of subterranean subversive behaviors develop that aren't necessarily visible. This means all the singular, unwritten, unpredicted, unsaid, invisible behaviors of individuals, groups, etc., that subvert what they have been prescribed. Ultimately, there are all kinds of non-prescribed, more or less subterranean possibilities of using what is offered differently from what a supposedly all powerful prescription intended. This is to say that it is always possible and effective for, not only individuals, but also small groups, micro-societies, sub-associations, all sorts of marginal networks and sub-networks, to respond to the hypertext with a hypotext. To return to what Claus Rath was saying about the two kinds of language, poetic and informational, I think that there are always a certain number of possibilities, of pathways, to reinvest informational language with poetic capacity. You can always resubvert, at a certain cost. I do think it is possible to reinvest very powerful, dominant languages, with poetic capacity. It isn't always easy, and involves sometimes lots of losses and sacrifices, etc. From that perspective, the term "psychotization," which was used earlier does seem to have a dual aspect. It is true that there is a latent psychotization in our societies. But I do think that sometimes psychotization is also a response, a defense, against the powers, the forces that alienate us, the way that, in Freud's time, hysteria had a kind of sufficient response capacity it no longer has today. The forms of hysteria, which could do so in Freud's time, can no longer defend us against power in the long term, and in effect, more extreme forms are now needed, deeper and more shatter-

■   BEING HUMAN:

The Technological Extensions of the Senses   137

Claus-Dieter Rath and Richard Teitelbaum – Being Human Colloquium, New York, 1996

Seth Shostak and Claude Rabant – Being Human Colloquium, New York, 1996

THE TECHNOLOGICAL EXTENSIONS OF THE BODY ∎

ing. In a certain sense, they have more to do with psychosis. But they are ways as well, and psychotics are people who have a lot to teach us about the perception of these undergrounds, these resistances, these shapeless creations that are also everything that is a creation without code, without shape, without object, another type of creation than strictly formed and coded creation.

PAOLA MIELI: I would like to return to the issue of the subject relation to the other from a different angle, asking Seth Shostak, director of SETI, Search for Extraterrestrial Intelligence, to make some comments on the project *Phoenix*, initiated in Australia in 1995, the most sophisticated program for the detection of extra-terrestrial life. Of course this project raises all sorts of interesting questions, for instance, how would the scientists respond if they got a sign from extra-terrestrial life.

SETH SHOSTAK: I'm not going to answer the question, that's up to everyone else, of course, but maybe just take one minute to set the stage. What SETI is about is just using big radio telescopes. They're nothing more than large antennas to try and eavesdrop on any radio traffic that may be coming to us from other thinking critters in the Milky Way galaxy. That's the experiment, very simple. We haven't heard anything yet. I know that many of you believe that the aliens are not only out there, but they've also come here, and they may be abducting your neighbors for salacious purposes. Many of you may wish that this will happen to yourselves. However, from a scientific point of view, there's no evidence for any of that. I have no doubt that they're out there, but we have not heard them yet. But the experiment that we're running today is about 100 million, million times more sensitive than the first experiment that was run thirty years ago, and that gives me personally some optimism that in the lifetime of everybody in this room, unless some of you are suffering from a malady that's especially aggressive, we will hear a signal from some company in the galaxy. So, you may wish to address the question, if we do hear something, what might we say back, if anything, so that's the thing.

■ BEING HUMAN:

PAOLA MIELI: Could you describe a little bit the proposals for such an answer from the different scientists, researchers involved in your project?

SETH SHOSTAK: There have been suggestions – this will prejudice, of course, the answers, I'm afraid. Some people say we shouldn't reply anything, why tell them that we're here? They might come down and take our chlorophyll, or our women, or whatever. I wouldn't mind if they removed the chlorophyll. No, I like salads. But we are already letting them know about us. *I Love Lucy,* after all, has reached approximately 500 nearby star systems already. So they're probably judging our civilizations on the basis of that, or maybe *Mr. Ed,* I don't know. There's no point in saying we won't say anything, unless you just want to clam up. Some people have said, look, we ought to have a very carefully orchestrated response. Let's get the United Nations together, and have them decide. Well, you all know that the United Nations is a terribly slow and inefficient organization, I'm not really sure that they should speak for humanity, but that's an idea. Another idea, which I think is more likely, is that yes, the United Nations will be called into session, but of course everyone who has a satellite dish in their backyard is going to swing it around in the direction of this star, and start transmitting their own personal philosophies, or pictures of their dog, or whatever. So, that's a possibility. It has also been suggested, somewhat seriously, that we, the astronomers who find the signal should immediately reply, saying: "We hear you, and by the way, here's the password, here's the secret password, 'bagel' or something like that. And if any message you get in the future does not have this secret password, ignore it, okay, because we're the only ones you really ought to be listening to." If Columbus heard from one guy, "We have a password, don't listen to anybody else," the first thing he would do is say: "I want to hear somebody else." But, those are some suggestions, I'm sure you have better ones.

PAOLA MIELI: Do the panelists have any suggestions?

DOUGLAS TRUMBULL: I don't know how to exactly answer that,

but *2001* is a really neat example which relates to what Seth is talking about. The model of *2001* was an omniscient, omnipotent, solid state representation of where the Internet will get to in the end. At the end of the film, a final transformation takes place. Keir Dullea goes through a completely abstract experience and gets transformed into the star child to return to Earth. Maybe Kubrick was suggesting that this would be the transformation of mankind to another state, some higher level, some other plane, still being a human being, still being physical, but improved in some substantial way. All of this talks about where we're going, our transformation, whether it's our psychological, emotional transformation, or getting rid of our pathologies or neuroses or psychoses, or whatever it is, is ultimately leading to some more enlightened state of existence. And I don't know where that is, but I see a lot of avoidance of actually coming to grips with truly changing our nature.

DANY-ROBERT DUFOUR: About the message and the answer, we are assuming here that we are the ones who should answer. Whereas what is going on here is that they are the ones who don't want to answer us. Maybe they are everywhere, hearing all the messages we are constantly sending, and under no circumstances do they want to answer.

CLAUDE RABANT: I think this is a magnificent question, because it is almost, not quite, the ultimate question. It has been asked for a very long time because people have been asking what angels were for a very long time. Angels are *angelos*, which is message, already. The message question and the question of the extent of the message are there from the outset. How can you hear what you don't hear, and how can the other know what he doesn't know? I'll answer this with a non-answer. I'll answer with a very beautiful text by the German writer Jean-Paul, entitled "The Death of an Angel." It begins this way: "The angel of the last hour, we so harshly call 'death,' is the sweetest angel, the best of angels, chosen to gently and delicately pick the submissive human heart, removing it from life, removing it from our breast without bruising it in its warm hands. His brother is the angel of the first hour, who kisses man twice, the first time to give him life, the second, so that he awakens in the next life, unblemished and smiling, not crying, as he came into this life..." That is the beginning of the text, and the rest, which concerns us directly, is that the angel of death, the angel of the last hour, who brings death to humanity, in fact doesn't know what he's doing. He doesn't know what death is, so he asks his brother to allow him to know what death is, and he is granted his wish. At the last moment, he enters a dying human body, and by that very fact he begins to experience human life. The Other, the one who brings us the ultimate message, doesn't know what he is bringing. He doesn't know what he is bringing us, because, in order to know it, he would have to have made the same journey as us before we had. We again find the division mentioned earlier between opacity and transparency. The ultimate message we are waiting for is, in fact, the message of death. But this message cannot be known directly, because it can only be known at the end of a human life; when we are about to receive it ...we are no longer there to receive it. There is a kind of fundamental, absolute otherness. True otherness is otherness in relation to death itself. This could be expressed by something Freud says about death. Death is recognized by its denial. You can't distinguish the message itself. You can't distinguish creation, life. You can't distinguish how we recognize death from how we deny it. What message could put us in contact with the radically other?

PAOLA MIELI: I would like to know if there are more comments from the panelists.

RICHARD TEITELBAUM: I hesitate to raise this after this very profound and poetic discourse, but maybe I'll risk it anyway. Something that's been kind of disturbing to me about the whole event, maybe this is a question for you, too. It was touched on a bit about techno-elitism. On the one hand, we've talked very much about total communication, global village, and the benefits of technology and achieving this kind of thing, but to look at the makeup of the panels, one would have to assume that being human means being a white male, largely, with you being the main exception, and I'm a little disturbed by the fact that other cultures, societies, races, etc., are not represented, and it seems to me that there is a real danger in this – well, I just made up this term, techno-elitism – that you know the universal language is becoming English, that the universal music is becoming American pop music, and I'm just surprised, really. I mean, I would like to see an Australian aborigine here to discuss the relationship of bush telegraphy to the Internet, or something like that. That may be a little far-flown, but to have no representatives from Japan here when you're talking about digital technology, or no black people when you're talking about these issues. I'm a little surprised, and I'd be curious to know what your response is to that.

PAOLA MIELI: When we first conceived this project and invited various persons to be part of it, we encountered a significant and revealing resistance. It was a difficult task to convince people, technicians from different fields, to come to this forum and have discussions, to look for a common language and common ground, stepping out of the specificity of their domains. They couldn't understand why. They couldn't understand, for instance, why psychoanalysts would be interested in new technologies and why would they want to organize such an interdisciplinary encounter. We were looked at as if we were the "aliens." We contacted many people, women and men, from different countries, of different races and from different fields. The ones who accepted to be part of these working days were, simply, you. Of course, one obstacle was economic. Après-Coup is a nonprofit organization and this colloquium was made possible by small, individual contributions. We couldn't afford major expenses. But *Being Human* is a project. It is a work in progress that doesn't exhaust itself with this colloquium. We hope that in the future more people will join our discussion.

PAOLA MIELI: A question from the floor, yes sir.

From the audience: I welcome the comments that were made by Richard, because for me it brought to the surface a concern that I had listening to these panels. You have three basic themes that you've been working with. The technological extensions of the mind, of the structure of the body, of the senses. I think Richard's latest comment sort of brings to the floor the importance of dealing with the technological extensions of democracy. As I sit here as a minority member of our culture, I feel a concern that maybe technology as we're experiencing it in our culture, is not going to be available to the underprivileged, the non-elitists within our society, people who cannot afford accessing it through institutions, hardware, or software. I just got my master's degree two years ago, and I basically spent $50,000 of my own money to derive the benefit of an extended education. Two years later, I looked back at what I learned, and I said to myself, if I really want to stay ahead of what's happening in the technological field, I really should be going back to school and starting all over again, and picking up the tools and the information that I will need in order to benefit from many of the things that were discussed here yesterday and today. My question to the panel is this: where does democracy really fit in if technology is only going to be accessible to the few and not the masses?

PAOLA MIELI: Thank you for your question that summarizes some of our concerns; the frustration, for instance, provoked in this domain by the gap existing between the offer and the true availability of the product.

SETH SHOSTAK: I'm not nearly as pessimistic as you are. I admit that there's a serious problem there, it's at least recognized. In 1950, there were only ten million television sets in the United States, and today there are sufficient numbers of television sets that people have them as furniture. I'm not sure they even turn them on. It's probably better that they don't. I think the same is going to be true of access to the Internet and other technologies because, as Doug would tell you, there is in fact a market among the entire populace for this sort of access, so it will be driven down in price. People are working on that in Silicon Valley right now. Computers that don't run spreadsheets, they just tune you into the Internet, may cost just a few hundred dollars, the same price as your TV set.

DANY-ROBERT DUFOUR: Faced with this optimism, I feel the need to play the role of pessimist. I think that there is a struggle that began long ago, as I have already pointed out, and in order to pick up on this struggle, yesterday I brought up Nietzschian prophecies made a little over a century ago. I think there is a struggle between what are called the last men and others who are following another path. I think this struggle will become increasingly bitter. I've chosen my side. I'm a last man and I want to remain a man who dies. So, I'll have to fight. It's a fight I've already lost against those who want to break free of limits in time and space, that seem to me to be the sad foundation of history, of art, of our common space, of our common humanity, I would even say. I think that this idea of a common humanity is imperiled now. Too many accumulated inequalities mean that we are in a period of very high risk and at a turning point in civilization. I think we are in a crucial period. And I think that what has happened here over the past two days must absolutely be continued if we want to shed some light on the critical choices faced by our civilization.

RICHARD LEDES, *from the audience*: Psychosis has been mentioned a couple of times. Couldn't we say at the very least that the expectations surrounding future technology have significant similarities to Schreber's fantasies of being able to replicate other people, of being able to create other worlds? I have in mind here the etymological link of the ancient Greek word *techne* to its cognate *tikto*, meaning "I give birth." This etymological link is actually emphasized by Aeschylus at one point in the *Orestia*.

CLAUDE RABANT: In any case, I agree with your reference to this landmark. I think that it is one of those historical or textual sites that can be referred to again and again...Let's say that today we need something beside pure pathology or individual pathology, something that may be endowed with prophetic power, analogous to what was said yesterday about Nietszche or Spinoza. To some extent our own destiny is expressed in these texts. What seems clear is that we are in an era when, whatever positive or negative meaning we give it, psychosis does not occupy the same place it did when psychoanalysis was invented. This requires us to reexamine certain patterns of psychoanalysis itself.

From the audience: I have three questions. The last two I would like to address to Mr. Trumbull. The first, which is very short, is to Mr. Shostak. When we receive a signal from one of our cosmic companions – when, since you say it's going to happen definitely – how can we be assured that we, as a global community, will know this? I mean, can you assure me that SETI's going to release it to the world immediately, or is there a possibility of a global conspiracy, that we better not tell the public?

SETH SHOSTAK: Well, the answer is, you will know. There will be no conspiracy. The American public, which is very fond of conspiracies, I might say, is indeed convinced that there would be some sort of cover-up, that the CIA or FBI would sweep down into the radio telescope, assuming they could find it, and shut us all up. But, let me tell you what's going to happen. To begin with, there's no policy of secrecy. It would be practically impossible to shut up the hundreds of people who would know about it. Secondly, this is what's really going to happen. If I'm sitting at the telescope, or one of my coworkers is sitting at the telescope, and they get a signal from ET, it hasn't been confirmed yet. They're not supposed to say anything until it's confirmed. You don't want to spread a false alarm. But they will send an E-mail to their girlfriends Matilda, or boyfriends. Matilda, don't tell anyone, but we've got the big one here. Why? Because they want their picture on the cover of *Time* magazine, not their boss's. Matilda has a brother who's always been interested in astronomy, so she sends an E-mail to him. He sends 10,000 E-mails to all his buddies. Within three hours, the entire world knows and it hasn't even been confirmed yet. I have seen this in action, I'm sure of it. There is no way it could be kept secret if you wanted to, and we don't want to.

From the audience: Okay, thanks. Mr. Trumbull, we've all known about the subliminal messages, and the technology, film, video, photography. How much of it is actually going on now, and is there a possibility of political, I don't know, brainwashing, or influence on a global scale? Are subliminal messages also possible and are they being used now in an audio sense, not just video?

DOUGLAS TRUMBULL: Well, I don't know the inside word on this, but I've read a little bit about subliminal programming, and I have been led to believe that subliminal programming is underway on a regular basis in a lot of supermarkets, and a lot of retail stores where, in the Muzak that is playing in the stores, are embedded messages. "I will not steal today. I am not a shoplifter. I'm a nice person. I won't do this." I don't know if this is going on, but I've read that this is going on. I don't know of any subliminal material that is going on in any kind of regular feature motion pictures or television shows or anything like that. I know it's very possible. I know it works. I've done it myself as an experiment. Fortunately, my experience is that in the entertainment-media business, the people who are in charge, the managers, the studio owners, the agents, the attorneys, the vice-presidents, don't have a clue about technology, thank God. It's also frustrating because it's very hard for me to do my work, but they don't have any way to manipulate it or understand it. A great thing about the Internet is that the common man can get the message out without much control from higher-ups, or political forces.

DANIÈLE LÉVY: I wanted to ask a question about supermarkets. I have to recognize, or to finally realize that when I am in a hypermarket, which does, after all, have something to do with contemporary technology, I experience some anxiety and a kind of overstimulation. Lacan said in the '50s, that contemporary buildings, housing projects, could only promote aggressivity, because everyone's house is so similar, with the same TV on the same cabinet. Identities or specificities must be all the more violently asserted as they are not written down anywhere. So I tell myself that there is an obvious effect of the new technologies. It is the reaction to standardization, which can only engender aggressivity. Now, that is the mass technology side of it. It doesn't answer the question of the other aspect, which is the frenzy of representation men-

■ BEING HUMAN:

tioned earlier. But I would like to suggest a relationship to something else that was brought up this morning – money. Sir, you said, as did others, that it was absolutely impossible to produce anything that you weren't reasonably sure was going to earn a lot of money. It is always "a lot of money" that is said, never "money" or "enough money," money must increase unconditionally.

DOUGLAS TRUMBULL: Your statement answers its own question. I think you're absolutely right. I do think it's all about money. I find that incredibly depressing and demoralizing. It pisses me off every day and I wish it weren't. And that's why I keep talking about, well, what's the next situation? What's our next level of evolution? And when I talk about that myself, I want to know what's the next political system that's bigger and better and more personally rewarding to me than capitalism and democracy, because I'm very disappointed with many aspects of a money-driven society. We have a society that's based upon images that shampoo will bring sexual gratification, and you know what I mean by that, and I'm just disgusted with it, and I don't think my own sense of joy in life will ever come from a product, or a car, or a social status thing. It comes from some other area and, unfortunately, that's not an area that has profit in it, and so it's not a big part of our culture. It's only part of a subculture.

PAOLA MIELI: Marcos, would you like to make a closing statement?

MARCOS EINIS: I simply want to try to be the least depressed possible when I leave here. Speaking for the French, I want to thank all the Americans for their hospitality, and particularly, for making it possible for us to talk about things we aren't used to talking about. I want to point out that I found this an absolutely exceptional place to be, in terms of how difficult it is to find a place to speak about themes that aren't our cup of tea, that aren't in our habits, and force us to rethink how such things should be talked about. Leclaire used to say that something changes through analysis. It isn't language. What changes is the relationship to speech. I don't want to dwell on the depression caused

Colloquium Proceedings

by our separation. I would like us to say that there will be a next time, for instance, across the Atlantic. A special thanks to Paola.

PAOLA MIELI: I want to thank you all for your participation. Thanks to the panelists and the public. I don't want to repeat myself. I hope you will give us your suggestions for the pursuit of this work. I wish to conclude with a special thanks to our colleagues from Paris, for their irreplaceable moral, economic and symbolic support, in particular the colleagues from L'Association APUI of Paris, and Le Cercle Freudien of Paris. Thank you very much indeed.

Afterthoughts

Jacques Leclaire

On November 9th and 10th, the *Being Human* colloquium was held, in a very informal manner, on the grounds of New York City's Union Theological Seminary.

The organization and management of the colloquium corresponded exactly to my father's vision. He was convinced that a pure academic meeting would be insufficient. It was certainly not such a meeting. My father would have appreciated this. In addition, the different sessions had brilliant moderators, able to listen and add a "crescendo" to sometimes intense discussions.

The gathering was, as my father imagined it would be, a "work in progress" activity, where moderators with a knowledge of each session's dossier guided a discussion between panelists.

He would have been delighted to experience this atmosphere, and he certainly would have made friends while discussing original ideas in the huge upper dining hall, or while taking a walk in the November sunlight.

— JACQUES LECLAIRE

Jacques Houis and Serge Leclaire – New York, 1991. Après-Coup workshop "Myth and Witness."

Elements for a Unifying Thread

Serge Leclaire

>   To Paola Mieli, Dany-Robert Dufour and Marcos Einis

In order to refresh what is at stake in *Being Human,* here are, on the one hand, in "A," notes expressing what I think should be our project's unifying thread. On the other, in "B," is the outline of a possible public document.

A. *Elements for a Unifying Thread*

Today's most acute problem appears to be that of the mutations underway in the thought processes.

Be they modalities of representation, symbolization, or formalization; be they speech or language (*parole, langage, langue*), writing or grammar (subject, object, noun, pronoun, and conjunction); be they logical models, referents, and "invariants" of all sorts; be they ideas and concepts especially, it must be allowed that they are all undergoing a process of insidious but irreversible metamorphosis.

To think through this mutation of the elementary functions that make up the activity of thought is the current priority of all those who, whatever their itinerary, have actually experienced overstepping the boundary (*limite*) of any closed field (whatever that boundary's legitimacy), of those who, in short, are in a position to recognize, and hence to regard critically, the acts of "transgression" that have been theirs. The most inexorably qualified among them are the psychoanalysts, because they have, with guilty innocence, "chosen" to work at lifting the boundaries [of "repression"] and calling into question what it is that constitutes a border (*dé-limitation*), a barrier, and a frontier.

In this case, thought does not really need to be done over, nor undone, but in a conjunction as critical as it is urgent, restored to the sovereign power (not to be confused with any omnipotence) it derives from the transgressive movement that constitutes its essence.

Thought does not come up against an "object," but rather against the difficulty of finding a boundary to cross, since that is the purpose of its function. Now, most of the "boundaries" (*limites*) it has played-off of, and to which we have been accustomed, have, *de facto*, lost the greater part of their consistency.

However, if what structurally constitutes a supporting edge (*appui, butée*), and thus a limit of sorts to thought, is the inexhaustible "real" in the Lacanian sense, we must admit that it is some sort of major crisis in the relationship of thought to the real that constitutes the current floating and drifting of thought.

A question arises here that seems crucial to me: what is the index of consistency of the real of the body?

Can it be said and thought that the body is the natural interface between the symbolic and the real? If, as I believe, the "body" is primordially ruled by a unary type of logic, one would parse its function as interface from the standpoint of this hypothesis.

What then is meant by "mutation of the thought processes"?

First, an irreversible transformation of the most common referents such as the concepts of life/death, nature/culture, empty/full, matter/anti-matter, free energy/bound energy, space/time, etc...but especially of concepts related to the "person": body/mind, and "psychology": conscious/unconscious. Note how these concepts are coupled in a binary system! Also, a transformation of the conception of speech and language, originating in psychoanalytic practice, and marked by the advent of Lacanian concepts, to mention the most eminent: that the signifier is what represents the subject (subjective function) for another signifier;

■ BEING HUMAN:

or the concept of "the real"(this will need to be explained for our purpose) as remainder of the processes of symbolization; or, also, the very concept of subject.

In short, it is the idea of the concept itself that seems to be in the process of imposing its mutational, if not transgressive nature, and is prey to a mutation effect returning from objects and their mutations: thus, for instance, the mutation of the consistency of matter, which is no longer able to assure the function of supporting referential edge (*butée référentielle*), requires that thought finally produce other models of referents (in the manner of the Lacanian signifier for speech and language).

In other words, the concept can only assure its referential function by effectively assuring the mutational process of transformation/transgression that promotes and specifies it as concept. (As such, it differs from the idea conceived as "eidos," image or representation, which tends to temper or contradict the transformative function of the concept). A concept can only belong to an open system referring to other concepts or conceptual functions, similar to a "biological," living system, in constant transformation.

The question remains of the "place" of the constant, of the durable, outside of any reference to ideas of immobility. As well as the question of how to understand the manner in which the figure of "repetition" intervenes as symptom and/or necessity.

The constant or durable could then be, not what can repeat itself, but, rather, what can reproduce itself, with the notions of creation or re-creation the term implies. In short, what can engender itself, even if it appears to do so identically, something temporality would seem to contradict.

The identical is sustainable only through a process of symbolic identification, of nomination (which brings up the question of genealogy, of transmission of the name.) If we could make it understood that the identical only occurs through repetition of the name or, more generally, through an operation of symbolization, then we would produce an actual act of thinking and a mutation with it.

THE TECHNOLOGICAL EXTENSIONS OF THE BODY

Can identity, as the basis of the one, be considered the matrix of the unary?

The prospect of thinking through the mutations of the thought processes partakes of unary logic. It also and thereby constitutes a supporting edge for the project.

Just as the concept of a "fold"(*pli*) is transversal, so, eminently, is that of a transgression-overstepping (*transgression-franchissement*).

To think through the overstepping (*franchissement*) involves using the disproportions of unary logic as the primary supporting edge (*butée*), the boundary always yet to be crossed, in order that thought which assumes its own mutations may legitimately unfold.

This is probably where one would pose the question of how the advent of writing tends to shift the concept toward the idea (*eidos*), by producing a sort of image-stop (*image-arrêt*), that makes a sacred effigy of the text yet to come.

B. *First Brief Sketch of a Position*

Considering the transformations or, better, the mutations occuring in the different fields of science (from astrophysics to molecular biology, from the exploration of elementary particles to that of the human genome), we should:

1. Recognize the correlative mutations occurring in the "psychic" processes of thought (the advent of new forms of logic, the taking into account of the hypothesis of the unconscious).
2. Acquire the means of pertinently thinking through the, perhaps implicit but surely decisive, mutations of the thought processes, in particular in order to restore to the conceptual function its specific role as operator of mutations, of overstepping, of transgression.

■ BEING HUMAN:

3 Take into account the function of the body, conceived of as an interface, not able to be circumvented, between the world of so-called "objective," exterior reality, and the world called interior, "subjective," calling its boundaries into question in the process...

— SERGE LECLAIRE
June 12, 1994

Comments on "Elements for a Unifying Thread"

Dany-Robert Dufour and Paola Mieli

As part of a working correspondence shared by Leclaire, Dufour, Einis and Mieli, these notes should be understood in the context of an ongoing discussion. For this reason, they contain no reference to the thoughts that preceded them, nor to the reflections that followed and that were interrupted by the death of Serge Leclaire.

Despite their brief and fragmentary nature, these notes constitute an important document, showing what was at stake for Serge Leclaire in the *Being Human* project.

In the main, we would say that they make up a working document seeking to introduce the difference between binary and unary paradigms as it relates to thought, to "the idea of the concept itself that seems to be on the way to imposing its mutational nature." This involves attempting to replace the binary paradigm, which sustains the conscious articulation of the concept in language, with the unary paradigm, which seems to govern the body, by virtue of the fact that the very idea of "thinking through the mutations of the thought processes partakes of unary logic." According to Leclaire, this approach would form the project's "supporting edge."[1]

---

[1] The unary world is built on definitions that have *one* term (hence the expression *unary*) with repetition of the subject in the predicate (examples: "Who says *I* is *I*" and "the *signifier* is what represents a subject for another *signifier*"). These definitions engender spaces where inside and outside are inverted (example: the "Mobius strip"), circular temporalities where before and after are inverted, rationalities where cause and effect, yes and no, are reversed.

The binary world is built on definitions that have *two* terms. The space of science is a construction based on binary definitions distinguishing an effect and a cause, an element *a* from an element *b*. These definitions lead to classical rationality – a space of differences between inside and outside; linear temporalities based on distinctions between before and after; rationalities founded on the difference of cause and effect, yes and no.

Comments on "Elements for a Unifying Thread"

In these notes Leclaire does not explicitly discuss the trinary paradigm. Previously examined in detail, the trinary is an implicit reference here. The accent is placed on the binary/unary opposition. But the role of the trinary in the symbolic is presupposed. The symbolic is created with the trinary, the basis of the very structure of language; basis, as well, of the subjective division that makes it possible to think according to the binary logic that guides conscious thought.

Leclaire refers here to the relation of the identical to identity. The transmission of the name, of the identical, is based on the existence of a third party from which all genealogical counting can begin. The Name-of-the-Father, to use Lacan's term, refers to this third party that allows the founding of the "one" of the identical. Genealogy can be understood as what happens when this point of reference becomes fixed. It is because of this "one," which establishes an origin, that the trinary system cannot be understood without referring to the binary. It is only through the binary – the number, the relationship between two magnitudes – that there can be counting, adding or subtracting.

While the "one" of the identical sustains symbolic transmission, identity supposes the use of the "one" as generator of difference. There is, as Lacan puts it, an "affinity" of the proper name with the mark, and the mark registers the "one" of identity as difference. Leclaire wonders whether identity, basis of the "one," could be conceived of as the "matrix of the unary."

Adopting the same format Leclaire uses to summarize his thought, we shall highlight the following points as central to his project:

The trinary world is derived from definitions having *three* terms that cannot be reduced to two (examples: the triad of enunciation – "I," "you," "he/she/it"; the Borromean knot; the Holy Trinity...). These definitions produce origin, provide points of reference and support, supply the bearings upon which symbolic systems are based. It is on the basis of the ties formed by these definitions with three terms, that we are able to count, speak and write. See: Dany-Robert Dufour, *Les bégaiements des maîtres,* Edition François Bourin, Paris 1988, treatise on the unary; and Dany-Robert Dufour, *Les mystères de la trinité*, Gallimard, Paris 1990, treatise on the trinary.

■

1. We should be aware that the new objects presented by science require and produce the development of new forms of logic – forms that respect the hypothesis of the unconscious (a place where the unary manifests itself since it ignores the difference between yes and no).
2. Thought itself is in the process of reflecting the mutations and transgressions these objects imply. Certain concepts produced by the practice of analysis, some of the Lacanian concepts in particular – the definition of the signifier or the concept of the real, for example – point to the mutations taking place. We should follow this path, the path of the unary, and try to free the concept from the binary structure that characterizes oppositions such as: life/death; natural/cultural; empty/full; matter/antimatter; free energy/bound energy; space/time; body/spirit; conscious/unconscious; etc. The concept, in its referential function, shall belong to an open system, itself in a state of constant transformation. By reflecting mutation, the concept shall lose its status as "eidos," as fixed representation. Leclaire suggests that the "place" of the constancy necessary to the concept's role as reference, can be found in repetition, conceived of as a generative reproduction within temporal difference.
3. Of the mutations that have occurred in the different fields of science, several involve the body, about which Leclaire forms the hypothesis that it is governed by a unary type of logic. These mutations, for example those produced by the latest discoveries of genetics or molecular biology, clearly show how the boundary between the symbolic and the real is moving. Thanks to the implementation of new systems of signs, of new decodifications, the symbolic is "gaining" on the real. This raises the question of the index of consistency of the real of the body. The hypothesis of the body as interface between the symbolic and the real belongs to the Freudian/Lacanian tradition: the psychic work, the unconscious structured like a language, is the deciphering of the body's

ciphering, etched by the real of *jouissance*.* The interface between the real and the symbolic is an interface between impression and deciphering. The unary image appears of a body as pure edge, like the edge between the two sides of a Mobius strip. The binary conception of the body as the site of the separation between interior and exterior, subjective and objective is found to be obsolete.

— DANY-ROBERT DUFOUR *and* PAOLA MIELI

---

* See translator's note, *Being Human*, 94.

An Introduction

Brief Preliminary Considerations on Sameness, Otherness, Idiocy, and Transformation

An Introduction

Paola Mieli

> *The human discovers himself first as different, then as mortal.*[1]
> – JACQUES HASSOUN

> *L'ethique est relative au discours.*[2]
> – JACQUES LACAN

From a television program we learn that the American Government is in the process of creating a special site in the New Mexican desert, 2000 feet underground, to collect nuclear waste. It is meant to last ten thousand years.[3] How do you make a nuclear waste site safe for eternity? With the help of technologies old and new, specialized scientists, geologists, and engineers are guaranteeing the efficiency, safety, and durability of the construction. Once such waste is disposed of, we shall fear no leakage nor any other kind of dispersal of the buried material. Buried indeed for eternity. Still, a problem remains, and it's not exactly a "technical" one: how does one signal to future generations that an extremely dangerous radioactive material is buried in this particular area? *Discover Magazine* reports that a special committee of scientists, scholars, and technicians has been formed to study the question: what sort of signal shall we use to warn future civilizations not to dig or tap in any way into the indicated place?

---

1 *Being Human*, 59.
2 *Télévisions* (Paris: Seuil, 1973), 65.
3 *Discover Magazine*, "Immortality," November 1994, Binball Productions, Marc Etkind and Robert Kirwan editors. All related quotations come from the same source.

All the eminent scholars agree that they cannot envisage those future civilizations, much less the languages they will be speaking. It is very likely that, in a thousand years or so, current English will have about the same effect on people that Egyptian hieroglyphs have today. Archaeologist Maureen Kaplan advises against the employment of precious materials because, as history has shown, they act as an inducement for all sorts of explorations and sample-gatherings. She recommends instead a stone marking system, made up of fairly massive elements. She emphasizes that, as was the case with the Rosetta Stone, it is necessary to provide an interpretative key: warning signs are to be written in several languages.

What if future societies no longer know how to read? For this reason the committee includes an expert in communicating with "non-readers": Frank Drake, an astronomer and a specialist in sign production for creatures unfamiliar with our planet. It was Drake who came up with the message for aliens carried on the Twin Voyager spacecrafts launched in 1977, on a mission to photograph Jupiter, Saturn, Neptune, and Uranus. Since these ships are intended to navigate through space until some sufficiently advanced civilization appropriates them, they provide us with an opportunity "to immortalize ourselves and to make our existence and our nature known." The message, recorded on gold disks, is composed of a set of the planet Earth's characteristic sounds, and by 120 encoded photographs, each "carefully chosen to convey information about who we are."

For the site containing nuclear waste, Drake suggests a "pictorial" signal "to spell out" one thing: dig here and you die. For instance, the whole site could be covered with a giant skull. Architect Michael Brill objects that nothing guarantees that a skull be understood as an index of danger and death: a civilization different from ours might even take it to be a sign of life. Brill suggests therefore that we communicate not with signs but with "feelings": by devising, for example, a monument of forms that will arouse terror, made of sharp objects such as knives, teeth, etc. A proposal to which physicist and science-fiction writer Greg

Benford responds that, if it's true that a landscape of terror may put off some, it will probably attract others. He concludes that the smartest course of action would be not to signal the presence of nuclear waste site at all, so that no one will think it's anything special, but just another thing to be forgotten.

Discover Magazine informs us that a plan, put forward by a government contractor, has been chosen, and it gives us its general outline: the whole area will be covered by a huge hearthwork supporting broad granite monuments, each carved with warning signals in the world's seven most-spoken languages. This perennial structure is slated to be built in the year 2083.

The need to signal danger leads to the building of a monument; *monumentum* comes from the Latin *monere*, to remind, to warn. How could a "signal to posterity" not be transformed into a monument? This time, however, it is the need to ward off that turns the signal into a monument. A monument that reminds people that they cannot touch it. It is the creation of a taboo.

What a peculiar legacy to posterity. What an edifying representation of the values that animate our society: building in order to alert people of the presence of the death we prepare for them. As Michael Brill points out, it is through such a structure that future generations will appreciate our interest in them: "What was important to us is you ten thousand years away."

What was important to us....

*

New Leaf Superior potatoes can be bought at many American supermarkets. They are clones of clones of genetically engineered seeds from the chemical company Monsanto. They've been genetically engineered for the plant to produce its own insecticide, a bacterial toxin – BT toxin – that automatically kills the beatle that craves them. "This year, the fourth year that genetically altered seed has been on the market, some 45 million acres of American farmland have been planted with biotech crops, most of it corn, soybeans, cotton and potatoes that have been engineered to either produce their own pesticide or withstand herbicides."[4] Michael Pollan, author of an interesting article on the subject, informs us that we find no warning against the toxin in question on the bags of potatoes we buy; the Food and Drug Administration sees no reason to point it out, since BT toxin is not an additive but a pesticide, and as such exempt from regulation or labeling. Needless to say, the development of these crops, which reduce some of the problems of natural farming, assists large-scale producers, allowing for the agricultural industry, which is mainly a monoculture, to be concentrated "in a shrinking number of corporate hands."[5] The firm Monsanto obviously reaps the benefits of its patent; cyclical inspections carried out in the fields will soon be rendered obsolete by "terminator[s]," a complex of genes that will sterilize every new seed the plant generates.

"One way to look at biotechnology is that it allows a larger portion of human intelligence to be incorporated into the plant itself."[6] The true taste of intelligence.

*

"Cloning humans was not so hi-tech,"[7] declares Lee Bo-yon upon the first public announcement of human cloning that took place in Seoul on December 16, 1998. The team of South Korean scientists, as is well known, obtained four embryo cells by cloning the cell of an infertile woman, without, however, implanting them in the donor's

---

4 Michael Pollan, "Playing God in the Garden," *The New York Times Magazine*, October 25, 1998, 45.
5 Ibid., 51.
6 Ibid., 48.
7 Sheryl Wu Dunn, "South Korean Scientists Say They Cloned a Human Cell," *The New York Times*, December 17, 1998, A 12.

■   BEING HUMAN:

womb – which would potentially have given place to a genetic replica of the woman in question. It may have also raised a few ethical questions.

The news did not stir much controversy, and no one denied the likelihood that human cloning had been tried out "secretly in other countries."[8] As Doctor Michael Roberts reports, "Cloning is becoming routine"[9] in the animal world, where, unhampered by ethical standards, it is now a fact of life. Just think of the very recent success of the team led by Dr. Tsunoda at Nara, Japan: they have managed to carry out a successful cloning of eight calves from a single cow – four of which died at birth.

Meanwhile, the recent separation and cultivation of human embryonic stem cells – which, given their transitory nature, are hard to capture – promise extraordinary advances in the realm of organ transplant and gene therapy. It appears that such cells may be developed in many if not in all corporeal cells; someday they may well function as universal spare parts. "The cells may also offer effective routes to human cloning, although both the researchers and their sponsor deny any interest in this application."[10] Research in the area of the cultivation of stem cells, conducted at the University of Wisconsin and the Johns Hopkins University School of Medicine, are financed by Geron Corporation, a company specializing in anti-aging processes. The world of bio-technology and the stock market have paid a lot of attention to this Corporation after the discovery, it announced last year, of how to keep human cells alive past their natural term.

"From the start, Geron has been aggressive in financing academic scientists in exchange for the exclusive commercial right to their work."[11]

---

8   Ibid.
9   Gina Kolata, "Japanese Scientists Clone a Cow, Making Eight Copies," *The New York Times*, December 9, 1998, A8.
10 Nicholas Wade, "Scientists Cultivate Cells at Root of Human Life," *The New York Times*, November 6, 1998, first page.
11 Andrew Pollack, "Small Company Gains High Profile in the Scientific World," *The New York Times*, November 6, 1998, A24.

THE TECHNOLOGICAL EXTENSIONS OF THE BODY

The man who conceived and started the Geron Corporation, Michael West, is currently chief executive of Advanced Cell Technology, the company that announced last November the successful creation of a mixed stem cell from the fusion of a human cell with a cow's egg stripped of its nucleus: an operation said to have humanized the bovine cell. In the wake of the concern President Clinton expressed about such a hybrid and its possible fate if it were to be transplanted to a human womb (but why not also a cow's womb?), various members of the National Bioethics Advisory commission have expressed a lack of alarm. As Dr. Carol W. Greider, a Johns Hopkins biologist puts it, "There are some ethical issues there but they are much less worrisome."[12]

A bill in the United States Senate against human cloning was defeated in February 1998, mainly, it is rumored, because of the restrictions it would have placed on stem-cell research – a step forward for industry and research, and, to be sure, for the art of making people accustomed to the idea. As Dorothy C. Wertz, a bioethicist at the Shriver Center, declares, "Human cloning will likely also be accepted once it becomes a reality. Most of today's ethical arguments against it were previously used against in-vitro fertilization and turned out to be false."[13]

*

Current capitalist discourse confronts us with a peculiar notion of otherness and transmission. The story of the nuclear waste site is a fact of reality: a very real reality that engages current and crucial concerns of our civilization. It illustrates one radical aspect of humans' technological extension beyond death. No need to point out that nuclear waste and the issues regarding its burial are the outcome of powerful tech-

---

[12] Nicholas Wade, "Ethics Panel is Guarded about Hybrid of Cow Cells," *The New York Times*, November 21, 1998, A 9.
[13] Nicholas Wade, "Scientists cultivate Cells at Root of Human Life," *The New York Times*, November 6, 1998, A 24.

■ BEING HUMAN:

nologies and political interests. Let's observe instead that this story is emblematic of a certain human position in today's historical context. The various proposals and the sense of uncertainty and confusion they generate, shed light on a particular embarrassment. When experts and scholars are asked to deal with future individuals or unknown creatures, tecno-scientific planning comes up against a limit. *Alterity* carries a contrast, an unknown. For example, it brings in the "desire" factor. Who can tell if what frightens certain people may not attract others, as Benford has pointed out. Who knows what kind of desires will guide future individuals…. Given the unpredictability of the human being, calculability is transformed into speculation; and speculation reveals the fantasies of the speculators.

In the example of the nuclear site, what shakes the apple cart is a problem of "communication"; and this is all the more significant today that the discourse on communication seems to be endless, and both computer science and the media allow for the globalization of information. Some people have begun to mythologize the advent of "total" communication. Claus-Dieter Rath recalls that according to Lasswell an act of communication between two individuals is complete when they understand the same sign in the same way. But "complete" communication finds its limit in the human being; in the fact, that is, that nothing guarantees that man, as a being of language and desire, will not receive the message sent to him in a reversed form, as the experts' concerns about the nuclear waste site clearly show. The fact that language may signify different things from those it says and that there is an insuperable difference between enunciation and statement, is part of the very nature of human language; the subjective division illustrated by it and deriving from it, cannot be overlooked.

The act of speech, the enunciation, does not coincide with what is said, what we call the statement. This split bears witness to the presence of a border: the human being is perennially exiling himself in the act of speaking. Dany-Robert Dufour remarks that the dream of total communication can be achieved, it's true, but on one condition: that the subject ceases to exist. The plurality and abundance of today's communication media, the quantity of information whose access they facilitate, come to enrich a subject who, no matter how endowed with new technological prostheses, new information, and new creative spurs, will not refrain from misunderstanding others as well as himself.

Communication, by its very nature, implies an otherness with which to communicate. Such an *alterity,* however, is assumed to be sufficiently similar for the message to be received. When we are faced with the idea of different cultures or future civilizations, or alien creatures, then the ever-latent question of diversity comes again to the fore. To this question, as Seth Shostak shows in his close examination of the behavioral traits of extra-terrestrials, researchers try to respond "by holding a mirror up to themselves."[14] Difference is filled in by the imagination, even if the imagination, faced with such radical alterity, reveals its limits and permeates the unknown with the logic of the known.

Fantasies, according to Freud, constitute a reservoir of the pleasure principle in a world in which the individual has had to adapt himself to reality. They correct reality, its discontents, in favor of an imaginary universe in which desire can find its satisfaction. And if imagining is not necessarily the same thing as daydreaming, it becomes one with it in the face of the unknown. The relation between imagination and technology follows a circuitous path. Production of new technologies is anticipated by the imagination. Science fiction anticipates the discoveries of science, as, for instance, NASA technicians attest in creating Virtual World Stations along guidelines provided by William Gibson's books. And yet, the possibilities opened up by the application of new technologies, once turned into reality, exceed the imagination, denounce its limits and force it to go further still. As Salvatore Guido remarks, machines occasionally demonstrate "a propensity to do things that neither their builders or programmers had anticipated."[15]

---

14   *Being Human,* 484.
15   *Being Human,* 398.

Brief Preliminary Considerations

Faced with the actualization of technical possibilities and unprecedented operations, we find ourselves practicing the unimaginable, and the body's technological extensions force us to produce new visions.

The human need to cull pleasure from one's own mental processes finds its special expression, Freud remarks, in the pleasure we derive from rediscovering the already known, in the tendency of the psychic apparatus to spare the psychic expense procured from working out the new. The attraction, the seduction of the new, is counterbalanced by the work it implies. Sometimes we respond to the effort derived from technological transformations by resisting; sometimes with a refusal. Yet, "At this stage of the game...the future has already happened."[16]

The perplexities raised by the new bio-genetic potatoes reminds us that food introduces an essential aspect of the relation between sameness and otherness. The original form of judgment is, in Freud's view, a judgment of quality. It manifests itself in a structuring affirmation – a *Bejahung*, a symbolic yea-saying: what is "good" is introjected, and what is "bad" is expelled, in a selective movement that differentiates the ego and its essential and existential otherness, its world. The relation to food is never neutral, charged as it is with vestiges of this archaic and mythical distinction. Being at the mercy of external care for survival, the infant's body becomes the site of an exchange with the other which, inscribing libidinal excitations, design the map of infantile sexuality. The satisfaction of the child's needs provokes the emergence of desire, as well as the longing for its fulfilment. The relation to food will always be trapped in the difference between need and desire, in the declensions of the demand for love.

Identification is the first form of an affective bond, Freud remarks, while incorporation is the primordial form of identification. We generally eat alterity to convert it into ipseity. However, there are various ways in which alterity persists, as in the case of forms of poisoning,

---

16 William Gibson, from the movie *Cyberpunk*, Mystic Fire Video, New York 1991.

real or imaginary. From this standpoint, one can't help but note how American New Age has turned food-related concerns into a widespread fad. Care for organically grown food, however, doesn't necessarily issue from biogenetic manipulation in agriculture or breeding. It often anticipates them, or else it is completely unaware of them. Such care is nourished by the idea of the clean, by a counter-discourse of "freshness," as Margaret Morse calls it, which often institutionalize a phobic relation to the other, to an "external" implicitly perceived as poisoning or poisoned. Here we encounter the paradox of how the spread of bio-genetic foods, whose harmful effects over time are unknown or unforeseeable, goes hand in hand with the fashion for the most exaggerated hygienicism, without these two tendencies necessarily entering into conflict. Both, by the way, yield generous profits.

According to the primordial dialectic between good and bad, ego and world, the bad is equivalent to what is *foreign* and may threaten the subject. The term "foreign" shows how the extraneous is etymologically linked to the external (*foraneus*, from the Latin *foras*, out of doors). The binary logic of introjection and expulsion soon reckons with an interior that is presented, occasionally but insistently, as *foreign*; illnesses, symptoms, anxieties, dreams, and so on, attest to a subjective division in which we find ourselves necessarily inhabited by what is presented as *other*. The body itself – at the mercy of physiological transformations, needs, drives that elude rational control – reveals its essential alterity.

In western medicine, illness is, for the most part, implicitly understood as a difference, an anomaly that is not an integral part of the body and that overpowers health. One is assailed by it. A pathological anatomist has recently described the horror that seized him during an autopsy: in the disease-ravaged corpse, cancer appeared in the guise of white excrescences spread throughout the organs, as though, as he put it, some extraterrestrial had maliciously played havoc with the body's natural balance and beauty.

This is one of the many perspectives in which death comes to represent absolute otherness.

■ BEING HUMAN:

No doubt today's conquests in medicine, pharmacology, bio-genetics, represent extraordinary advances in the struggle for health and survival. One need only think of the possibility of producing organic replacements less susceptible to rejection for reconstructing parts of the organism; or think of the possible interventions, in gene therapy, into hereditary illnesses. And if illness is represented as alterity, let there be new possibilities to combat it.

It's interesting, though, to observe how today's bio-genetic applications nurture a certain spirit of *"sameness."* The other side of the coin in stem-cell research, for example, is cloning. The fact that animal and vegetal cloning research is animated, among other things, by the principle of the identical, is something researchers and industry openly declare. Dr. Tsunoda, for instance, explains that the reason for cloning cows is: "To reproduce exact copies of animals that are superb producers of meat and milk."[17] In this sense, reproduction functions like a new type of selection, though in a very different sense from that of Darwinian selection. But industry is not concerned with the consequence of this. Moreover, as Robert Pollack remarks, our efforts to decode human DNA go hand in hand with our methodic and careless destruction of various animal species. "This is on the scale of evolution a catastrophe equal to any asteroid that ever struck the planet."[18]

Dr. Dorothy C. Wertz makes a revealing statement about today's resistance to human cloning: that the ethical arguments against it are the same as those formerly used against artificial insemination – and discredited, she adds. Whether or not assisted-reproduction industry takes an interest in cloning, it must be noted that certain practices within that industry have for years now been oriented toward a particular type of

---

17 Gina Kolata, ibidem, A8.
18 *Being Human*, 61. *A propos* of transgenic plants, research conducted by a team from Cornell University, provided the first evidence that pollen from bio genetically altered corn harms monarch butterflies, a non-pest species – an indication of the unforeseen effects on the environment caused by genetically engineered crops. See Carol Kaesuk Yoon, "Altered Corn May Imperil Butterfly," *The New York Times*, May 20th 1999, first page.

selection: selection of sperm, for instance, or of eggs for implantations, depending on various chosen somatic characteristics – color, race, height, etc. – but also on chosen psychic characteristics and cultural background. The possibilities opened up in this area by new technologies clearly demonstrate how, when it comes to reproduction, human beings measure themselves against a certain ideal. One tends to hand down an idealized self-image. Freud has remarked that love for one's children is a form of secondary narcissism.

Applied to the body, the new technologies permit a pursuit of the ideal, the implementation of the imaginary through an intervention on the real. In this light, cloning merely brings to completion a tendency that's been brewing for some time. The fact that technicians see no great logical difference between in-vitro fertilization and cloning, says a lot about the spirit that has always guided assisted-reproduction practices.

In cloning, reproduction is self-reproduction, parthenogenesis. How does all this weigh on the question of transmission?

One can't help but note that cloning carries to its extreme consequences the reality of many practices of assisted reproduction: that is, reproduction without a partner. In the case of cloning, the function of the other, of the *third,* has been even further expelled. Filiation by cloning, whether we're dealing with reality or fantasy, is binary self-reproduction, duplication that bypasses, at least in principle, the Oepidal structure, by its nature trinary. What the Oedipal structure ushers in is precisely the notion of *difference*, above all sexual difference. A difference between the sexes that opens up the question of the Other's desire. Difference implies otherness.

Cloning can be understood as a further step in medical science's attempt to *eliminate* sexual difference. Sex-change operations are just one example of this, performed for years now, as if it were implicit that to change sex all you need is surgical intervention. We see how applied medical science nurtures and incites the acting-out, when, in the face of transsexual discomfort, or the fantasy of self-duplication, of elimination of the other in reproduction, it responds with an intervention on the real of the body. Reducing subjectivity to pure mechanical materiality, biomedical deterministic ideology reinforces the symptom in order to actualize it in a practice. "They say brain so you will think even thought, good, is limited and inside, a muscle."[19]

Going back to the questions raised by the construction of the nuclear waste site, we can now take stock of the extent of the paradox in which we are enmeshed: we are part of a generation at once concerned with warning future beings of our own destructiveness and unconcerned with the immediate or future side-effects of some of our own technological applications. Perhaps it will suffice to point out that the logic of immediate profit holds sway here, and that it is absolutely oblivious to the implications of its own operation. We are dealing with a way of thinking that, however much it claims to be motivated by the struggle for well-being, for the prolongation of life, the elimination of suffering,[20] never ceases to manifest a downright suicidal drive. The case of the mad cow is exemplary. We'll see what the consequences of biogenetic pollution and of cloning will be. Love of sameness, narcissistic reproduction of the ideal, are part and parcel of self-destruction.

What can one say about a historical generation of parents who jeopardize their own children for the sake of immediate profit? A fundamental aspect of parentage is being called into question through an attack upon the figure of transmission. Transmission implies implementation of the symbolic function, the assumption of one's own mortality. Chances are that when a parent refuses to think of himself as mortal (and this quite apart from his real death) the child will be exposed to psychosis. In fact, people speak of the psychoticization of present-day social reality.

---

19 Amiri Baraka, *Being Human*, 283.
20 *A propos* of suffering: How can a medical program that purports to be dedicated to well-being and longevity, not do its utmost to alleviate physical pain? Is this one more expression of the puritan fundamentalism animating American public life? Or are we dealing with one of the many fruits of a judicial system that for the sake of the citizen's rights, denies him adequate assistance and fosters the fear of possible malpractice suits?

THE TECHNOLOGICAL EXTENSIONS OF THE BODY

While love for one's child manifests an aspect of the parent's narcissism, if not the hope of realizing one's ideal ego in one's offspring, transmission implies the recognition of the child in its structural alterity. One characteristic of identification is precisely the assumption of traits that, in resemblance, embodies above all a difference. If the identical is carried through the symbolic dimension of the name, identity marks unicity, singleness.

The myth of self-reproduction reveals a disdain for filial difference. In confusing transmission with perpetuation of the equal, in privileging one's own ego as the ideal, love of sameness bestows upon filial difference a status that is, if not diabolical, threatening. [21]

The refusal of transmission shows the ascendency of the death-drive. It's interesting to note how in the case of the monument to mark nuclear waste, the scholars' questions revolve around a presupposition: namely, that of a break in transmission with future generations, an interruption implying a difficulty in passing and circulating the message from one generation to the next. It's true that the history of civilization, and thus the history of technology, is a history marked by interruptions – interruptions caused, however, by destructions for the most part man-made. In this sense there is nothing inherently progressive about the history of technology. Just think of the break created in the history of knowledge – technological knowledge included – by the destruction of the library of Alexandria. To this day, we don't know how the Egyptian pyramids were built. Yet, organizing transmission around the idea of *interrupted* transmission says something not only about the will to take past history and possible interruptions into account, but also about the definition of our present-day history.

Jean Pierre Lebrun points out that the supremacy of modern science parallels the supremacy of the statement over the enunciation. The internal coherence of the statement, based on the principle of conformity

---

21 In fact, transmission as reproduction of the equal, involves a notion of engenderment that echoes the Nazi ideal of the "natural" and "pure" reproduction of the race. In this view, purity is guaranteed by a transmission that occurs integrally and leaves no remainder.

between word and thing, gains the upper hand and the enunciator gets ousted. The model of mathematical propositions is generalized, the statement dealt with as though it were a truth without a subject, and its relevance extended. In this debate, however, it's necessary to distinguish *science*, its evolution, its conquests, its advancement from the attendant *discourse of science*, which expresses itself differently depending upon various historical contexts, social ties or particular ideology. To think, say, that science is 'true' because it is mathematically transmissible is, as Lacan says, "sheer delirium."[22] The very program of science is articulated upon transformation, reformulation, and the refutability of its statements.

The growing faith in the authority of the statement, as it is expressed every day in commonplaces about scientific discoveries and in the current parlance of the media, should be understood as the fruit of a particular ideology. Retracing, step by step, the evolution of scientific rationality introduced into Western thought by modern science, Koyré clearly illustrates the various scientific discourses that science gives rise to in various historical contexts. Newton's energy-god, the divine maker of a perfect world-clock, is progressively revealed to be a *"fainéant,"* a slacker; a new discourse announces that this watch works so well that there's no need even to wind it and, as Laplace put it when Napoleon asked him what role God played in his *System of the World*, "Sir, I had no need of that hypothesis."[23] If scientific rationality articulates the vision of a world which, in inheriting the ontological attributes of divinity, slowly rids itself of God, the following question arises: to what extent does the myth of scientific rationality produced by the current discourse of science *tend to rid itself of man*?

Nevertheless, the Freudian unconscious is unthinkable without the birth of modern science; it carries with it a radical modification of the position of the subject, inaugurated by the "rejection of all knowledge"

---

22 J. Lacan, "Note Italienne," *Ornicar*, No. 25, 9, Paris, Edition du Seuil, 1982.
23 "Sire, je n'ai pas eu besoin de cette hypothèse." Alexandre Koyré, *From the Closed World to the Infinite Universe*, The Johns Hopkins University Press, Baltimore, 276.

THE TECHNOLOGICAL EXTENSIONS OF THE BODY   ■

of the Cartesian cogito.[24] The conception of the world as rational introduces a new notion of error, which the human being is subjected to; subjected to the difference, for instance, between enunciation and statement. A new relation to knowledge opens the road to the idea of modification and relativity: "In no field of knowledge can one have the last word, any more than in any human relation."[25]

Today's fetishization of the authority of the statement is a symptom of current scientific discourse; yet, as various political, legal, economic situations show, enunciation maintains its authority. The ups and downs of the stock-market, for instance, are clearly the outcome of a political-economic mechanism at the mercy of the effects of enunciation. One might ask just how these forms of authority partake of the social discourse in which we are currently immersed.

*

A discourse is that which determines a form of social bond. Lacan defined five figures of discourse.[26] Among those figures, it seems fruitful to our purpose to pursue here aspects of the *capitalistic discourse*. The coordinates of a discourse are designated by four places: that of the agent or dominant place, that of the other, that of truth, that of the product or production.

---

24 Jacques Lacan, "La science et la vérité," in *Ecrits* (Paris: Editions du Seuil, 1966).
25 Serge Leclaire, *Being Human*, 50.
26 In his 1969/1970 seminar *L'envers de la psychanalyse*, Lacan fully elaborated the *discourse of the master*, the *discourse of the university*, the *discourse of the hysteric*, the *discourse of the analyst*. In 1972, he proposed a fifth figure, the *capitalistic discourse*, which he briefly sketched in a talk he gave in Milan, entiteld "Du discours psychanalytique." See: Jacques Lacan, *Le seminaire, Livre XVII, L'envers de la psychanalyse* (Paris: Editions du Seuil, 1991); and Jacques Lacan, "Du discours psychanalytique," in *Lacan in Italia, 1953–1978* (Milano: La Salamandra, 1978).

■ BEING HUMAN:

Brief Preliminary Considerations

$$\frac{\text{agent}}{\text{truth}} \longrightarrow \frac{\text{other}}{\text{product}}$$

A discourse is characterized by what comes to occupy each of these places. S_1 is defined as the master or primal signifier, the signifier that marks subjective identity; S_2, the repressed signifier, represents the set of signifiers that constitute knowledge; $\emptyset$ stands for the divided subject; *a* represents the object cause of desire.[27]

$$\frac{S_1}{\emptyset} \longrightarrow \frac{S_2}{a}$$

To access capitalistic discourse, let's first look at the *discourse of the master*. In the discourse of the master, the dominant position, occupied by the primal signifier, is defined in relation to a knowledge that stands in the place of the other, of the servant. The fact that it is the servant who possesses the master's knowledge, founds the master's authority. What emerges from this is a *jouissance*,[28] an enjoyment, a *plus-de-jouir*,[29] as Lacan puts it, which Marx identifies as surplus-value (*a*). Here, the divided subject occurs in the place of truth.

---

[27] For the articulation of these concepts see Jacques Lacan, *Le seminaire Livre XVII*, op. cit. See also: Jacques Lacan *Ecrits* (Paris: Editions du Seuil, 1966); *Ecrits, A selection* (New York: Norton and Company, 1977). For an English introduction to the work of Lacan, see Joël Dor, *Introduction to the Reading of Lacan*, Edited by Judith Feher Gurewich (New York: The Other Press, 1998).
[28] See translator's note, *Being Human*, 94.
[29] The expression "*plus-de-jouir*" means at the same time "more-enjoyment" and "no-more-enjoyment."

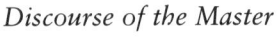

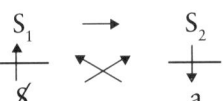

In capitalistic discourse, the relation between $\cancel{S}$ and S_1 is reversed. The divided subject is in the position of agent, the dominant position, while the primal signifier, that of identity, occurs in the place of truth. Knowledge is in the place of the other, giving rise to labor through *jouissance*. The product remains enjoyment, a *jouissance* entropic by its very nature, a *"more enjoyment"* that's never enough, that is sheer dispersion.

The divided subject in dominant position represents the figure of the symptom.[30] The signifier of identity in the place of truth, on the other hand – what capitalistic and university discourse share – marks the triumph of the *"I-cracy"* (*"Je-cracy"*).

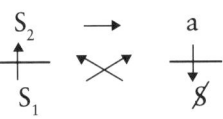

The transcendental and illusory *I* is, as Lacan points out, the irreducible operation of the *university discourse*; therefore also of the capitalistic's.

The *I* (S_1) occupying the place of truth expresses itself as a pure imperative. The myth of the mastering ego, the *I* of the enunciator who assumes to be identical to itself, is what the university and capi-

---

30 The symptom in the dominant position is what the discourse of the hysteric illustrates. See Jacques Lacan, *L'envers de la psychanalyse*, op. cit.

talistic discourses cannot eliminate from the place of truth. And, by the very fact that the primal signifier, the signifier of identity, occupies the place of truth, "the whole question of truth is, strictly speaking, crushed."[31]

The discourse of science, in its fundamental arrangement, anchors itself to the university discourse, coincides with it. This is evident from the place occupied here by knowledge, that is the dominant place, that of order, that of the command "Keep on knowing." Yet scientific subjectivity, as it is expressed today, for instance in the discourse of the media, seems rather to latch onto capitalistic discourse. What capitalistic and university discourse do share, however, the *I-cracy,* is what feeds a love for sameness.

But who is the subject of a discourse, such as the capitalistic, that manifests the symptom in the dominant position, and the self-identical *I* in the place of truth? A subject, we might say, that *never ceases being an idiot* in producing enjoyment from exploitation. *Idiot*: a term to trace back to its etymology in *idiotes*, that is, a private person, individual, "one in a private station," – from *idios*, one's own, separate, removed from social responsibility. *Ta idia prattein*, to mind one's own business. A separate individual is one who keeps to his own affairs and, if he embarks upon any public cause, does so as a private person *(idios en koino staleis)*. His ignorance needn't be pointed out, since, being separate in his sameness, he entrusts his knowledge to the place of the other, and experiences his own reality as a divided subject only in the form of the symptom.

The subject of capitalistic discourse and of today's scientific subjectivity finds himself in the forced position of *doing nothing but acting,* producing. In submitting to the imperative of his own ideal ego, to mastery, to production of surplus enjoyment, he often pursues as idiot the possibilities opened up to him by techno-scientific innovation; symp-

---

31 Jacques Lacan, *L'envers de la psychanalyse,* 120.

tomatically, he attends to profit and immediate enjoyment heedless of its descent, or thoughtlessly dismisses side-effects of technological applications performed in good faith and to good purpose. Today, the position of many researchers is, from this standpoint, exemplary; we witness them oscillating between university and capitalistic discourses, between the imperative for knowledge, the all-justifying "Knowledge Go Foward," on the one hand, and, on the other, a symptomatic stupidity, the ignoring attitude of the idiot. We need only think of some of the arguments with which medical researchers approach their own operations, for instance, their claim that there's no difference between adoption and in-vitro reproduction, or between in-vitro fertilization and cloning.

Technicians, moreover, are themselves part of the work force, the knowledge, used by capitalistic discourse to produce surplus enjoyment, *plus-de-jouir*. They are its tools. Ultimately, they belong to the category of *Bestand*, the "standing-reserve," the "ground," that for Heidegger characterizes the manner in which everything that relates to modern technique is present, understood as a challenging (*Herausfordern*) of nature, of the real, towards discovering and transforming. Only, what appears to us as standing-reserve (*Bestand*) loses its character as object (*Gegenstand*). We are dealing here, according to Heidegger, with the danger of modern *techné*: if what it brings to light is no longer present to man as an object but concerns him solely as ground, as standing-reserve, then man himself can be taken as a standing reverse, as pure background.[32]

Man is confused with the product whose tool he is. "The consumer society derives its sense from this: what makes for the so-called 'human' element is given the homogeneous equivalent of any *plus-de-jouir*, sur-

---

[32] It is, then, when man "comes to the very brink of a precipitous fall…[that] he exalts himself to the posture of lord of the earth." Martin Heidegger, "The Question Concerning Technology," in *The Question Concerning Technology and Other Essays* (New York: Harper Torchbooks, 1977), 27.

plus enjoyment that is the product of our industry."³³ An example of how the human translates into a commodity is, say, the drug addiction promoted by the pharmaceutical industry, the success of drugs like Prozac or Ritalin, which exploit *jouissance* to reinforce behaviors "highly valued in this particular culture." ³⁴ Sponsoring a biomedical ideology that identifies human subjectivity with mechanical materiality, the capitalistic discourse profits from the confinement of a person's *jouissance* to the solipsistic, masturbatory enjoyment of the idiot.

Capitalistic discourse sketches a particular figure: the self-identical *I* in the place of truth is situated between the object *a* cause of desire, in the place of production, and the divided subject is in the dominant position. The relation between the subject and the object that is the cause of its desire defines the *fantasm*,³⁵ and the *fantasm* guides the subject in its relation to the world. According to Freud, nothing is more variable than the object of the drive, since any object whatsoever that may assume its function, will always be other than the mythic primal object of satisfaction, lost by definition. Having inaugurated *jouissance*, such an object has also inaugurated the movement tending to its repetition. The notion of *plus-de-jouir* takes account of the loss necessarily inscribed in the repetition of satisfaction. It is never *it* – this is why it calls out for *more*.

Capitalistic discourse fosters the metonymic substitution of objects; for the sake of satisfaction, it institutionalizes dissatisfaction, the entropy inherent in *jouissance*. Propped up against the figure of the *fantasm*, it carves out of it the spirit of production. The difference that the quest for identity carries into repetition, is exploited. Once one object is used up, there's always another handy to take on the same function.

---

33 Lacan, op. cit., 92–93. [t14]
34 Eric Parens, *Being Human*, 61.
35 We use the word "fantasm" to indicate the Lacanian notion of *phantasme*, which doesn't translate fully with the English "fantasy." The concept of *fantasm* specifically defines the relation of the subject to its object of desire.

THE TECHNOLOGICAL EXTENSIONS OF THE BODY

For this reason too it works so well. "But in fact, it works too fast, it wears down, it wears down so well that it wears out."[36]

If the imperative of production and instant gratification implies consumption, the new technologies seem to accelerate this process. Is it this relationship between consumption, *jouissance* and the death drive, that leads experts to say that technologically superadvanced cultures are headed for self-destruction?

*

Transforming the world on the basis of need and desire is an activity as old as the world itself. Fantasies can be various, as can the technologies placed in their service. If today capitalistic discourse indulges the fantasy of reproduction of a self-identical ideal ego, compliant with the creation of a world ideally purged of otherness and difference, the least we can do is point out that such an ideal world is essentially *totalitarian,* a world implicitly promoting segregation. The movie *Gattaca* offers a fine example for this realm of biogenetic dreams: achieved racial perfection occurs in tandem with radical alienation, with the attack of the "natural," of the "technology-free" as alien and terrifying.

The reproduction of the ideal ego in the real doesn't turn reality into an ideal; rather, it stages the consequences of ideological absolutism. We see how totalitarianism, which brings about segregation, neatly represents a dominant aspect of today's so-called democratic society, oriented as it is toward a global economy that makes a mockery of socio-economic differences and local cultures. Biogenetic potatoes and Bill Gates's products are but two examples. In a world prey to uniformity, the Macintosh slogan "Think different" appeals to the narcissism of distinction.

---

36 "Mais justement ça marche trop vite, ça se consomme, ça se consomme si bien que ça se consume"Jacques Lacan, in *Lacan in Italia*, op. cit., 48.

■ BEING HUMAN:

However lavishly capitalistic, university, and scientific discourse manufacture a sutured subject, an ideal master of himself, the effects of the divided subject proliferate. Difference remains in the world – biological difference, sexual difference, economic and discursive difference. And capitalistic discourse is but one of the possible discourses, albeit the prevalent one. If anything, the new technologies *accentuate* the relation to the alien and extraneous, the contact with difference. Is it this accelerated traffic with one's own alterity that prompts the call for the identical?

The debate stirred by the idea of how to answer ET when and if he will give us a sign[37] – after all the messages sent by SETI technicians funded by our contributions – clearly shows how difference and otherness are above all part of our own world, and how, in the face of a desired absolute otherness, people do not know how to conceptualize it. It is here that the symbolic finds its limit in the real. New technologies can only accentuate the contrast.

Nothing keeps the new technologies from becoming vital, creative resources, except the constraints of some discourses. It's clear that technological advances in medicine allow for extraordinary leaps forward in the struggle with disease, and that the new technologies could be used for the preserving of species, as Jim Yount suggests, for the fight against pollution, and so on. They are useful in reconstructing history, as attested to by the DNA analysis of young people who, as babies, were torn from the hands of "disappeared" political prisoners and adopted by families close to their parents' persecutors. Technology can be placed in the service of the right to know, of unveiling rather than concealing.

If the subject of capitalistic discourse is an idiot, such a role does not suit everyone. Some will take other positions and question the consequences of the current political situation, if not the implications of the

---

37 *Being Human*, 139.

new scientific discoveries and the technological revolution that has been our fate. Some will use their knowledge not for doing and not knowing, for shutting their eyes and ears, but to question its effects and uses, assuming ethical responsibility as subjects and citizens, members of a *polis* that implies their participation in the commonweal.

As Michel De Certeau points out,[38] the strategies put in motion by the dominant discourse in view of centralized, expansionist, and totalitarian production are countered with consumers' tactics, the myriad practices of daily life that, in metaphorizing the dominant order, make it run on a different register, transforming it from within. However framed by prescribed syntaxes, the vocabulary received by a system can be used to produce other codes, non-prescribed acts that pursue indeterminate courses and gradually erode preestablished schemes.

The idea animating the present book falls within this framework: in opposition to compulsory action, we are trying to create occasions for reflection, trying to think about aspects of present-day technological reality in a dialectical interdisciplinary perspective, one which includes impasses, differences, contradictions. We are trying to listen to and learn from differences. These records of the *Being Human* colloquium clearly reveal the Babel of languages produced by a meeting of scholars for the most part unaccustomed to dealing with points of view radically different from those of their own field of operation and research. We see how many questions were raised and can still be raised, how confused some answers are, how unforthcoming others have remained; but we also see how interesting such a confrontation is, and the multiplicity of the tasks it instigates.

We may not yet fully realize the scope of today's technological acceleration, nor quite what to do with this world rife with so many new dazzling toys. Since many traditional points of reference are lost, new ones will have to be created. Present-day technologies introduce differ-

---

38 See also Claude Rabant, *Being Human*, 136.

■ BEING HUMAN:

ence – for instance, the very difference constituted by the possibility of pursuing a love of sameness through an intervention on the body. And one may ask, as Leclaire did shortly before his death, if the transformations we're experiencing reflect changes in our thought process.

In certain instances, the technological extension of the limits of the body makes a myth of difference. In the realm of computers and virtual reality, the fascination with devising and using new codes, new sign systems, is accompanied by the wonder of gaining some brand-new experience. In obtaining an unaccustomed relation to the world and to itself, it enables the subject to confront and accentuate elements of the human that had, in fact, always existed.

A new experience of time and space, ubiquity and immediacy, display a subjective relation to the body for the most part relegated to second place in Western culture: in the traditional binary opposition of mind/body, internal/external, reason/instinct, the body's otherness is mostly reduced to a mechanical materiality. Some new technologies, on the contrary, reinforce the body in its essentially *creative* alterity. We store our own knowledge in a computer, travel virtually – and marvel at this. The body manifests its extension, its malleability. It's confirmed, reinforced in its *being mind* – and in its capability of being elsewhere. This very malleability, which readings, dreams, meditation, had always borne witness to, is thrown into new relief; it becomes a familiar, commonly shared notion.

When Douglas Trumbull describes the success of his virtual film *Back to the Future* as the fruit of "a huge physical manifestation of an out of the body experience,"[39] he is expressing the paradox of a certain vision of the body: that of a body that limits experience, when in fact it is what makes it possible. Thanks to the senses the body perceives itself to be at once here and there, inside and outside. In shattering a restrictive, closed image of the body, the new technologies nurture an experi-

---

39 *Being Human*, 125.

ence of the body as *edge*, as an "interface"[40] between inner and outer world.

Richard Serra explains how using a computer in planning his *Torqued Ellipses* simplified the conception of hitherto nonexistent forms, allowing him to work out an architecture in which the body finds itself to be simultaneously centered and decentered, according to a real space that seems virtual. Many of Merce Cunningham's recent productions, designed by computer, lead the body into otherwise unthinkable articulations and new rhythms, new stases, new movements sketch unprecedented visions. As Richard Teitelbaum remarks of his work in musical composition, using new technologies allows him to discover his own fantasies; and to create harmonic material from them.

"It's some sort of major crisis in the relation of thought to the real that constitutes the current floating and drifting of thought," remarks Leclaire.[41] The implementation of new reading and decoding systems, from the realm of biology to that of chemistry, physics, or computer science, the multiplication of the symbolic hold over reality through the production of new signs, moves the old boundaries between the symbolic and the real. In transforming aspects of the relation subject/world, contemporary technologies force man to confront new products of his own subjectivity, and a different limit experience returns to the human being the potentially transgressive quality of his thought processes.

To consider the scope of present transformations avoiding irresponsible indifference and conservative moralism, may be challenging. Yet, we want to learn from our products, learn to make use of their creative and transgressive range, learn to take measure of the new referents they introduce.

---

40 *Being Human*, 155.
41 *Being Human*, 152.

"The natural interface between the symbolic and the real."

The Biological Truth Criterion: A Shaky Foundation[1]

Serge Leclaire

I would like to sincerely thank, on the one hand, Madame Piera Aulagnier for having recalled so clearly and in such striking fashion, the crux of the psychoanalytic position on the subject of identity and desire; and on the other, Monsieur Patrick Guyomard who so brilliantly illustrated the modalities and possibilities of psychoanalytic work as applied to situations of adoption and artificial insemination.[2]

These two contributions will allow me to limit myself to an attempt to analyze a point that seems to have posed a problem for group v of the High Council, the biological truth criterion.

The current state of work, as Madame Françoise Héritier-Auge reminded us, shows that, among the four criteria which determine filiation in our tradition, namely: the criterion of legitimate parenthood legitimized (and even naturalized) by marriage; the common law criterion; the criterion of will; and the biological truth criterion, the tendency is to exaggerate the importance of the latter by granting a dominant truth-value to "biological truth."

Here, from the outset, we witness the return of the question of the opposition between a "natural" order, biological in this case, and a "social" order about which we have just been reminded "there is nothing natural about social organization."

---

[1] "Le critère de vérité biologique: un appui bancal" first published in *Topique*, no. 44 (Paris: Editions Dumod, 1989).
[2] At the time of my talk I had not yet heard Guy Rosalato's contribution which substantially broadened the psychoanalytic point of view.

I will therefore pause to examine "order," the common denominator of the terms which are in opposition. Whatever predicate specifies it, order consists of a symbolic organization. The present state of science and of biology in particular, bears adequate witness to the fact that biological order, if not biological "truth," consists, above all, in an organization of symbols (names, numbers or letters) assigned to elements made visible by experimentation: cells, molecules, proteins, antibodies, etc., which can thus be identified.

Natural order in general and living (biological) order in particular, can today be read as a symbolic order, a system of laws and, as such, does not intrinsically differ from social order: symbolic reality, "human nature" as, essentially, a domain of speech, language and writing, is at work there.

To put it simply and briefly, I would say that the opposition between nature and culture today takes the shape of an opposition between two different types of symbolic activity.

The first type, to which scientific research and production belong, has as its top priority to discover and account for the order of things so as to deduce a practice, a utilization, an exploitation of "the riches of nature." The second type, to which social practice belongs in all of its aspects, political and especially ethical (regulation of the relationships among subjects, men and women, individuals and groups), should have as its priority to institute the determining function of the symbolic order in the ordering of the relationships among subjects. The legislator's work belongs to this second type of symbolic activity.

While there seems to be today no difficulty or obstacle to the deployment of a symbolic order that "accounts for," as witnessed by the successes of science and its production of the semblance of a universal language, we find it very difficult, in the current state of our civilization, to implement, other than through archaic forms, the work of the symbolic which "orders" human activity and relationships. Not without reason is this primordial function of the symbolic mistrusted. It is because it continues to be perceived and experienced as something partaking of a supernatural, transcendent order which in fact deprives "human nature" of its essential quality of speaking being. Under the pretext of legitimately challenging the tenacious belief that the word can only come "from above" (from a sky, a God, a church or a State, a prophet or a master), we go so far as to challenge, in the same motion, any power to order the symbolic order.

Since its foundation, psychoanalysis has asserted itself as a practice that seeks to rehabilitate the virtues of the word. Within each cure it strives to give [back to] the subject the opportunity to speak.[3] Psychoanalysis thus challenges any belief or ideology that attributes the origin of ownership of language to some supreme being. The psychoanalyst works with speech and language as primordial constitutive elements of human nature, just as a biochemical therapist brings into play molecular interactions thanks to modern pharmaceuticals. This is how the psychoanalyst is able to testify to an experience which, in more ways than one, partakes of a chemistry of signifiers (or an alchemy of words) rather than of a "magic" of language. It is also why he is today probably the only practitioner to systematically implement a rational technique based on a "natural" theory of speech and language. Of course, the logic which governs it can be surprising because it considers the meaning-effects of speech to be secondary, granting a privileged status to the elementary nature of words, their "signifying" values, their molecular functions, to put it metaphorically, which have no "meaning" other than the qualities representative of their belonging to the hypercomplex system of speech and language, the only one able to account for the fundamental nature of the subject (of the subject-effect, strictly speaking) other than in the mode of a pure intuition or an ideological construct. In brief, the subject, in the psychoanalytical sense of the term, is the irreducible difference which maintains the "tension" of the for-

---

3 Translator's note: "rendre la parole au sujet" – parole as speech but also as a psychoanalytical subject's ability to turn the symptom back into an utterance.

THE TECHNOLOGICAL EXTENSIONS OF THE BODY

midable energetic system of desire, the one no machine will ever be able to duplicate.

In any human procreation, natural or artificial, blind or farsighted, this energy is at work. In our work, we should consider it a parameter not to be circumvented, which would usefully specify the "criterion of will." It will be understood, or at least intuited, that this approach to the living human allows for a productive re-thinking of the obstacle encountered by the High Council and which consists in, let us recall, the tendency to "exaggerate the importance of the biological truth criterion and give it a dominant value."

How? First of all, by relativizing the "weight of truth" of the biological criterion whose power of attraction and even fascination owes as much to the mastery of a symbolic order it puts into play, as it does to the reality of the processes it accounts for. The technique this procedure allows and the practice it produces engender a truth effect in the strongest sense of the term (*adequatio rei intellectu*) but it is a truth that relates to objects which certainly do determine the process of reproduction. However, it leaves out the truth of "subjects," the hard nuclei of the system of desire at work in reproduction and veritable "objects" of the symbolic order, the one that regulates and orders relationships among people. Here it is possible to grasp the crux of the problem we encounter in our work, a difficulty which arises moreover each time the question of ethics is considered (all too rarely, it is true!). The ethical order, as I see it, consists in a body of laws and rules which govern the "hypercomplex system" of relationships among subjects, and therefore tells "how to live with the other." It is based on a conception of human nature which differs according to civilizations and eras, always influenced by the current state of beliefs and knowledge, affected by cultural experience and interactions. No matter its foundations, its substance is essentially symbolic, written or spoken laws or rules. These laws concern an object (human nature or subject) whose reality is substantially conditioned or determined by the ideas held about this object, if not by the laws themselves. It is commonplace nowadays to point out that the

■   BEING HUMAN:

great systems of belief or thought (religious, philosophical, or even political) which, in the course of history, have assured the remarkable cohesiveness of important human groups (civilizations, societies) through the "view of man and the world" they proposed or imposed, have now lost, if not their dogmatic cohesiveness and nostalgic claim to universality, at least a great part of their power and credibility: "a crisis of ethics" which, for my part, I would call a crisis of the power of the symbolic.

It is only natural then that the ethical demand should be addressed to a type of symbolic system with a proven coherence and effectiveness: scientific discourse. For lack of an answer to the question, "how do we live with the other?," at least some echo can be expected to the question, "how do we live in good health?" if it is addressed to medical discourse's imposing system of knowledge! To my mind, it is from this displacement of the demand toward the place where the power of the symbolic has proven itself – the power of accounting for experimental data – that the excessive weight of the biological truth criterion we face, derives.

How to acquire the means to overturn this momentum, to rehabilitate a power of the symbolic in its legitimate place, the subject not the object, is the question to which the psychoanalyst can contribute.

It simply involves recognizing that the object of the human sciences in general and of psychoanalysis in particular has its own consistency which depends on a non-ideological conceptual elaboration which is neither religious nor metaphysical or scientific; that is to say, an elaboration which is pertinent to the realities it addresses; realities which essentially consist of relationships – fantasies, drives, desires – among terms which can be conceptualized but not objectified: subject (of desire), representation (unconscious), object (of the drive).

To nevertheless make use of "scientific" metaphors, I would say that our physics of human nature (physics of *phusis*: nature) would belong to the order of speech and language, its atoms, words or phonemes and its elementary particles, "signifiers." Psychoanalysis is a systematic approach to a space made up of dimensions other than those which account for the geometry of a three or four dimensional world. The reality of this no less "natural" space is made of "dimensions" (memory, forgetting, drives, desires) for which the categories of quantity and measure are not pertinent. The concepts derive from phenomena such as repetitions, resistances, memory lapses, reminiscences, mutations, spotted *in vivo* in the unfolding of the cure, which allow the extraction of structural constants, of specific conditions of interactions, and more generally of the elaboration (keeping to my scientific metaphors) of a physiology of the signifier which more closely resembles a hypercomplex and unstable immunological system than a pyramidal and unifying explanatory system.

But here, at least in this type of practice and elaboration, the symbolic is at work in its primary aim of affirming or causing the ever-present re-emergence of the human thing.

Even in the procedures of artificial fertilization, desire is at work, not only in the project of procreating in the face of certain physiological or accidental handicaps, but also in scientific elaboration and the techniques it is able to perfect. The components of desire's movement may vary but its conditions remain constant.

To introduce among the determining criteria of filiation a "desire criterion" seems to me a simple way to concretize, in "recommendations" or even in a regulation if not a law, the weight of a symbolic truth considerably more determining for the human thing than a biological reality whose truth function is but relative and only becomes exalted, as I think I have shown, by default.

It remains to be seen how to account for the force of desire in terms that do not surreptitiously reintroduce religious, metaphysical or even scientific hypotheses. The vocabulary of psychoanalysis or, better yet, the conceptual framework it has been forging with Freud, Lacan and many others, may help, if only this undertaking is remembered by our interlocutors...and by ourselves. As for me, it is what I do wish for.
I would like, in any case, to thank Madame Francoise Heritier-Auge and the High Council for having initiated this meeting, and express the hope that it will have an impact.

Human Individuality in the Age of DNA Diagnosis

Robert Pollack

> Science is the criticism of myth. There would be no Darwin had there been no Genesis.
> — W.B. YEATS

Among the many myths that we seem to be unable to escape, despite all the evidence of our science, is the notion that there once was, or once may be an ideal person. Life comes to us in species. From Plato and Aristotle until today, the notion of an ideal version of each species existing somewhere in the natural world, has underlain the view of life held by many otherwise sensible, modern people, some of them scientists and doctors.

Whether they liberally imagine the exemplar of a species to be an average of the individuals in it, or conservatively conjure up an abstract ideal to which all the individuals should aspire, hardly matters. In either case, they have made the same mistake. By considering the individual variations that distinguish one person from another to be deviations from an ideal, they ignore the existence of natural selection, a mindless mechanism that has been producing new species from old for the past four billion years, since life on this planet began.

Since the 1860s – the decade that saw the freeing of America's slaves, the liberation of Russia's serfs, and the publication of Darwin's *Origin of the Species* – natural selection has haunted the dreams of every romantic. It is an adequate and testable explanation for our origins that in one stroke eliminates the possibility of an ideal human, and that celebrates human individuality and the differences among us as essential to our birth, and to our continued viability as a species. Since that book was published, the evidence anyone has been able to accumulate remains consistent with its basic conclusions, especially as they apply to our own species. Natural selection can create new *species* only by choosing from among the individual, inherited variants born within every living species. This leaves a reasonable person with only one choice: to abandon the notion that an ideal person could ever be found or even described, or to abandon the astonishing, interlocking accumulation of evidence for our own emergence as a species by the operation of natural selection. That choice was clear to Darwin himself, and it left him incapacitated by various psychosomatic illnesses for decades, until Wallace's competing manuscript obliged him to disgorge his own notes; Darwin never considered *Origin* a finished work, even when it sold out its first five printings.

In the past decade or so, the resistance of even the most sophisticated of Western societies to the realities of natural selection raises a new problem at the intersect of medical and legal ethics: defining the propriety and the utility – if any – of DNA-based testing for inborn differences. The ethical presumption of medicine, extending to medical science, is the overriding desirability of creating and extending optimal lives for individuals, one individual at a time. DNA-testing may do this in some cases, but in many cases it accomplishes the opposite, by revealing a fact about a person's future fate, without offering anything for that person to do about it. In those cases it reduces the time of optimal life for that person, by revealing information that shrinks opportunity, destroys happiness, and threatens well-being and future medical care.

When this may happen, and no matter whether society as a whole would benefit from the information, the test would be an unethical event unless it were wholly voluntary and wholly confidential. But how to protect the privacy of such information? That is impossible, so long as insurance is predicated on informed risk, and so long as companies can legally ask for the information in order to set the membership of their various medically insured pools. Thus the first prerequisite of ethical application of DNA-testing, would be to make insurance a wholly-pooled enterprise. In other words, the demand for ethical DNA testing should spell the end of private insurance for health or life.

■ BEING HUMAN:

Beyond the immediate need to change the laws so that they either end insurance as we know it, or protect an individual's DNA from being analyzed as a requirement for insurance, is the need to clarify the meaning of these tests in biological terms. There is no "gene for" anything. To think so, is to confuse a complex and very long chain – the way the many genes of a person interact to make the body and the brain – with any link in that chain. Break any link, and the chain is broken: that is the meaning of a single-gene disease like Huntington's chorea. But when the chain is whole, that does not mean that any single link at risk is the sole source of all the brain-functions damaged when one such link is damaged. The existence of a "gene for Huntington's" failed brain function does imply the existence of a "gene for normal" brain function in anyone else.

Yet that is the presumptive meaning many people put to the test: they inherit one version of one gene and they are sick, another version of the same single gene and they have a normal brain. There, of course, the Platonic ideal peeks from behind the curtain of unspoken presumptions: "Normal brain" in that context can only mean singularly, ideally normal; variations can only be diseases. This is manifestly not so, and our species's grand triumph of robust variation – our astonishing outbred difference from person to person – is lost in the confusion.

The problem of what a DNA test means, gets even more sticky when the platonic ideal of Normal is applied not to DNA-based inheritance, but to a mimic of genetic inheritance, the passing down of habits of the mind from parent to child. Some single-gene diseases are inherited in a recessive pattern. This means that the difference shows itself only in people inheriting a damaged version of the gene from both their parents; persons receiving one functional copy from one parent are not effected by the presence of the non-functional version from the other person.

Genetic recessiveness has its mimic in the passage by parents to their children of experiences in their own family's past, experiences, they may not themselves show in their own lives. As psychoanalytic researchers have shown in abundant and depressing detail, sudden disaster for which there can be no planning, and no possible genetic content – near-death in a concentration camp for instance – can be passed on to a third or even fourth generation as a precise, reproducible series of social and clinical difficulties. These are not resolvable by DNA-data, but they are by solving the traumatism of the past, through psychoanalysis.

The two kinds of inheritance – familial and genetic – come together in the moment when a person learns that the family history of fear of a disease has been resolved in their own case by a DNA test. At that moment an "inherited" family history is derailed by new information that requires the entire family story to be re-understood: the psychic strains are likely to be as great whether the DNA data are "good" or "bad." The implication here is simple: not only should DNA-based diagnosis be voluntary, restricted, and private; it should also take place in a psychotherapeutic context, rather than a medical one.

When one examines the current American legal situation in these contexts, its absurd, backwards dysfunctionality becomes evident. While a psychoanalyst or psychologist may keep potent and therapeutically useful knowledge of a person's family-inherited past wholly confidential, no one can keep similar DNA-data – another form of potent knowledge of the past – from anyone else's scrutiny. Yet once it is known, DNA's news of the past cannot be disentangled from context in any one person's life. I conclude that there can be no ethical, clinically-sound reason to obtain either form of information, unless both forms of revelation of the past are equally protected from abusive scrutiny. That is where the law should take us.

■

Psychoanalysis and Genetics: Clinical Considerations and Practical Suggestions

Andrée Lehmann

Consultation with a medical specialist in genetics has a clearly defined purpose: "to evaluate the subject's risk of genetic predisposition" and to inform him or her about "possible avenues of medical care." The distinctive feature of consultations of this kind is that they do not fall under the rubric of therapeutic practice: they produce a predictive knowledge which as often as not has no immediate implications for cure.

This acknowledged gap between knowledge and its effective application places patient and physician in a novel situation as compared with traditional medicine. No doubt it is the unease, not to say the alarm, that this situation causes geneticists that has led them to turn to mental health professionals of one kind or another. Occasionally, a psychoanalyst is called upon to assist patients who may be troubled by the kind of information they receive in this context. Naturally, confronted by such a task, no psychoanalyst could fail to consider what the likely effects on the patient of gaining such knowledge might be.

THE CONSULTATION ITSELF

Consultation with a geneticist has two stages: genealogical research and physical testing. The goal of the first stage is to ascertain whether there is a chance that the illness under consideration – female cancer, in the case of the consultation that I observed – may have been transmitted genetically. With this in mind, the patient is asked to seek out any family members who can help supply the information needed to reconstruct the family's medical history. On the basis of the data thus obtained, working with the patient (and perhaps with members of the patient's family), the geneticist will establish the genealogy of whatever illnesses have occurred in the family. The tree that emerges should make it possible to establish whether occurrences of the same illness may have – or cannot have – been transmitted genetically.

This finding will be immediately conveyed to the patient, along with such advice on monitoring and screening as is consistent with medical knowledge of the condition concerned. This moment is especially delicate, not just because of the nature of the information imparted, but also because genetic transmission is by no means – far from it – the sole factor that heightens the risk for cancer. Thus, closer observation of a patient (as compared with that recommended in any case based on the risk profile of the population to which he or she belongs) may even be ordered at the very same time as a negative result comes in as to that patient's genetic risk.

Where the genetic risk result is positive, the geneticist will request that the patient take a blood test in order to ascertain whether the suspect gene for the illness in question (in the case of breast and ovarian cancers: $BRCA_1$ or $BRCA_2$) is in fact present, and, if so, whether or not it has undergone mutation. The blood samples obtained at this time may also be stored for later testing.

The search for the gene can take several months, at the end of which period a letter will be sent to patients advising them that the investigation results are at their disposition. Patients decide for themselves whether or not to obtain the findings. They also decide whether to inform other family members, and should they opt to do so, the task itself falls to them also.

*

Obviously, no part of this whole investigative process can be a matter of indifference to patients. The inquiries that they make following the initial consultation oblige them to reflect on both their individual

BEING HUMAN:

and their family histories, to renew contact with relations of whom they have lost sight, perhaps for many years, and so on. In the course of the consultation the geneticist-physician may well ask unexpected questions, press for long-forgotten details, and urge that no one at all be left out. He will inevitably register remarks made about this or that family member, and note the attitudes, reactions and doubts that his questioning evokes in the patient. Physician and patient are liable to collaborate, so to speak, in the creation of a delicately nuanced dialogue.

When the findings are communicated to the patient, the physician will strive to give as comprehensible an explanation as possible – always bearing in mind the specificity of each individual – of the genetic predisposition that has been discovered. In all cases, the patient should receive a nuts-and-bolts account of the procedures that have been followed and be clearly apprised (as appropriate) of their experimental character.

The protocols governing these consultations have been drawn up with the utmost care and thoughtfulness by doctors much concerned with the ethical aspects of the situation. But no matter how many precautions these specialists take, no matter how much empathy they may manifest, there is no getting around the fact that at the end of the consultation process patients find themselves alone with the new information they have received, confronted by an absence of certainty, with no prospect of an adequate therapeutic response, and faced with a new set of responsibilities.

*

Research on the aftermath of consultation, carried out for the most part by social psychologists or epidemiologists, has revealed that patients experience heightened anxiety and paradoxical reactions. Anxiety has been found to increase as much amongst women confirmed as being at no genetic risk as amongst those for whom a problematic gene has been identified. Similarly, recommendations to the patient regarding preventive measures may be followed or not, there being no apparent correlation in this regard with positive versus negative consultation results. Patients seem to have difficulty grasping the notion of risk.

For the psychoanalyst there are many questions in need of answers here. How is it that these women are so willing to undertake consultations, and so cooperative when they do so? Can this be attributed solely to the persuasiveness of the doctors? What is the nature of the *demand* that these patients make upon a knowledge which is liable to take the form of a verdict? Why do these women seem overall to be less affected by anxiety than might be expected? Why do some of them appear satisfied with what they are told, even when it is bad news?

It was only after realizing that these consultations were designed for women from families *already considered at risk* that I was able fully to grasp their true purpose and to outline a clinical attitude and a way of thinking about them based on fresh criteria. The fact is that the women who consult the geneticist, whether or not they have developed cancer, already have questions about their future, about the origin and transmissibility of cancer. They are already full of doubt and uncertainty, hence already subject to anxiety.

*

These women patients consult the geneticist to get answers to their questions.[1] True, geneticists are not doctors in the customary sense – but nor are their clients patients in the customary sense. They do not come to consult because they are ill, even if they happen to be; they come to consult about a possible risk that they may fall ill sometime in the future.

What they get is a reply. Even though this reply may bring no absolute certainty, even though it may be complicated and hard to under-

---

[1] This text was written in part to address the concerns of physicians who wonder what type of psychological support should be provided to patients engaged in oncogenetic consultation. This is particularly true of the remarks which now follow.

BEING HUMAN:

stand, it does supply a piece of discerning and objective knowledge about familial and personal predisposition.

Reactions to such replies present striking differences. Some women seem satisfied, evince feelings of relief, proffer their thanks and resolve to follow the preventative recommendations. Others, by contrast, seem just as anxious as before, if not more so.

What is very often noted (and all the studies stress this) is that the information given is not easily absorbed. The notions of risk and probability, and the distinction between genetic and general risk factors, are poorly grasped. Even when they appear to have grasped them, patients are liable to say things or have reactions betraying the fact that they have not really understood. Two logical systems seem to coexist, so that even where an intellectual comprehension would seem to have been achieved, it fails to override pre-existing anxieties or beliefs.

*

Consultation with a geneticist thus poses two main problems, that of the disjunction between objective knowledge and the mental state of the patient, and that of the anxiety that it seems to occasion. These two problems warrant a deeper analysis of the psychological responses to such consultations.

THE MENTAL IMPACT OF CONSULTATION

All the studies carried out to date underline the necessity for more thoroughgoing investigation of the mental impact of oncogenetic consultation.[2] From the psychoanalytic point of view, such investigation cannot

---

2 For more information on oncogenetics, see the collective work *Hérédité et cancers; vers une médecine présomptive* (Paris: Editions Lavoisier, in press); and the collective study *Prise en charge des risques génétiques des cancers du sein et l'ovaire* (Paris: INSERM, in press).

be confined to an evaluation of anxiety and the behavioral disturbances that are liable to accompany it, for anxiety is a normal defense mechanism when a subject is confronted by a painful situation that he is able neither to combat nor to escape. Disturbed behavior is merely an expression or manifestation of an internal state with its own dynamics. Depression, anxiety or confusion as reactions to such situations are not necessarily pathological. They simply reactivate the subject's relationship to his family or to the outside world.

In order to assess the need for optimal psychological *support* and *follow-up*, and to decide what forms these should take, the psychological effect of consultation must be examined in depth. In our experience this effect has three aspects: the doubt the subject suffers from, his or her fear of illness, and the disturbance of family equilibrium.

As I have already noted, *doubt* antedates the consultation itself. It torments and overwhelms the subject, destroying all his firm points of reference and affecting everything in his world. Release can only be achieved by ruses, and is always transient. Doubt consequently gives rise to a diffuse, permanent anxiety.

Anxiety is the feeling that one is threatened by an imminent danger from which it is impossible to escape. The fear of cancer can therefore generate anxiety. And genetic consultation, inasmuch as it is liable to reinforce the feeling that the threat is imminent, is bound to increase such anxiety, especially if it fails to relieve doubt.

The subject's fear of cancer certainly has a basis in reality (all the women I observed had had to deal with illness, either personally or in their families). But it also has a basis in the psyche, and this in two respects: in the first place, it builds on the fantasies that cancer evokes (physical and mental decline, abandonment, loss, and mourning); secondly, it is sustained by the place that the idea of cancer has taken or may take within the family constellation.

This last aspect is particularly relevant to genetic consultation, in the course of which *family relationships* are both thoroughly revisited and thrown into turmoil. The reconstruction of the family tree leads patients and their families to dredge up particular past events, particular long-forgotten or long-eluded family traditions, or particular sets of beliefs acknowledged by family members themselves to be irrational. A return to family history is set in motion, complete not only with its discoveries or rediscoveries but also with its inevitable revisions. The tree technique provides this return with an anchor, a set of markers, a reference grid that facilitates identification. The subject's connection with the family line is rethought and oftentimes in consequence more effectively assimilated.

All the same, the fact that this re-evaluation of family ties takes place under the sign of illness and death is bound to take its toll. For one thing, it encourages the bringing up of the most carefully concealed issues, everything to do with the family that is felt to be baneful (secrets, unforgivable acts, unresolved guilt, abandonments, failures, and so on). Any suspicion of a family curse will naturally be reactivated by the idea of a shortcoming in the genetic heritage, although objective knowledge may also serve to relieve such fears.

In addition, the discovery – even the hint – of the existence of a destructive gene plunges the patient into a state of malaise, into a web of internal conflicts. As a child of her parents, for example, she may feel that she is a victim of the family group, hating her parents for having transmitted the bad gene to her and hating herself for hating them. At the same time, as a potential or real parent herself, she may worry about what she might be passing on to her offspring, and feel guilty. According to the subjects themselves, such concern about genetic transmission is their chief motive for consulting the geneticist.

In some cases the reactivation of the most worrisome of ideas and feelings, or the putting into question of equilibriums that have been achieved only at great cost, may produce negative reactions – anxiety, sadness, balking, rebellion, etc. Such reactions can only impede the comprehension of the information, at once new and complex, that the geneticist has to convey to the patient.

Two surprising phenomena repeatedly mentioned in the literature and observable in the wake of consultation may also be attributable to reactivations of this kind. The first is that individuals defined as "at risk" do not ask *if* they will develop a cancer, but *when*: they interpret the notion of risk as implying an ineluctable fate. The second phenomenon is that anxiety does not invariably disappear in subjects with negative test results: their anxiety about cancer is such that they cannot draw a distinction between genetic risk and general risk.

Otherwise, and more encouragingly, the developments sparked by the patient's decision to consult tend to produce a psychological working-through that leads to the establishment of a new balance. This process of working-through is the precondition of acceptance by patients of the message they are being sent, and of their ability to draw the necessary conclusions for themselves.

*

Let me conclude this discussion of the psychological impact of oncogenetic consultation with a word about *knowledge*. Knowledge is an object of exchange; it exists only to the extent that is shared. A knowledge exclusive to one person is in effect a form of delusion and turns into morbid rumination. Incommunicability foreshadows delusional thinking.

In order to incorporate a piece of knowledge in this sense, it must be possible to discuss it with other people. Failing this, it remains abstract, neither real nor unreal, as when people speak of what they know nothing about.

At the same time, knowledge has a relationship to desire. To acquire knowledge, one must feel an interest therein. One must also have authorization to know something, and be capable of comprehending it. Knowledge that is threatening is even harder to incorporate.

The question of *psychological support* should thus be framed as follows:

Women who have had to deal personally or in their family situation with cancer cannot help but be worried about their future and that of their loved ones. To have it confirmed that one is potentially at risk, or carrying a destructive gene, inevitably increases this anxiety and gives new urgency to the questions surrounding illness, death and genetic transmission. The problem for the subject now becomes knowing how to face this new situation, how to assimilate this new information, how to live with these without being taken over by a sense of doom, and how, despite this sword of Damocles, to rediscover a measure of freedom in his or her life.

Once the initial shock has passed, subjects must have time to take stock of their changed situation. Time by itself does not suffice, however. The new information has to be successfully worked out, that is to say, the subject must absorb it, assess its consequences, and find a way to live with it.

Implicit here is the idea that the subject will independently find her own way forward, helped if possible by a family entourage just as implicated in this turn of events as she is; of course, that entourage may sometimes be incapable of dealing with the situation.

Generally speaking, and most of all in cases where the person is isolated, alone with her suffering and questions, it is desirable that one or more interlocutors be found who are capable of listening to her, of encouraging her to voice her concerns while at the same time allowing her to discover her own way ahead.

The form such support should take (whether on an individual or group basis), and the right moments for it to be proposed (whether before or after the patient is notified of risk) are matters that have to be carefully weighed on a case-by-case basis.

In all cases, however, the form of support offered must allow patients to find their own way. It should be designed to help them bring their subjective problems into conjunction with the objective knowledge that they have just acquired. Only by means of such a conjunction will they be able to reach a rational assessment of their situation, and hence make a decision that can be maintained in the long term.

Appearances to the contrary notwithstanding, a rational evaluation from the subject's standpoint is not identical to the one that the geneticist is likely to propose, even though they may sometimes produce the same practical conclusions. The patient need not necessarily espouse the physician's arguments, but she does need to understand them and take them into account.

This implies a work of transference, for the hope placed in the doctor – that is, the demand that is directed to the doctor – is commensurate with the unfamiliarity of the knowledge that the doctor imparts, as also with the level of anxiety of the patient. The gap between scientific knowledge and the ability to cure – a dramatic one in the case of genetic medicine – makes such a work of transference unavoidable for patients in this kind of situation. The genetecists are worried by it also, which is doubtless the reason why they have sought help from mental health professionals. In the event, they have chosen to call on clinical psychologists. As for psychoanalysts, the vast majority of them are either diffident or ignorant on this issue.

The work of transference here consists, as always, in the substitution of elements symbolizing subjective desire for the hopes placed in an omnipotent other. Where scientific knowledge is successfully incorporated *complete with its limitations*, it becomes what it really is, namely an element of reality which acts as a law in the constitution of a "we" at the very site where the you/me struggle had formerly been enacted. The doctor thus resumes his or her proper position as the possessor of a relative knowledge of, and a relative power over the functioning of bodies. The physician's subjective position in the doctor-patient dialog may once more be distinguished from his knowledge in its scientificity.

■ BEING HUMAN:

CONCLUSION

As scientific knowledge advances, it penetrates into fields whose exploration once seemed impossible, and was often forbidden. The consequent emergence of new realities produces powerful reactions: enthusiasm or anxiety, blind faith or, on the contrary, mistrust and even feelings of helplessness. Individuals and social groups are obliged to transform themselves in order to assimilate new ideas and new practices. Little by little they must work out new bases of identity and new ethical criteria.

These considerations clearly apply in the case of genetics. The initiation of the Human Genome Project in the late 1980s still smacked of the utopian. The speed with which its concrete applications have taken root has produced not so much surprise as a feeling of unreality. Personally and ethically, this can only worsen our perplexity. Meanwhile, the gruesome historical antecedents of the word "genetics" hardly inspire confidence.

At the level of the imaginary, the intrusion of the observable and the quantifiable into realms that were once the exclusive preserve of religious belief can encourage the conflation of the biological individual and the human subject. For science, the only thing involved here is another step towards the realization of its project, which is to subject everything to analysis and so arrive at a grand synthesis. And it seems that nothing can arrest science's progress along this path. Our best course, therefore, is probably to devise new equilibriums that will help us assimilate science's impact rather than be put at panic stations by it.

Still, for human subjects another boundary has been shattered. A cognitive adjustment has to be made. In other words, beliefs that are often unconscious, and essential to the sense of identity, are now confronted by an unchallengeable certainty buttressed by scientific research. At the same time, however, this objective certainty applies only in part to the patient herself. In the first place, it has an arbitrary aspect: its edicts are expressed solely as statistical likelihoods. Secondly, it is impossible, on the basis of the current state of knowledge, to recommend appropriate curative measures, for genetic therapy is still in its infancy.

A distinction has therefore to be drawn between the general meaning of this new knowledge – which is bound to spawn changes in humanity's ways of viewing itself – and its particular meaning to a human subject learning that she is (or is not) the bearer of a destructive gene, that she has inherited this gene from this parent or that, that there is such and such a degree of probability of her passing it on to her progeny, etc. Such a subject will be confronted by the necessity of carrying out a work of assimilation which differs radically from the parallel task which faces the community as a whole. (Much will be gained, moreover, when we get a clearer idea of the relationship between these two levels.)

Genetics presents us with a complex discourse, and it behooves us not merely to grasp its meaning but also gradually to invent a new way of living with knowledge that is at present undergoing a permanent revision. We need an ethics for the scientific age. And, as Serge Leclaire put it, the psychoanalytic conception of the subject is undoubtedly the most appropriate one upon which to found such an ethics.

"I don't think it matters to anyone where their eggs and their sperm come from."

Reading, Writing, and the Discourse of DNA, or The Mind of a Molecule [1]

Ona Nierenberg

> And now the announcement of Watson and Crick about DNA. This is for me the real proof of God.[2]
> — SALVADOR DALI

Molecular biology has been the source of significant knowledge about inherited diseases in recent years. Single genes have been identified as the causes of such diseases as Huntington's chorea, sickle-cell anemia, cystic fibrosis and early-onset breast cancer. The promise is that such discoveries will lead beyond screening to therapies which will be able to ameliorate the pernicious effects of these debilitating and life-threatening conditions. However, a gap between such knowledge and the production of cures has remained, in part, because the more molecular biologists discover about genes, the more complex they are found to be. As one journalist surveying contemporary genetics writes, "Every day we do more research on genes, and every day we know less about them."[3]

Yet, at precisely the same moment that many geneticists are grappling with the problem that most genes do not function according to the established linear, predictable, mechanistic model, the notion that

---

1 This paper is greatly indebted to the works of Evelyn Fox Keller (1990; 1995) and Lily Kay (1995), two historians/philosophers of science, who have independently traced the genealogy the discourse of DNA. Each of these authors reveals how absolute power became embedded in the discourse of DNA well before the material discovery of the double helix. Following their lead, it is my aim to show how situating the supposed cause of human behaviors in the genome is itself *an effect* of social and historical events, practices, and paradoxes which have been rendered nearly invisible.
2 This statement was chosen by Francis Crick as the epigraph for *Of Mice and Men*.
3 C. Burr, *A Separate Creation: The Search for the Biological Origins of Sexual Orientation* (New York: Hyperion, 1996), 209.

"we, and all other animals, are machines created by our genes"[4] still dominates the search for biological explanations of so-called "human behaviors." Criminality, homosexuality, alcoholism, manic depression, racial links to intelligence, the propensity to divorce, and homelessness are just a sample of the myriad human phenomena which recent investigations have claimed as genetically determined. Gene-of-the-week media reports demonstrate a distinct lack of limits when it comes to the supposition of genetic causality: "Is a Gene Making You Read This?" inquires a *New York Times* article monitoring human behavioral genetic research.[5] Here it is significant to note that many assertions of genetic determinism are made on the basis of studies which have been published – and publicized – despite frank admissions by their authors that the evidence they contain is inconclusive. Nevertheless, these same authors express indefatigable confidence that these genes will ultimately be found. Of course, it is this conviction which animates such research to begin with – but then, we are left with the question of what animates this conviction?

Eschewing all doubt, scientists working on the Human Genome Project have promised that "once we have found all the genes in the human body and learned their genetic sequences, we will have a blueprint, an instruction book for people."[6] A *blueprint*. An *instruction book for people*. DNA is defined as "the coiled strands in which the key lies to all that an organism is and does."[7] It "carries invisibly within itself all the information necessary to build everything alive. The instructions are written in a code."[8] *Information. Instructions. Code.*

4   R. Dawkins, *The Selfish Gene* (New York: Oxford, 1976), 2.
5   G. Kolata, "Is a Gene Making You Read This?" (*The New York Times*, January 7, 1996), 4.
6   G. Kolata, "Unlocking the Secrets of the Human Genome." (*The New York Times*, November 30, 1993), C8.
7   F.T. Lee, *The Human Genome Project: Cracking the Code of Genetic Life* (New York: Plenum, 1991), 2.
8   J. Levine & D. Suzuki (Writers); Corporation for Public Broadcasting (Producers) (1993). *The Secret of Life: The Immortal Thread* [Television documentary].

■   BEING HUMAN:

While this terminology may seem rather benign and even instructive, the problem is its effect: When DNA is the blueprint, the instruction book, the code, or the key, human existence all too quickly becomes reduced to "follow[ing] the orders of DNA. We have no choice. We are prisoners of our genes."[9] We are prisoners and *machines*: "Robot vehicles blindly programmed to preserve the selfish molecules known as genes"[10] to quote a prominent author on the subject. However unpleasant our fate, we can at least enjoy the certainty of having already been written.

By now, the trope of DNA-as-instruction, code, etc., has become so ubiquitous as to be rendered nearly invisible as metaphor – it is, in fact, DNA's *sine qua non*. But representing life and heredity in this language was not the result of discovering either the "inner logic" or the material structure of DNA.[11] This particular vocabulary has its source in information theory, a discipline which originated in the realm of engineering in the late 1930s, quickly gained prominence as a result of its usefulness in formulating and solving problems of military significance during World War II, and rapidly dispersed throughout the physical, social and life sciences during the 1940s and '50s. The history of the discipline of molecular biology is inseparable from the influence of information theory, for better and worse. In the 1953 article in which Watson and Crick identified the structure and confirmed DNA as the genetic material, they defined it as "the *code* which carries the genetical *information*."[12] This definition has been described as a "stroke of genius"[13] beyond the material discovery of the double helix, for

---

9 Lee, *op. cit.*, 6.
10 R. Dawkins, *The Selfish Gene* (New York: Oxford, 1976), p.v.
11 L. Kay, "Who Wrote the Book of Life? Information and the Transformation of Molecular Biology, 1944–55," *Science in Context*, 8.4, 611.
12 J.D. Watson & F. Crick, "Genetical Implications of the Structure of Deoxyribose Nucleic Acid." *Nature*, 171 (1953): 967 – emphasis mine.
13 E.F. Keller, *Refiguring Life: Metaphors of Twentieth Century Biology* (New York: Columbia University, 1995), 18.

with this publication Watson and Crick institutionalized the deployment of information theory terminology into the domain of biology.

Why is this so important? Because it was only the use of the language of information theory that made it appear that the "secret of life" was a secret no more:

> The kind of pre-established design that is manifest in every living organism is not found among inanimate objects. It was therefore long considered to result from a special agent, some vital force escaping the laws of physics. *Only in the last decades has a mechanistic interpretation of the activities observed in a living organism been considered to be compatible with its properties and behavior. In particular, the paradox came to an end when molecular biology borrowed from the theory of information the concept and term of "program" to describe the genetic information of an organism.* Accordingly, the chromosomes of the fertilized egg are assumed to contain, coded in DNA, the genetic blueprint that is considered to direct the development of the future organism, its activities and behavior.[14]

Here biologist Francois Jacob claims that the use of information theory terminology resolved the problem of life's escape from the mechanistic world view, saving science from having to admit a "special agent, some vital force escaping the laws of physics." This would be no small feat, given that three centuries of modern science had been unable to decipher the "mystery" of reproduction, life's most unique feature, its most inpenetrable "secret." Yet, when we note the omnipotent powers granted DNA in deterministic discourse, it becomes apparent that *this problem has not really been resolved, but, rather, has been occluded through the use of the language of information theory*. The claim that

---

14 F. Jacob, *The Possible and the Actual* (New York: Pantheon, 1982), 13 – emphasis mine.

■ BEING HUMAN:

mechanism has triumphed can be seen as dubious given the identification of DNA, an inert molecule, with a "code," "blueprint," "message," or "instruction book." All of these require *writing* and *reading*. And such activities require *a mind*. Thus, *"written in"* to the discourse of DNA is a *mind* that is reading and writing. The question of whose mind, however, is not only left unanswered by the discourse of DNA, but has remained *unasked*. While explicit references to "vital forces," "special agents," or God do not usually appear in articles regarding molecular biology or behavioral genetics, DNA is conceptualized as "the cause of itself and of all other things,"[15] and the "agency, autonomy and causal responsibility"[16] that were once the exclusive purview of the Divine now belong to DNA. Thanks in large part to the rhetoric of information theory, such dominion has shifted from the heavens to the molecular level.[17] While this particular shift in the universal alignment has been the source of numerous achievements, among its most notable must certainly be the uncanny ability to efface the invisible disembodied intelligence which is fundamental to the discourse of DNA. And although this discourse is a uniquely 20th century response to the age-old question of life, the effort to answer it by deploying a rational, disembodied mind (God) can be traced to the origins of modern science in the 17th century.

*

The theory of Preformation arose in the mid-17th century, during the early stages of the development of the microscope, and remained viable for some biologists until 19th century advances (i.e., cell theory, dating the age of the Earth) rendered it unsustainable. Preformation

---

15 R.C. Lewontin, "The Dream of the Human Genome" (*The New York Review of Books* [May 28, 1992]): 32–33.
16 E.F. Keller *op.cit.*, 9.
17 Ibid., 55.

accounts for the formless appearance of the embryo through the belief that this amorphous quality is an "illusion" of our perception. In fact, the complete set of adult organs was supposed by Preformationists to be always already present in a miniaturized form. These organs were presumed to be minute and "liquid" in character, which prevented them from being seen. What appeared to be a gradual process of formation via development, they believed, is actually solidification and increase in size. These "preformed germs" were believed to have been put into place by God at the beginning of Creation. Thus, all descendants of an original set of parents of all living organisms – all creatures existing at any given time – were supposedly already present at the origins of the world. In the case of human beings this was illustrated as an infinite regression of homunculi present inside Eve, or in later versions, inside the "head" of the sperm. One tiny human being was thought to be nested inside another like an endless series of Russian dolls. Supposing that adult organisms were always already present, and mature organs always already formed, allowed development to be explained mechanistically. According to Preformationist theory, "there was no true formation of any new individuals and the development of new offspring was simply a process of solidification and expansion, which was assumed to be the result of the same simple mechanical laws that also controlled physical processes."[18] Preformationism, with its appeal to God the rational creator who put matter and the laws of motion into place, appeared to solve the problem of explaining development mechanistically – at least, for a time.

Substantial improvements in the microscope by the mid-18th century allowed for finer observations of embryonic organisms. Using this new technology, biologist Kaspar Friedrich Wolff observed that the miniature forms posited by the Preformationists were simply not there from the start. In his view, organs were actually *formed* during development,

---

18 E. Gasking, *The Rise of Experimental Biology* (New York: Random House), 38.

■ BEING HUMAN:

rather than being already present. Since this could not be accounted for mechanistically (*transformation* is not a simple case of matter in motion),[19] Wolff declared that the organism goes from a simple form to a more complex one through the active ongoing intervention of God, who acts as a vital force shaping the organism in its various stages. This school of thought, which believed in actual *development* shepherded by God, was called Vitalist/Epigeneticist.

But even as the qualitative changes taking place during development became easier to observe, the Preformationist point of view that the miniature organs were present, though not visible, remained dominant. Many natural philosophers believed that the Epigenetic theory relied too heavily on the ongoing intervention of God, a supernatural vital force. They believed that Preformation was ultimately a more properly scientific, or mechanistic theory, despite its shortcomings, because it de-mystified development through the assumption of already-present organs and assumed that God created life to follow the rational mechanistic laws of matter in motion.

The canonical history of science holds that preformationism went down in "flaming defeat,"[20] and so it might seem, given the thorough confirmation that has been provided for the epigenetic thesis that true *qualitative* change takes place during development. However, this view of scientific progress as the march of truth ignores the rather obvious and insistent preformationist principle at the core of our concept of the

---

[19] Mechanism does best at explaining phenomena which are repetitive and ongoing, for example, the beating of the heart. Biological processes which involve historical change and irreversible transformations, like development, do not lend themselves to mechanistic theory: "The problem is that the machine metaphor leaves something out...The problems of biology are not only the problem of an accurate description of the structure and function of the machines, but also the problem of their history...Of course, machines too have histories, but a knowledge [of them] is not an essential part of the understanding of their workings" R.C. Lewontin, "Genes, Environment and Organisms," in *Hidden Histories of Science*, R.B. Silvers, ed. (New York: New York Review, 1995), 118.

[20] S.J. Gould, "Forward," in C. Pinto-Correria's *The Ovary of Eve: Egg and Sperm and Preformation* (Chicago: University of Chicago, 1997), xiv.

gene: "The gene is...a 'message,' and the truth of preformationism is that what is 'preformed' is the information for making an organism."[21] Thus, it can be said that the discourse of DNA is "the final successor to preformation,"[22] lacking only the acknowledgment that for something to be preformed, there must be an agency *per*forming the *pre*forming.

*

Preformationism was originally known by the name "evolution," from the Latin *evolvere*, which means to disclose or unfold. As a noun, *evolutio* is the Latin word which refers to the unfolding or reading of a scroll.[23] "Evolution" was apt terminology for the theory which described mini-adults as "unfolding" from their cramped positions in the embryos and increasing in size until they were born.[24] But by the late 18th century, epigeneticism, which had become the dominant theory of embryological development, became known as "evolution." (In fact, the logical similarity of epigeneticism and preformationism, once viewed as opposing and competing theories, can be detected by the way in which "evolution" slid from naming one to naming the other.) Gradually, "evolution" was broadened to include species through the idea of "recapitulation," which held that the development of the embryo begins in a general, undifferentiated state common to all organisms of its general type and increasingly differentiates to a more distinct state linked to its particular species. This was largely taken to mean that organisms progress from a less complex state to a more complex state as they develop.

21 J. Mazzeo, "Introduction," in O. Hertwig's *The Biological Problem of Today: Preformation or Epigenesis? The Basis of a Theory of Organic Development* (Oceanside, NJ: Dabor Science Publications, 1977).
22 Pinto-Correia op. cit., 309.
23 R. Richards, "Evolution," in *Key Words in Evolutionary Biology*, E. F. Keller and E.A. Lloyd, eds. (Cambridge, MA: Harvard,1992), 95.
24 S.J. Gould, *Ever Since Darwin: Reflections in Natural History* (New York: Norton, 1977), 35.

Of note, Charles Darwin did not refer to his theory as "evolution" in the original edition of *The Origin of Species*; he used the phrase "descent with modification." But social theorist Herbert Spencer (who coined "survival of the fittest") persuaded Darwin to use the term "evolution," and its teleological and progressive implications did not take long to resurface.

Unlike the theory of special creation, with its God-given goal of perfection, Darwin's evolution posits "no perfecting principles; no guarantee of general improvement."[25] Natural selection only pertains to very specific, local conditions in which organisms live, and operates only on those variations which are produced by chance. Darwin's theory posits that organisms will increasingly adapt to their own particular environment over time, and that this process also leads to the diversity of species. Thus, Darwin's theory provides a theory of causality without any innate purposiveness. It explains the present in terms of the past, but offers no predictions for the future, given that the circumstances an organism may have to adapt to are random and unpredictable.

However, the ideal of progress as natural law permeated Victorian social, legal, historical, and scientific thought. Darwin was forced to explicitly refute the attribution of progressive and purposive ideas to his theory by his opponents and adherents alike. But despite Darwin's protests, the popularization of evolution as a perfecting principle proliferated in the late 1800s. According to Darwinism (as opposed to Darwin himself), all attributes and actions of organisms and species can be causally attributed to "survival of the fittest" and the progression to complexity (from single-cell organisms to *homo sapiens*). By thus raising natural selection to the *sole* mechanism of evolutionary change, Darwin's interpreters transformed evolution into a process whereby organs and organisms could be perceived as ideal relative to

---

25 Ibid., 31.

the environmental challenges present (as in today's sociobiology).[26] This ideal is, of course, Nature's *purpose*. And when evolution is imbued with purpose, it is difficult to distinguish from God's grand design. It was only by maintaining the enormous significance of chance in evolutionary processes that Darwin was able to avoid the presumption of a mind at work, a "ghost in the machine." Once chance disappears, we are confronted with the mind of God: though evolution according to Darwinism may challenge the Biblical account of Creation, "God" as omnipotent "planner" is no less essential to its logic.[27] Thus, the Pope can embrace evolution, as he did in 1996, while in no way renouncing God.

*

When Darwin created his theory of "descent with modification," there was as yet no established scientific understanding of intergenerational transmission. Darwin created his own theory, Pangenesis, which can be seen as significant largely by virtue of its inadequacy. Inspired by Darwin's flawed notion of a unit of transmission which is based on the living body, biologist August Weissman discovered that the germ cells were the only source of inheritance, and that

---

[26] Darwin's own emphasis on natural selection as opposed to the other means of evolutionary change he posited (i.e. correlation of growth) contributed to the metamorphosis of his theory into Darwinism, as did certain remarks in *The Origin of Species* in which he attributed purpose to natural selection. "Of Darwin's thought, as of that of a great many thinkers, it could thus be said that it did not succeed in fully keeping pace with its own novelty," D. Lecourt, Introduction to F. Gros' *The Gene Civilization* (New York: McGraw-Hill, 1989), 7–17.

[27] As Lacan writes, "In evolutionist thought, although God goes unnamed throughout, he is literally omnipresent. An evolution that insists on deducing from continuous processes the ascending movement which reaches the summit of consciousness and thought necessarily implies that consciousness and thought were there from the beginning." Jacques Lacan, *The Ethics of Psychoanalysis,* Book VII, 1959–60 (New York: Norton, 1992), 213.

■ BEING HUMAN:

the germ plasm itself was isolated, immortal and inviolate. Weissman's work, along with Darwin's theory of evolution and advances in biochemistry, allowed for a new approach to the application of mechanistic laws to the question of heredity. But it took the rediscovery of Mendel's work at the turn of the century to establish the discipline of genetics. Mendel's genius was not just finding the hereditary material, but discovering a mathematical technique for quantifying the transmission of traits from generation to generation.

Mendel provided the conceptual framework and experimental techniques which permitted the discipline of genetics to be established, while also circumscribing a domain that excluded from consideration numerous questions which had formerly been considered central to understanding reproduction. Notably, differentiation and growth were not accessible to the same statistics, experiments and methods that yielded data on the heredity of simple, quantifiable traits. These more complicated, multidimensional (rather than linear) aspects of the origins of life became severed from questions of heredity. Genetics and embryology, once inseparable, became distinct disciplines by the early 20th century, with genetics assuming ultimate power and prestige as the science of the basis of life itself.[28]

The structure of institutional science in the 1930s fully established the split between embryology and genetics. Warren Weaver, who coined the term "molecular biology" and was head of grant dispersal at the Rocke-feller Foundation in the 1930s and '40s, played a pivotal role in this division. A mathematical physicist and dedicated mechanist, Weaver deployed the enormous resources at his disposal specifically toward applying the techniques of physics, chemistry and mathematics to the study of life. This amounted to a virtual "colonization" of biology by physicists, for according to Weaver's analysis, biology had the problems

---

28 E.F. Keller op.cit., 5–14.

and physics had the technology.[29] Physics also had the power and credibility, due to well-known developments in quantum mechanics. Weaver also brought to the new science of molecular biology the explicitly formulated wish to bring "dangerous" human sexuality into the domain of the rational.[30] In his Rockefeller Foundation progress report of 1934, he wrote:

> The challenge…is obvious. Can man gain an intelligent control of his own power? Can we develop so sound and extensive a genetics that we can hope to breed, in the future, superior men? …Can we obtain enough knowledge of the physiology and psychobiology of sex so that man can bring this pervasive, highly important, and dangerous aspect of life under control?[31]

Warren Weaver's work through the Rockefeller Foundation contributed to the solidification of a mechanistic logic of life, the concept of the gene as both origin and outcome, and a synonymy between knowledge, intervention and control which established the discipline of molecular biology. Yet, early molecular biologists did not use the language of messages, codes, and information to describe their discoveries. This vocabulary emerged in the 1940s "as the technosciences of cybernetics, information theory, and electronic computing began to define the discursive space of the early postwar era."[32]

---

29 P. Abir-Am, "The Discourse of Physical Power and Biological Knowledge in the 1930s: A Reappraisal of the Rockefeller Foundation's 'Policy' in Molecular Biology." *Social Studies of Science,* 12, 343.
30 Genetic determinists have addressed this wish (in theory, at least) by turning genes into the motive force behind sexuality (Dawkins, 1976; Wilson, 1978). According to such thinkers, all species are subject to the imperative to propagate genes: in this, humans are viewed as no different from other animals or even single-cell organisms. Consequently, sexuality is transformed into a rational, predictable force and human sexual reproduction can be more easily situated within a mechanistic paradigm.
31 Cited in R. Kohler, "The Management of Science: The Experience of World War and the Rockefeller Programme in Molecular Biology." (*Minerva* 14 [1976]): 300.
32 L. Kay, "Who Wrote the Book of Life? Information and the Transformation of Molecular Biology, 1945–55." *Science in Context,* 8.4, 616.

■ BEING HUMAN:

Austrian theoretical physicist Erwin Schrödinger co-won the 1933 Nobel Prize in physics, and invented the quantum wave equation that bears his name. In 1944, Schrödinger published his work, *What is Life?*, a book that has been called the "*Uncle Tom's Cabin* of the revolution in biology that...left molecular biology as its legacy"[33] and the "ideological manifesto of the new biology."[34] Numerous physicists lauded this text as instrumental in their leap to biology after World War II (among them Francis Crick and Maurice Wilkins), and many biologists (including James D. Watson, Salvador Luria, and Francois Jacob) also heralded it as a seminal piece of work. In *What is Life?*, Schrödinger posed the question which scientists had struggled with throughout history: what is it that differentiates living from nonliving things? What is the essence of the characteristic of being alive? Schrödinger decided that *continuity of order* was the "secret of life": the fact that individual organisms can maintain life over time, and species can reproduce themselves.

Both staying alive and reproducing seemed remarkable to Schrödinger in light of the second law of thermodynamics, *the* most important law of physics, which held that increasing entropy was the rule of all matter in motion. As the science of heat/energy, thermodynamics posited that action/work required loss of energy, which ultimately led to a running down of any system into complete disorder. Schrödinger, a physicist, found it incredible that living organisms appeared to defy this rule throughout the life span, as well as through the generations. After all, no inanimate matter could escape this inevitable degradation. Whatever capacity or "device" allowed living systems to do so must be the secret of vitality. And Schrödinger decided that this secret lay in the contents of the chromosomes (genes), which possess the quality of be-

---

33 G. Stent, "That Was the Molecular Biology That Was." (*Science* 160 [1968]): 392.
34 R.C. Lewontin, "The Dream of the Human Genome" (*The New York Review of Books* [May 28, 1992]): 31.

ing like *written information*: "It is these chromosomes...that contain some kind of hereditary code-script, the entire pattern of the individual's future development and of its functioning in a mature state." But in addition to being a "code-script," the chromosomes act as an *animating force*: "The chromosome structures [genes] are at the same time instrumental in bringing about the development they foreshadow. They are law-code and executive power...architect's plan and builder's craft all in one."[35] The gene is not only the plan, but is that agent which activates and realizes the plan. It is a "code-script" that reads and writes itself.

Logically, Schrödinger's idea of the DNA as both "law code and executive action" is problematic: "If one takes the notion of a genetic program literally, one falls into a strange loop: one has a program that needs its own product in order to be executed...To carry on the program, it must already have been executed."[36] Keller characterizes this problem by describing this concept of the gene as "Janus-faced: it is part physicist's atom and part Platonic soul – at one and the same time a fundamental building block and an animating force."[37] And though many scientists, as early as the late 1950s, recognized this contradiction (as well as other limitations of Schrödinger's model), it nevertheless maintained its power for many geneticists and became the basis for the discourse of DNA which has proliferated in popular culture.

A profound and sustaining ambiguity was created through the use of the term "information" in relation to DNA, a term which had two directly opposite connotations, one in the colloquial sense, involving *meaning*, and the other in the context of information theory, that referred to a *quantity* from which semantic content was excluded. This ambiguity allowed for a discursive hegemony that shows few signs of abating.

---

35 E. Schroedinger, *What is Life?* (Cambridge, UK: Cambridge University, 1944), 22–3.
36 F. Varela & J.-P.Dupuy, eds. *Understanding Origins: Contemporary Views on the Origin of Life, Mind and Society* (Boston: Kluwer, 1992), 4.
37 E.F. Keller, *Refiguring Life: Metaphors of Twentieth Century Biology* (New York: Columbia University, 1995), 8.

■ BEING HUMAN:

The scientific conceptualization of information grew out of technological problems facing scientists and engineers during World War II, i.e. the control and deployment of missiles; the development of radar systems; the need for secure communications; and the necessity for automatic computing machines in the development of weaponry. In the postwar years, "information theory" became a legitimate discipline that also spawned related endeavors like molecular biology, artificial intelligence, mathematical modeling of the brain, cybernetics, systems analysis; in short, a plethora of disciplines which re-conceptualized their objects as information processors. It did not matter if the object was living or not. In fact, one of the main accomplishments of information theory was consistent with one of the main goals of mechanism: universality. Information theory became the interdisciplinary schema which equated cells, molecules, animals, humans and machines, as psychologists, neuropsychologists, biologists, and engineers employed a common vocabulary of messages, communication, information, codes, control and feedback to describe their respective domains in terms of data input/output devices and processes.

Though the scientific concept of information goes back to the late 19th century and thermodynamics, its modern incarnation is usually dated to 1948, with the development of a specific technical definition laid out by Claude Shannon, an engineer with a background in mathematical logic. In his work at Bell Laboratories, Shannon was concerned with the probability of the accurate transmission of messages between sender and receiver over channels such as telegraph wire, telephone cable, or radio waves. A "message" in the context of this theory has a very specific meaning: it is "a series of symbols taken from a certain repertory – signs, letters, sounds, phonemes, etc. A given message thus represents a particular selection among all the arrangements possible. It is a particular order among all those permitted by the combinative system of symbols."[38] Shannon's contribution was to mathematize what he called the *amount* or *quantity of information* based on the probability that a certain message would be selected amongst all others, and he specified that a particular message (as a combination of elements) must be distinguished from all other possible messages to be recognizable. His definition of information necessitates a closed system, like a particular language, with a finite (even if quite large) amount of possible combinations of symbols (like letters). The "rules" of the system, i.e. the probabilities of particular combinations of symbols, must be known in advance.[39] Shannon's quantification allowed for the statistical understanding and manipulation of the *codification* of messages. This is crucial for communications systems, because in order to be transmitted through their channels, messages must be "coded": sounds, for example, must be transmitted into electrical impulses to be transmitted over telephone or telegraph. Finding a way to quantify, measure, and manipulate "information" would allow Bell Laboratories to calculate and engineer the maximum efficiency and reliability of telephone transmissions. Thus, the greatest number of possible signals could be effectively transmitted over a particular channel during a given amount of time.

One of the difficulties in understanding Shannon's technical theory of information is that we are so used to thinking of "information" and

---

38 F. Jacob, *The Logic of Life: A History of Heredity* (Princeton: Princeton University, 1973), 250.
39 The more highly probable a combination of symbols is, the less information is needed to transmit it. For example, in the English language, there is a high probability that the letter "u" will follow the letter "q." Thus, according to Shannon's rules, there is very little information necessary to communicate "u" following a "q." Information is quantified in *bits*, and Shannon based his system on the idea of binary digits whereby one unit of information, (a bit), designates the choice between two equally probable outcomes. That is, the bit is "a universal measure of amount of information in terms of choice or uncertainty. Specifying or learning the choice between two equally probable alternatives, which might be messages or numbers to be transmitted, involves one bit of information." (Pierce, 8).

■ BEING HUMAN:

"messages" in terms of their meaningful content. However, Shannon's concept of information has to do with probabilities of the combinations of symbolic units, and *does not involve the concept of meaning at all*.[40] In fact, Shannon explicitly severs his definition of information from any notion of semantic content:

> The word *information*, in this theory, is used in a special sense that must not be confused with its ordinary usage. In particular, *information* must not be confused with meaning...
>
> In fact, two messages, one of which is heavily loaded with meaning, and the other which is pure nonsense, can be exactly equivalent, from the present viewpoint, as regards information.[41]

Shannon's concept of information was quickly appropriated by the field of genetics because it was considered to be "isomorphic" with the idea of "negative entropy" in that Shannon's definition distinguished "information" from "noise," defined as the total background of all possible messages from which a particular one must be set apart.[42] Thus, Shannon's definition was seen as providing a measure of "orderliness" (information) against a background of chaos or entropy (the potential interference of other possible messages). Given Schrödinger's thesis that the chromosome "saved" the organism from entropic decay because of

---

40 Another way to conceptualize Shannon's novel use of the everyday term "information" is to think about the common practice of word processing with computers. Computers measure the amount of information in a file in "bytes" (which are series of eight consecutive bits). The amount of information in a particular file, as measured in bytes, is irrespective of whether the file contains mathematical equations, family secrets, or random sequences of letters. The computer is completely indifferent to what a file contains in terms of its *meaning*; nevertheless, what is in the file (the combinations of letters, numbers, symbols) is quantified as *information*. The technical definition of information cannot discriminate between meaning and nonsense.
41 C.E. Shannon and W. Weaver, *The Mathematical Theory of Communication* (Urbana: University of Illinois, 1949), 8–9.
42 F. Jacob, *op. cit.*, 250–251.

its "orderliness," it could now be formally stated that the secret of Schrödinger's code-script, the secret of life itself, was "information" as the source of that order.[43]

Nevertheless, Shannon's definition of information posed a problem some geneticists recognized immediately: a mutation would not impact the amount of genetic information, no matter how radical or devastating its effects.[44] In spite of this significant limitation, the term "information" became the *sine qua non* of the discourse of molecular biology. However, this convention could only be sustained by ignoring Shannon and Weaver's injunction to ignore the question of meaning. While many molecular biologists continued using "information" in the colloquial sense to indicate that DNA "contained" meaningful content like human behaviors and attributes, Shannon's technical definition, which specifically excluded any reference to meaning, lent the term scientific legitimacy and authority. The opposing colloquial and scientific usages of the term "information" have been able to coexist in the discourse of DNA, establishing an ambiguity which allows the "information" supposed to inhabit the gene to have both scientific legitimacy *and* content. Thus, Keller's reference to Schrödinger's gene as "Janus-faced"[45] refers not only to its ability to incarnate both "law code and executive power," but also can be seen as a reference to the two faces of "information." No specific qualities could be determined by the gene's "information" if the word were being used in the scientific/engineering sense. However, since the discourse of DNA is predicated on the idea of *information as meaningful content*, "information" is being used by geneticists in the colloquial sense even as the specific, technical definition provides scientific authority. Consequently, questions raised by the anthropomorphization implicit in the colloquial use of "information," the presumption that the DNA molecule can somehow "read"

---

43 E.F. Keller, *op. cit*, 94–95
44 Ibid., 19.
45 Ibid., 8.

■   being human:

and "comprehend," are masked by the technical definition of information which would presume no such thing. On the other hand, questions which might be provoked by use of a precise quantitative term like "information" in relation to the gene (i.e., quantity of what?) are effaced by the colloquial meaning of information. This essential contradiction lies at the core of the enormous power the discourse of DNA is able to sustain.

*

But how did "information" become bound to determinism? To understand this trajectory further, we need to look toward another important figure in the history of molecular biology, Norbert Wiener. Wiener was a brilliant mathematician who named and promoted the science of *Cybernetics*, a system of thought he saw as a new lens through which to interpret problems in every discipline. He viewed his work, and that of his contemporaries, as contributing to the birth of an epoch which he believed would become known as the age of "communication and control."[46]

During World War II, Wiener worked in the gun control section of the United States Office for Scientific Research and Development under the leadership of [the omnipresent] Warren Weaver, specifically on self-guiding, goal-seeking weaponry. As a result of this work, Wiener came to his belief that *feedback* was essential to all purposive behavior, from that of missiles, animals, and cells to human beings. All "servomecha-nisms" were modeled on the idea of the homeostasis of living organisms, the delicate checks and balances by which organisms maintain vital functions, and which Wiener defined as a process of *communication control*. It is precisely this link between communication and control which has allowed the "information" contained in the DNA

---

46 L. Kay, *op. cit.*, 623.

molecule to be interpreted as an imperative, an *instruction* or a *command*.[47]

Consequently, the discourse of DNA has promulgated the idea that phenotype (the observable organism) is the product that is "intended" from the genotype (the genes themselves). Once DNA became an "instructor," the end result of development could easily be posited as the cause, obscuring the difficulties inherent in this logic, including the simple known fact that numerous, random factors can intervene between genotype and phenotype.

Just like Schrödinger, Wiener saw the living being in contrast to the entropic degradation of inanimate matter: "Organism is opposed to chaos, to disintegration, to death, as message is to noise," he wrote.[48] The characteristic feature of the individual organism isn't a particular quality of matter (as in earlier vitalistic theories) but the perpetuation of a message containing information about structure and process. He believed that the memory of *form* is the critical aspect of the living organism which is transmitted genetically and during cell division. As Wiener put it, "We are not stuff that abides, but patterns that perpetuate themselves."[49] He was sure that science would eventually be able to transmit a human being from place to place simply by transmitting his "pattern" – an idea he proposed well before Star Trek began "beaming" the crew of the Enterprise.

Transforming the model of homeostasis into a problem of control and communication allowed cybernetics to perform an ingenious feedback system of its own: a model of vital systems was applied to the inanimate world of weapons and machines, and then re-deployed to

---

47 The fact that Wiener himself did not fully identify "control" and "communication" with "*command*" did not prevent this equivalence from being made in his name. See Guido, this volume, for a close reading of how Wiener's writings explicitly problematize such an interpretation.
48 N. Wiener, *The Human Use of Human Beings: Cybernetics and Society* (London: Sphere, 1950/68), 35.
49 Ibid., 86.

■ being human:

describe vital systems. In the traversing of boundaries from human to machine to animal and back again, the linguistic metaphors were getting lost in the shuffle. In effect, cybernetics proved communication *is* control as its vocabulary soon gained hegemony in molecular biology, and the control of the flow of information became the "reality" of the gene.

*

The ideal of reproduction as a mechanistic, logical process lies at the core of molecular biology. Discovering this logic, and by extension, the logic of life itself, was the quest of John von Neumann. Von Neumann was a central figure in information theory; one of the foremost leaders in the realms of strategic military planning, electronic computing, automated control systems, and nuclear power (he worked on the Los Alamos atomic bomb project); and the creator of game theory.[50] He presumed a fundamental equivalence between living organisms and machines, and the premise underlying his work was "if you understood automata...you understood not only machinery better – you understood life."[51] Like his colleagues, he believed the characteristic phenomenon of life was connected to heredity: this is not just because organisms survive and reproduce, but also because he viewed the history of life as "a trajectory of increasing well-ordered complexity."[52]

---

50 Von Neumann had important relationships and collaborations with the key thinkers in this burgeoning field, and had a particularly close, ongoing relationship with Norbert Wiener, though they differed in many areas (for example, unlike von Neumann, Wiener made every effort to disengage himself from military science after World War II). In particular, "whereas Wiener attempted to unify cybernetics around the idea of feedback and control problems, von Neumann hoped to unify...both the biological and mechanical realms, around the concept of an information processor – which he called an 'automaton' (Aspray p.133).
51 S. Levy, *Artificial Life: A Report from the Frontier Where Computers Meet Biology* (New York: Vintage, 1992), 16.
52 Ibid., 29.

Von Neumann's Darwinism was in keeping with his ideology of rationality and his horror at the idea of chance occurrences: In a 1955 letter to a well-known geneticist, he wrote "I shudder at the thought that highly purposive organizational elements, like the protein,[53] should originate in a random process."[54] To reveal the logic of heredity, von Neumann set out to build machines that could "reproduce" themselves and "evolve," what he called "self-reproducing automata." It was a project "aimed at simultaneously explaining and simulating living systems, opening up a space of representation in which heredity was modeled on simulacra."[55] Furthermore, it was a project that might provide the ultimate proof that life and its origins are rational, saving von Neumann – and the rest of humanity – from the nightmare of random, unpredictable existence.

Von Neumann's model of the self-reproducing automata was based on Alan Turing's universal machine. Turing, an English mathematician with whom von Neumann had collaborated, is widely credited with having conceptualized the first modern computer. In 1936, Turing came up with the idea for a machine whose product was a "tape" that could describe another machine like (or unlike) itself. Expanding on this idea, von Neumann's machine was able to build another one just like itself from a *description*. In order to reproduce, a machine was placed in an environment rich in appropriate raw materials, the same resources from which it was made (batteries, tape, motors, wires, etc.). Following the coded instructions programmed into it, the machine (a kind of robot) would organize this material into a copy of itself. The machine could even be programmed to "mutate" and "evolve" through the incorporation of built-in random changes. Since information was the key to this self-replicating automaton, its key component became the mecha-

---

53 At that time, prior to the discovery of DNA, the gene was thought to be a protein.
54 Cited in Levy, *op. cit.*, 15.
55 L. Kay, *op cit.*, 618.

■ BEING HUMAN:

nism, called the "copier," which had the function of "reading" the instructions and copying them. The self-reproduction was, in fact, *initiated* by this act of reading. The deciphering of information animated the entire machine, "bringing it to life," so to speak. And this life perpetuated itself via these instructions which were passed along when the "parent" inserted a copy of the "tape" into its offspring. As von Neumann wrote:

> It is quite clear that the instruction [tape] is roughly effecting the functions of a gene. It is also clear that the copying mechanism B performs the fundamental act of reproduction, the duplication of the genetic material, which is clearly the fundamental operation in the multiplication of living cells. It is also easy to see how arbitrary alterations of the system...can exhibit certain typical traits which appear in connection with mutation, lethally, as a rule, but with a possibility of continuing reproduction with a modification of traits.[56]

If this description sounds familiar, it is because five years after von Neumann wrote it, Watson and Crick virtually replicated it in their description of their discovery of the structure of DNA in 1953. However, while it was obvious that a human being – John von Neumann – was the ultimate origin of the "instructions" given to the "self"-reproducing automata, the agency responsible for the inscription of genetic "information" was left uncredited. But the fact that this empty place did not become a *problem* can be attributed to the uncanny ability that information theory possesses to efface the implications of the agency it renders implicit. John von Neumann was able to provide what appeared to be a purely mechanistic and rational explanation of life be-

---

56 Cited in A.H. Taub, ed., *John Von Neumann: Collected Works* v. 5 (Elmsford, NY: Pergamon, 1963), 290.

cause information theory can make it appear that "it is the writing that writes."[57]

*

"Our difficulty is, to know in which direction to look for the myths of the 20th century...If we do think ourselves myth-free, when we are not, that is largely because the material from which we construct our myths is taken from the sciences themselves."[58]

What has allowed the discourse of DNA to remain so powerful, given that geneticists and other scientists have questioned its metaphors and reductionism since the very inception of the discipline of molecular biology? While the overdetermination of the success of the discourse of DNA includes significant political, economic, and ideological factors, I have narrowly focused here on how textual metaphors derived from information theory have enabled the discourse of DNA to veil its own exclusions, limitations and contradictions. To claim the "secret of life" is "written in" to our genes, what has had to be "written out" could fill a book in itself. As we have witnessed the disappearance of metaphor in the discourse of DNA (which has collapsed into an apparent and transparent "reality"), what has also disappeared from view is the specificity of *human* life: the peculiar ways that history, culture and language function as agents for the transmission of human desires, actions and attributes. These domains, with their complex rules, taboos and customs, have been reduced to the "environment," which can only facilitate or inhibit what is supposed as already inscribed. This generalized "environment" cannot incorporate the specificity of culture or historical time since biological determinism presumes that the human traits

---

57 L. Kay, *op cit.*, 615.
58 S. Toulmin, *The Return to Cosmology: Postmodern Science and the Theology of Nature* (Berkeley: University of California, 1982), 23.

■ BEING HUMAN:

it is trying to account for have always existed in the same way they do now (i.e. homosexuality, maternal and paternal roles, aggressivity). The theory of genetic continuity as an explanation for the current state of humanity leaves no room for the dialectical relationship between individual and collective existence which marks the discontinuity between human life and all other. The discourse of DNA smooths over such difficulties by giving the gene unlimited capacity to contain complex "information," thereby collapsing culture into nature, including the construction of the theory itself. In this instance, the effacement of the human construction of knowledge, which is part of the scientific method, has resulted in a peculiar paradox. While disavowing that language plays a part in structuring and representing their objects, scientists have attributed language use to an inert molecule. If nearly every week scientists proudly announce that they have been able to "read" a new "word," "sentence," or "chapter" in the "book of life," it is, in part, because what has been written out of this account are the ways in which they themselves are the authors.[59]

It is, in fact, the question of authorship that lies at the heart of the discourse of DNA, for it is very clear that "blueprints," "codes," "instruction books," or meaningful "information" require an author. We are left with a peculiar choice: either DNA writes itself, and is itself Divine, or it is a text written by God. Either the molecule itself has a mind, or it is evidence of the mind of God. In the former case, we are faced with explicit animism at the core of one our most sophisticated sciences. In the latter case, we are confronted by the link between theology and science which appeared severed by the Enlightenment.[60]

---

59 L. Kay, *op cit.*, 610
60 In *The Religion of Technology: The Divinity of Man and the Spirit of Invention* (New York: Knopf, 1997), David Noble points out that the idea of science and religion as distinct domains is a relatively recent contribution of the 18th c. Enlightenment. Noble argues that religion and science are "neither complements nor opposites, nor do they represent succeeding stages of development. They are merged, and always have been" (4). Noble also reveals that numerous scientists, including many geneticists as well as the

This leads us directly to the "dream of the human genome,"[61] the dream which has animated science since its inception: to unravel the mind of God and share in his creation. As one of the founders of the Human Genome Project put it, "For the first time in all time a living creature understands its origin and can undertake to design its future."[62] As we know very well, this is the dream shared by turn-of-the-century eugenicists and Nazi "scientists" alike. The problem is not that we need to establish more ethics committees to prevent repetition of the historical "abuses" of genetic determinism, but that the discourse of DNA makes it appear that this dream of controlling who and what we are is *realizable*:

> With life relocated in the genes and redefined in terms of their informational content, the project of "refashioning life," of redirecting the future course of evolution, is recast as a manageable, doable project...The kind of theory that would provide a workable guide for such a project is precisely a theory that focuses on causal relations between identifiable and controllable elements. Other processes, less controlled and less controllable, are bracketed in the double name of intellectual economy and technical efficiency.[63]

Grounded upon the bracketing of "less controlled and less controllable processes," the discourse of DNA allows us to partake of the dream

---

director of the Human Genome Project, are devoutly Christian and see in their work a deep religious significance. See also *From the Closed World to the Infinite Universe* (Koyre, A., New York: Harper, 1957) for an account of the development of modern science and its relationship to theology.

61 Lewontin, *op. cit.*
62 Sinsheimer, cited in D. Noble, *The Religion of Technology: The Divinity of Man and the Spirit of Invention* (New York: Knopf, 1997), 187.
63 E.F. Keller, "Physics and the Emergence of Molecular Biology: A History of Cognitive and Political Synergy." *Journal of the History of Biology*, 23 (3), 408.

which would transform human beings into both gods and machines subject neither to sex, nor to death, nor to incomplete knowledge. But there, where the limits of being human are ignored, we find an empty place concerning the question of responsibility for our individual and collective fate. This question cannot be addressed by gods or machines, but can only be taken up by human subjects. The wish to overcome the limits of being human is all too human, perhaps, but impossible, nevertheless. What is possible is to recognize this wish in its multifarious incarnations, and to transform the demand for comprehensiveness and certainty in ways that can accommodate the unknown – and the unknowable.

Interview with Renée Fox

Mark Stafford

Renée Fox, best known as a sociologist of medicine, is unique among America's social scientists in having studied the field of organ transplantation since its inception. In collaboration with historian of medicine and science Judith P. Swazey, she has written two ground-breaking works on this subject: *The Courage to Fail* and *Spare Parts*. The following conversation with Mark Stafford was based on a presentation that took place at the Après-Coup Psychoanalytic Association on May 15, 1998.

MARK STAFFORD: When did you begin to work in the field of organ transplantation?

RENÉE FOX: I have had a sustained relationship to organ transplantation and to the deployment of artificial organs – the artificial kidney and the artificial heart – for more than 45 years. I did my Ph.D. dissertation research in the Peter Bent Brigham Hospital in Boston, during the years 1951 to 1954, where the first successful transplantation of a human kidney from one identical twin to another was performed in 1954. The recipient lived for many years with the kidney that he received from his twin. This was not only a path-making case biomedically, but also ethically and legally. It involved the question of whether or not it was permissible to allow one brother to contribute to the other in this fashion – to undergo a major surgical procedure that would result in the loss of one of his kidneys, entailing pain, and both short- and long-term risks for him – in order to try to sustain the life of his brother. The youth of the twins also raised the question of whether they were sufficiently mature to give informed consent for the procedure. The Massachusetts Court was involved in the deliberations, and ruled that it would be psychologically injurious to the brother whose donated kidney might save the life of his sibling if he were deprived of the opportunity to do this for his brother.

MARK STAFFORD: How would you describe the current state of research on transplantation?

RENÉE FOX: Virtually all the solid organs of the human body are being transplanted with the exception of the brain. We are also doing multiple organ transplants: heart-and-lung transplants, for example. We are now in an era where in addition to transplanting organs, we are transplanting bone marrow, corneas, skin, and bones. And we are doing a considerable number of re-transplants and re-re-transplants. Since the end of the 1970s, with the advent of cyclosporine, an immunosuppressive drug with a greater capacity than its predecessors to attenuate or slow down the immune reactions responsible for the rejection of transplanted organs, there has been a veritable boom in the range of organ transplants undertaken, and in the number of transplants contemplated. There has also been an increasing preoccupation with what has come to be called the "shortfall" of organs. The language used is the language of supply-side economics, but the "resources" that we are talking about here are the living parts of human persons who contribute to others so that their lives might be sustained. The language of economics has been applied to this trans-economic phenomenon.

Ironically, one reason for the organ shortage is the increase in the number of transplants that we are trying to do in the United States, and an increase in the numbers of persons whom we define as being potentially eligible to benefit from organ transplants. Another reason for the shortage of organs is the number of re-transplants that are performed, despite the fact that persons who are the recipients of successive transplants become progressively less able to use the organs they receive well, and more and more likely to rapidly reject them. The supply of organs has been increased, but the number of organs that we aspire to transplant has also been greatly augmented, along with the number of medical centers in the United States that are engaged in organ transplantation.

■

MARK STAFFORD: A new source of organs are the organs of other species – genetically engineered "transgenic" pigs, for example. Can you tell us what kinds of ethical issues are involved in xenotransplantation?

RENÉE FOX: A phrase that comes to mind in this connection is an evocative statement made by Paul Ramsey, a renowned Protestant theologian, and an early contributor to the field of bioethics. In his book, *The Patient as Person,* published almost 30 years ago, he wrote about "the triumphalist temptation to slash and suture our way to immortality." What sometimes seems to me to be an effort to replace every worn-out part of the human body with human and organ transplants fits the paradigm about which Paul Ramsey had so much foreboding.

Some of the phenomena that this involves were visible, for example, at the Institute of Medicine's workshop on xenotransplantation held in June 1995, that I helped to organize, and in which I participated. Among the striking things about the workshop were the enthusiasm about organ transplants that prevailed, and the "we-are-on-the-brink-of-overcoming-them" optimism about the problems of xenograft rejection. These "accentuate the positive" attitudes were especially pronounced among the molecular biologists and immunologists doing cross-species animal studies, who were present. In contrast, the transplant surgeons who attended the meeting – including two who had conducted experimental clinical trials with baboon-to-human transplants – though still committed to the endeavor, were convinced that the time had not yet come when xenotransplants could be successfully and safely performed. The molecular biologists and immunologists, who were working on the level of cells and tissues rather than with solid organs, and who had less clinical experience and contact with patients than the transplant surgeons, were inclined to play down the many unknowns of xenotransplantation, and the degree of difficulty in understanding and controlling the hyperacute, and delayed graft reactions triggered by transplants between species. The atmosphere of positive expectancy was reinforced by the workshop participants from biotechnology and pharmaceutical companies in the United Kingdom and Switzerland, as well

■ BEING HUMAN:

as in the United States, who have invested heavily in xenotransplantation and in the development of transgenic pigs, and also by the presence and testimony of patients on waiting lists for transplants, who hoped that the development of xenotransplantation would make more organs available, and increase their chances of becoming recipients. The molecular biologists and immunologists agreed with the majority of the workshop participants that there was an uncertain, but real risk that xenotransplantation might unleash previously unknown or unrecognized pathogens into the human population; and they were mindful of the fact that the AIDS epidemic probably began through the transfer of a virus from chimpanzees to a human host. But they were among those present at the workshop who felt that this risk did not necessarily warrant the invention and implementation of special conditions of surveillance and regulation if xenotransplantations were to proceed.

MARK STAFFORD: Why is there so much willingness to take risks in this field?

RENÉE FOX: The current medical literature gives the impression that the major problem with which organ transplantation is confronted at this time is the shortage of organs. I find this surprising, because as I understand it, from the inception of the history of organ transplantation to this day, the chief source of biological and clinical difficulties and limitations associated with it is the rejection reaction – a mechanism that is so innate and unrelenting that in spite of all the advances in knowledge of the biology of organ and tissue rejection that have been made, and the development of immunosuppressive drugs that has occurred, it remains true that every transplanted organ will eventually be rejected, unless the donor and the recipient are genetically identical twins. By playing down the inexorable nature of the rejection reaction, and emphasizing the gravity of the shortage of organs, the medical literature, as well as the media, conveys the impression that once a person gets an organ that is a "good [immunological] match," the individual and the organ will live happily ever after in a problem-free way. I don't think that a deliberate attempt to hide the risks, limitations, and the downside of organ transplantation is involved here; but it is a reflection of the ardor that transplant professionals feel about this life-saving procedure, and as one thoughtful organ recipient put it, of the tendency of many of them to "discuss it only as a miracle."

MARK STAFFORD: What are the emotional and symbolic states which you feel are at work in the relationship between patient, donor, physician and nurse?

RENÉE FOX: From the beginning, organ transplantation has been defined by the medical profession and the public as a "gift of life." Initially, the notion of the gift was used metaphorically; but gradually it became apparent that what donors, recipients, their families, and the transplantation team actually experience are powerful psychological, social, cultural, and existential attributes of this extraordinary form of gift-exchange. To a remarkable degree, organ transplantation enmeshes its participants in what the renowned French sociologist, Marcel Mauss, identified as the constituent elements of gifts and their meaning in the classic essay, *The Gift*, that he wrote early in the twentieth century. No matter how spontaneous and expressive they may appear to be, Mauss pointed out, gifts are shaped by a triple set of "symmetrical and reciprocal" obligations: to offer and give, to receive and accept, and to seek and find an appropriate way to repay. Failure to live up to any of these obligations, Mauss pointed out, creates major social strains that affect the giver, the receiver, and those associated with them. He also emphasized that gifts have emotional and symbolic as well as material value and meaning. In this sense, he said, the gift and the obligations attached to it are not "inert." They create a spiritual bond between the donor and recipient.

It was with a mixture of awe and trepidation that some of the pioneers of organ transplantation thought about the gift of life for which they were intermediaries. Does the physician have the moral right to accept such a gift from a healthy person who wants to "amputate a vital portion of his[her] body," in order to prevent the death of another? they asked. And does the surgeon have the right to remove an organ from a deceased person for this purpose – engaging thereby in an act that could be termed "cannibalizing?" Today, some 45 years after the initiation of organ transplantation, this kind of moral questioning about the entwined acts of giving, taking, and receiving organs is no longer so acute or audible among transplanters. I don't think that it has disappeared, but it has "gone underground" – below the emotional and ethical surface of what transplant physicians characteristically claim is a relatively "routine" surgical act.

MARK STAFFORD: How is the emotional life of physicians impacted by their participation in this act of gift-giving?

RENÉE FOX: The title of the first book that Judith Swazey and I wrote together – *The Courage to Fail* – reflects one of the ways that transplantation was seen and felt by some of the members of the first generation of transplanters. The aphoristic phrase, "the courage to fail," was coined by Dr. Denton Cooley, a renowned cardiac surgeon, and a flamboyantly important figure in the early clinical history of heart transplants, during an interview that Judith Swazey and I conducted with him. What he dramatically invoked through this expression was an image of transplantation – especially of the highly symbolic heart – as a bold, uncertain, and often dangerous adventure in which physicians, and patients with end-stage disease were involved, the survival of patients undergoing transplantation constituted a "success," and their death, the archetype and the pinnacle of a "failure" that required "courage" for the physician as well as the patient to incur.

As I've already implied, transplant physicians today are inclined to speak of organ transplantation as a "routinely successful" part of modern medical practice, albeit a life-saving procedure. And yet, a transplant surgeon as prominent and intrepid as Dr. Thomas Starzl, who pioneered liver transplantation in this country, and trained most of the U.S. surgeons who perform it today, found it so worrisome and controversial to use healthy living donors for (kidney) transplants that he called a personal moratorium on doing them. Not only was he apprehensive about the catastrophic possibility that such a transplant might result in a donor's death – a possibility that he described as an event that would "almost stop the clock worldwide" – but even if the outcome was a positive one (which it usually is), surgically injuring well persons, and putting them in jeopardy by removing a vital organ from them was experienced by Dr. Starzl as an unjustifiable violation of the "first, do no harm" injunction that is a fundamental moral principle of medicine and physicianhood.

For years, transplant teams only considered persons biologically related to a recipient as prospective live donors – that is, the parents, the children or the siblings of a candidate-recipient. They stated that their policy was based on the biological assumption that a close tissue match between donor and recipient was more likely to exist if they were related in these ways, and that this would help to control the rejection reaction. I questioned transplanters about why they would not consider tissue-typing the spouse of a person awaiting a transplant, to ascertain if by chance alone he or she would be a good match; because even though we are not biologically related to our spouses, except through the children we mutually bring into the world, in the American family system, once we marry, they are our closest kin. I never received a satisfactory answer to this query, and could only surmise that for reasons that were more attitudinal than scientific, they felt less uneasy about taking a vital organ from a member of (what social scientists would term) a recipient's family of origin than of his/her family of procreation. Interestingly, as the shortage of organs has become a more and more preeminent issue for them, transplanters have created a new category of eligible live donors whom they call "emotionally related donors." These include spouses, and also "kin-like friends."

But your question about how the emotional life of physicians is affected by what Marcel Mauss called "the theme of the gift" that is a leitmotif in organ transplantation, carries us into the realm of how they react to what donors and recipients experience reciprocally when this special kind of gift-exchange takes place.

■ BEING HUMAN:

MARK STAFFORD: What are some of the concomitants of this extraordinary gift that the donor makes?

RENÉE FOX: To a striking degree, what happens closely parallels the gift-exchange paradigm that Mauss delineated. Although the American organ donation system is organized around the principles of voluntarism and freedom of choice, this does not eliminate the fact that no matter what efforts the medical team may make to protect the members of a prospective recipient's family from feeling coerced or self-coerced to become a live donor, the pressure to offer this gift of life is powerful. Cadaveric organ donations are also surrounded by inner and outer gift-giving pressures. For example, the "best" cadaver organs, from the point of view of their viability, are obtained from young, healthy persons, who have been fatally injured in a vehicular accident, who are victims of an act of violence, or who have taken their own lives. These sudden, unexpected, and premature deaths are especially tragic and fraught with problems of meaning. In these circumstances, the grief-stricken family may feel impelled to donate their young relative's organs by their intense need to make redeeming sense out of what they would otherwise experience as morally and existentially absurd.

Not only do donors feel such obligation-like pressures to give organs, but candidate-recipients feel complementary pressures to receive them. The recipient may not want a close relative to undergo the degree of discomfort, risk, or sacrifice that acting as live donor entails; or the recipient may feel that receiving such a "gift of life" from a particular relative might further complicate their already difficult relations. These reservations notwithstanding, refusing to accept a live organ transplant that is offered implies a rejection of the donor and the donor's relationship to the recipient (To use Mauss's somewhat arcane language: "Refusing to accept a gift is a refusal of friendship and intercourse.")

The prospect, and the actual experience of receiving a transplant – regardless of whether it is a live or a cadaveric organ that is transposed from donor to recipient – are likely to evoke buried, often animistic feelings that many people have about their vital organs and the integrity of their body. A widespread, preconscious sentiment that recipients seem to share is the belief that some of the psychic and social, as well as physical qualities of the donor are transferred with his or her organ into the body, the personhood, and the life of the individual in whom it is implanted. (Mauss identified this as a phenomenon inherent to gift-exchange more generally, writing that "the spirit of the thing given" and received is "alive and often personified.")

The psychic weight of these anthropomorphic connotations of the gift is augmented by the third set of obligations that organ recipients experience. They have received a gift of surpassing significance – one that philosophers would describe as "beyond duty and claim." It transcends what is ordinarily asked or expected of us. As Marcel Mauss could have foretold, because this is so, it is inherently unreciprocal. It has no physical or symbolic equivalent. As a consequence, the giver and receiver, and sometimes their families as well, may find themselves locked in a creditor-debtor vise that binds them one to another in a mutually fettering way. Judith Swazey and I have called these aspects of the gift-exchange dimensions of transplantation, "the tyranny of the gift."

In the case of a live organ transplant, this "tyranny" can take the form of a donor exhibiting a great deal of "proprietary" interest in the health and life of the person who has received his or her organ. In turn, the overpowering indebtedness that recipients may feel to the person whose life-saving organ they carry may make it difficult for them to maintain psychic distance and independence from the donor. This is vividly illustrated by a case with which I am familiar that involved a young woman who received a kidney from her donor-brother, and who had no contact with him for more than a year after that. However, when she self-reliantly achieved her goal of becoming Miss Pennsylvania and a top finalist in the Miss America contest, she contacted him, and she and he were tearfully reunited the night she sang an aria in the talent portion of the contest that she had invited him to attend. It appears that she felt that she had achieved enough on her own to reestab-

■ BEING HUMAN:

lish a relationship with her brother without being engulfed by her identification with him, by her indebtedness to him, and by the implications of the public statement that he made on this occasion, when he said that he was proud of his sister because he knew it was his kidney up on the stage that enabled her to sing!

Recipients of cadaver organs not only suffer from the unrequitable magnitude of the gift they have received, but also from a sense of the hovering presence of their unknown dead benefactor inside of them. In the words of professor of literature Richard McCann, who is the recipient of a cadaveric liver transplant, "The gift of life is saturated with death." In "The Resurrectionist," an essay that he wrote about his transplant experiences, he depicts the donor organ as "a bearer of its own cellular memories"; he describes the impact of his guilt-ridden realization that he had "participated in the pain and violence and grief of a human death" whose occurrence gave him the chance to live again; and he recounts that there have been nights when he has thought of the donor "with great tenderness," sometimes perceiving him as male, and sometimes her as female.

Transplant physicians were not prepared for the occurrence of what they were inclined to regard as these strange and disturbing phenomena. In my view, their own unease about "the gift of life" concomitants of transplantation, as well as their concern about donors, recipients, and their families led them to gradually develop a policy of anonymity around cadaver transplants. Rather than revealing the identities of the donors of cadaver organs, their recipients, and their families to one another, and providing them with details about each other's backgrounds and lives as they did in the early years of organ transplantation, transplant teams now only give a minimum of highly schematic information to those involved. They do not want to encourage donor families, recipients, and their families to become too entangled in each other's lives, they say.

Transplant surgeons rarely allude to the anthropomorphic aspects of organ transplantation. If asked directly about them, especially if

THE TECHNOLOGICAL EXTENSIONS OF THE BODY ■

they are questioned in a public forum, or within earshot of persons not intimately acquainted with the transplantation subculture, they may concede that such reactions occurred during the early days of transplants, when they were daringly innovative acts; but they are likely to claim that now that transplantation has become more "conventional," and "routine," responses like the reification of donated and received organs, survivor guilt, and the anguish of indebtedness are largely things of the past. In contrast, nurses and social workers who talk more freely about these phenomena attest that they are commonly experienced by recipients, donors, and their kin. I have found that transplant surgeons are more inclined to admit that this is the case when I talk to them in private, particularly if they define me as an old-time member of the transplant community because of all the years that I have been involved in its history as a participant observer.

MARK STAFFORD: What impact are recipients having on the politics of transplantation?

RENÉE FOX: The current preoccupation with the number of persons waiting for transplants has been driving transplanters to engage in acts that I think go beyond what we ought to be doing in the name of procuring organs. In my opinion, it is unrealistic, if not false, to assume that there are unlimited potential donors in this society, and that if everyone who was eligible to be a donor gave of himself or herself in this way, there would be enough organs for all those who might benefit from them. What is never said in the medical literature or in the media is that in the United States we already donate more organs for transplantation than in any other society in the world.

The criteria for accepting cadaver donors have been expanded to include people of advanced age (up to 80 years old in some programs), and also donors with diabetes, hypertension, some infections, some hemodynamic instability, or some chemical imbalances. These expanded criteria are said to have the potential of increasing the cadaver donor organ supply by 25 to 35 percent. We are pushing the envelope so much that we are now willingly accepting what the medical literature acknowledges are "marginal organs." An organ that you receive from an eighty-year old person, or from persons who have had diabetes or hypertension during their lifetimes are not likely to be of the same quality as those obtained from younger, healthier individuals. One of the issues that has arisen is whether a recipient who will be receiving a marginal organ should be told in advance of the transplant procedure about the status of the organ, and given the possibility of refusing it because it is of less than optimal quality. Still another question that the transplanting of marginal organs has posed is whether the recipient of such an organ should be given a higher priority for a re-transplant, because the initial organ transplanted was a less-than-ideal one.

A number of years ago, I was asked to make a presentation at the "Ethical Grand Rounds" of a well-known Children's Hospital. The case presented was that of a little girl who was born with biliary atresia, and who between the ages of five and seven years, had received four successive liver transplants for her congenital condition, each one of which she had rejected. She was awaiting a fifth transplant. My job was to discuss this heart-rending case. It quickly became apparent that the issues were not primarily biomedical. The patient around whom they turned was a little girl who had virtually grown up in this hospital. She was loved by everyone who had taken care of her, and they were determined to do whatever they could to sustain her life. They had managed to obtain four donated livers for her, and were now trying to procure a fifth. Given her prior history, it was highly likely that she would rapidly reject that organ. The shortage of donated cadaveric livers is especially severe, and livers small enough to be implanted in a child are even scarcer. This means that because the transplant team could not bear the thought of "abandoning" her (a phrase that one often hears in transplant circles when a re-transplantation is under consideration), they were "consuming" liver after liver on her behalf that might have benefited other youngsters on the waiting list for a liver transplant. What is more, if some of those waiting were "first-time" candidates, the outcome of their receiving an organ would be predictably better than the consequences of a re-transplant – and surely than those of a fourth re-transplant.

This brings us to the question of how donated organs are allocated in our society. The process takes place under the auspices of the United Network for Organ Sharing (UNOS), a private organization that was created by the U.S. government in 1986, when it passed the Organ Transplant Act. UNOS operates on a multifactorial point system for deciding who on local and regional waiting lists should receive a donated organ when it becomes available. The system was established with the aim of insuring that scarce, life-saving organs are distributed in a manner that is as fair and equitable as possible.

However, a close look at the multifactorial point system reveals that it is not totally value-free. One wonders whether any system for allocating such a precious and limited "resource" could be. For example, the system assigns a set number of points to a candidate, based on length of time on the waiting list, on the one hand, and medical urgency, on the other. The sickest people go to the top of the list in an attempt to rescue them, rather than giving more weight to those who might do better. If the potential recipient is a child, this increases the priority that the candidate recipient receives. For instance, a potential kidney recipient who is six to ten years of age is accorded one extra point in the UNOS system, while those who are zero-to-five years of age (0–5) receive two extra points. And being less than six months of age is one of the three defining criteria that affords a heart transplantation candidate top priority. So, we have built into the system our conviction about the special meaning of new life and of children, and about the added chances we believe they should be given.

Attempts are currently being made under the Clinton Administration to improve the effectiveness and fairness of the system of organ procurement and allocation by making the distribution of organs less local, so that organs are shared more broadly across city and state lines. A recent Institute of Medicine report has recommended that this be accomplished by breaking the country up into larger regions of about nine million people each, with organs distributed to the sickest patients first within each region. This recommendation has proven to be highly controversial, and it is still unclear whether or not the government will pass a regulation to this effect.

MARK STAFFORD: In light of the value you give to Ramsey's observation that we are attempting to "slash and suture our way to immortality," can you talk about how the limits of the body are being challenged by the new procedures?

RENÉE FOX: I think that some of the conditions that accompanied the re-introduction of so-called "non-heart-beating" organ transplantation on the American scene in the early 1990s are indicative of how the intense pressure to increase the number of organs available for transplant can lead to desperate, even dubious ways of treating the donor patients and their bodies.

A "non-heart-beating" donor is a person whose death is determined by the traditional criteria of pronouncing death – the "irreversible cessation of circulatory and respiratory functions" – rather than by the newer, "whole-brain death" criteria – the "irreversible cessation of all functions of the entire brain, including the brain-stem." Organs that are taken from brain-dead (heart-beating) donors are considered preferable to those recovered from individuals who are declared dead using cardiopulmonary criteria (the practice in the early days of transplantation), because the latter means that the organs will not continue to receive oxygenated blood up to the moment of their removal from the donor. The return to taking organs from "non-heart-beating" donors, who are either severely ill on life support, or who have suffered unexpected cardiac arrest has been propelled by what has been described as "the relentlessly increasing need for organs." Although they have been amended since, when protocols for the "modern" versions of non-heart-beating organ transplantation were first drawn up, they involved such conditions as transporting the patient-donor to the operating room while he or she was still alive; putting in a femoral line, and doing skin preparation and sterile draping for the procurement surgery prior to the individual's death; withdrawing life support in this enclave of the hospital inaccessible to his or her family or clinical caretakers; and determining that death had occurred, on the basis of only two minutes of electronic and arterial pulse pressure monitoring – all dictated by the driven need to shorten the time between the absence of circulation and the removal of organs in order to keep them viable for transplantation.

MARK STAFFORD: Finally, could you comment on the origin of your book *Spare Parts,* and the conclusions you reached in that particular work?

RENÉE FOX: The title *Spare Parts* had been suggested to me many years ago by the eminent sociologist, Erving Goffman. At the time that he proposed it, I thought it was too ironic. But as the enthusiasm about organ transplantation mounted, and the attempts to procure and implant organs escalated along with it, the concept of "spare parts" took on more cogency.

The book articulates our concern about the zeal that infuses the medical and social commitment to the endless perpetuation of life in this way, the willingness to obtain organs through questionable means, and the suffering and harm that such unexamined excess can bring. The last chapter of our book is entitled "Leaving the Field." It announces that we no longer intend to do first-hand research on organ transplantation (or other forms of organ replacement, such as the development and deployment of the artificial heart.). Our decision to distance ourselves from the field, we say, constitutes a moral statement on our part about where organ transplantation has been going, and the very slippery slope down which it seems to us to be sliding. It also expresses our conviction that the most important medically relevant social and moral problem in the United States today is not the shortage of transplantable organs. Rather, it is the fact that more than 44 million people in our society have no health insurance, and many more are underinsured, despite expenditures that exceed those of any other country in the world.

■

Some Reflections on Medically Assisted Reproduction[1]

Paola Mieli

> Al poco giorno e al gran cerchio d'ombra
> son giunto, lasso, ed al bianchir de colli,
> quando si perde lo color ne l'erba:
> e'l mio desio però non cangia il verde.[2]
> — DANTE,
> *Canzoniere*, 44 (CI)

In the depths of prostration – she can no longer sleep without waking up from a recurrent nightmare of suffocation – in the depths of a state of anxiety that is driving her nearly to suicide, Rachel seeks the help of an analyst after a series of failed attempts to become pregnant. She has already consulted four specialists, the last of whom urges her to let him attempt a second *in vitro* insemination. Rachel has no steady companion, and when she speaks of possible fathers for the baby she so desperately wants, she lauds only their physical endowments. These are not men you can count on, but then again none of them really counts in this quest she's on. The person she does trust is the medical specialist.

Her state of prostration is striking, and with it, a more insistent demand, a more insistent longing to experience motherhood – what she refers to as the miracle of birth. She visits one physician after another with the sense of approaching a final deadline. It takes five months of regular meetings for the taut wire of dreams that abruptly shatter

---

[1] The present is an English version of the paper "Verde: note sulle implicazioni attuali della riproduzione assistita," published in *La maternitá tra tecnica e desiderio*, edited by M. Fiumanò, Edizioni La Tartaruga, Milan, 1996.
[2] "I have come to a short day and a great arc of shadow, alas! and to the hill's whitening when colour vanishes from the grass; and my desire does not, for this, change its green..." *The Penguin Book of Italian Verse*, ed. George R. Kay, Penguin Books, Harmondsworth, 1957, 89.

her sleep to reveal the circumstances surrounding her own birth; only recently has she found out that she is not the biological daughter of the father who has raised her, but rather the sole fruit of a mother's transgression in a marriage that brought life to five children. It will be the course of analysis that impels her to discover that the family doctor, the very same doctor who helped her mother bring her into the world, is her biological father.

Let us merely point out here that Rachel, in her wish to be a mother, questions maternal desire, her mother's desire as a woman and as a mother, just as she questions in the doctor the desire of a man and a father. At the same time she questions the truth of which her coming into existence is the fruit, a truth that, though silenced, has continued to mark her entire life.

Needless to say, to hear out her demand by promising the success of an assisted fertilization is to evade the question underlying that demand and to strengthen the reasons for psychic infertility; it is to push Rachel to the brink of acting-out.

*

Marie-Magdeleine Chatel Lessana has shown in ample detail in her book *Malaise dans la Procréation* [Procreation and Its Discontents], that the occurrence of human fertility cannot be reduced to a pure biophysiological phenomenon, to an anonymous encounter of gametes.[3] Fertility is the result of a whole overdetermined set of elements that rest upon the materiality of the body, but that imply differentiated registers; a phenomenon of universal nature, it is inscribed in the specificity of individual history as the precipitate of a constellation of unconscious signifiers, of symbolic events, imaginary and real elements that constitute the uniqueness of subjective truth. The magical aspect of

---

3 M.M. Chatel Lessana, *Malaise dans la procréation*, Albin Michel, Paris, 1993.

conception condenses the contingent and accidental character of an encounter, an encounter that becomes a life-breeding occasion. In this sense, a conception is an event that traverses the generations; it localizes both a certain relation to motherhood that goes from daughter to mother and mother to grandmother, and sheds light on a precise relation between femininity and maternity, as well as the appearance of a paternal function entrusted with a symbolic transmission between generations.

It has been observed that at the very moment in which new medical discoveries are giving women the freedom to program and decide maternity, interrupt it, or defer it, we are facing a new problem: that of infertility. Various statistics show that often, when women find themselves wanting to conceive a baby they have put off having, they can't manage to have it. Obviously, we are not considering organic dysfunctions here, much less the physiological consequences of contraception or abortion.

Infertility as an effect of programming, let us say, interests us here insofar as it sheds light on a fundamental aspect of what we may define as the *temporality of motherhood*. By intervening in subjective temporality, the techniques of contraception reveal the gap that exists between will and desire; they expose the fact that the temporality of desire is logical, not a matter of linearity or decision-making. The possibility of controlling procreation gets confused with the idea of being able to schedule a child whenever one sees fit, with the illusion that one's own fecundity can be dictated by a rational decision. Let us emphasize at once that these statements in no way imply a critique of the use of contraception, much less suggest the return to some mythical reproductive "state of nature." But if it is worthwhile dwelling on the existing relation between control and fertility, this is because the application of new medical technologies to a woman's body radically demonstrates the truth of *subjective division*: the fact, that is, that human beings are inhabited by a knowledge that eludes conscious thought and whose effects no rational decision can avoid. The time of motherhood is such that the instant of conception compresses into one logical moment the falling into place of the set of signifiers that characterize a certain encounter, that inscribe such an encounter into a history spanning generations.

Psychoanalytical clinical practice, therefore, continually demonstrates the extent to which infertility may vanish in the course of a treatment.

*

The *New York Times* published a series of articles devoted to what has rightly been defined as "The Fertility Market."[4] As everyone more or less knows, assisted fertilization is first and foremost a booming business. It should be pointed out that in these articles, which are the source of the interesting data I'll be drawing upon in these notes, no mention is made of the possible psychological causes of what is defined as "infertility." The notion that subjective causes of a non-organic order may result in infertility is *a priori* ruled out. What is taken at face value, meanwhile, is the current definition of "infertility" given by the American medical establishment, namely, "An inability to conceive after a year of unprotected sexual relations."[5] This is taken as a fact; there is no discussion of either the correctness or the consequences of such a definition. In this view, a physician, at the end of the fateful year, should advise those patients eager to have a child either to undergo fertilizing treatments or to begin the course of assisted reproduction. The possibility of questioning the woman's personal circumstances, by sending her, for instance, to a psychoanalyst before going the route of medical intervention, isn't a consideration.

Psychological motivations for infertility are given no credence, then. And how convenient this is. In a heated debate on the new possibilities

---

[4] T. Gabriel, F.R. Lee, J. Hoffman, E. Rosenthal, *The New York Times*, January 7, 8, 9, 10, 1996.
[5] T. Gabriel, ibid., *The New York Times*, January 7, 1996, 18.

of genetic engineering – a different field from that of assisted fertilization but one that, ethically and socially, has rather similar implications – a doctor of immunology recently declared to me that, when all is said and done, the motive for scientific research is the desire for knowledge. Yes. But how is it then that such desire goes only in a certain direction, preferring to shut out a priori any other knowledge – psychoanalytical knowledge, say – that does not stop at the mere genetic, physiological, and environmental factors, but at the same time questions the complexity of subjective reality? People always answer this by leaning on the crutch of empirical, quantifiable data: on the famous formula of Lockean empiricism, *nihil erit in intellectu quod non prius fuerit in sensu* (Nothing will exist in the mind that was not first in the senses). But they fail to mention how Locke finished his sentence: *nisi intellectus ipse* (except mind itself) – which states unabashedly the idealist basis of such a doctrine of knowledge. It would be worthwhile to examine the basis of a certain contemporary scientific empiricism, which, clinging to the Truth of Statistics, spells Nature with a capital letter, thereby managing to rake in huge profits while giving the appearance of a clean conscience, free of the ethical implications of its own operations.

Obviously, we are dealing here with a position that can be so deftly upheld not only because it nurtures an enormously lucrative business, but also because in its own way it satisfies the demand of those who use medical science. It is in the interest of a certain economic reality for medicine to nurture ignorance and belief. Faced with a dead-end situation, it's only natural to look for a quick fix. The promise of a certain scientific discourse applied to the medical sphere is that of being able to *quickly* resolve not only illness but every least discomfort, every psychic or physical obstacle, by intervening in the patient's body – whether it's a matter of pharmacology, surgery, genetic engineering, or whatever. A solidarity is established between the logic of medical science and the logic of the symptom, which calls precisely for such direct intervention in the body. In the case of infertility we see assisted-reproduction practices reinforce the subjective conviction mentioned above

– namely, that to make a baby, you just have to want to; it doesn't matter, then, *who* wants to, whether it's the woman, her partner, the doctor, the clinic. Patients and doctors share the same faith in a totally rational, self-transparent subject; the response to the symptom, that is, is a symptom – and so its roots proliferate.[6]

Current medical discourse separates the sickness from the sick person; it makes the sickness its privileged object of study *at the expense* of the sick person. In reducing the symptom to the sickness of which it is thought to be a sign and treating it quite apart from the subjectivity that manifests it, this discourse relegates the individual to second place. This sort of schism between the sick person and the sickness, obviously operative in the logic that governs various practices of assisted reproduction or genetic engineering, is at its most blatant in the realm of current high-tech implementations of Life Extension. In the United States this situation is tragically paradoxical. The dying sick are too often kept alive thanks to intrusive machinery that spares neither physical nor moral suffering. The challenge a certain medical discourse throws out to death does not think twice about putting dignity and human suffering "on hold."[7]

Today, according to the National Center for Health Statistics, infer-

---

6 It should be mentioned here that psychoanalysts are not free from responsbility for the alliance that has arisen between individual malaise and collective denial, for instance, in increasingly popular areas like pharmacology or surgery, in response to various subjective problems, from ordinary unhappiness to so-called mental health. Traditionally trained on medical ideals, they often confuse treatment with ethical and behavioral prescriptions. No wonder such a form of psychoanalysis should be unsuccessful, and that patients should turn to other "doctors."

7 On the subject of the separation between sickness and the sick that medical discourse achieves, see J.-P. Lebrun, *De la maladie médicale*, Edition De Boeck, Bruxelles, 1995. On the sad paradox of extending life by means of the indiscriminate use of specialized machinery, see the article "Study Finds Doctors Refuse Patients' Requests on Death," *The New York Times*, November 22, 1995, 1 and C7. Despite the widespread practice in the United States of establishing a Living Will, a legal statement by which one safeguards oneself in a state of health against emergency and resuscitory medical treatment in the event of physical decline or terminal illness, the article shows how physicians persist in

---

■ BEING HUMAN:

Some Reflections on Medically Assisted Reproduction 263

tility is a condition that strikes 4.9 million American couples. According to the same source, in 40% of these cases it is the woman who is infertile, in 40% the man, and in 20% the infertility remains inexplicable. Medical specialists judge that approximately 50% of the couples in question can be aided by conventional systems of so-called low-tech treatments (drugs, intrauterine insemination, surgery) and 50% require so-called ART (assisted reproductive technologies), which include: in-vitro fertilization, gamete introfallopian transfer, zygote introfallopian transfer, micromanipulation, intracyctoplasmic sperm injection, crypreserved embryo transfer, egg donation, surrogacy.

From 1981 to January 1996 in the United States, 40,000 babies have been born by in-vitro fertilization and similar procedures; a number said to be five times lower than that of France, where assisted reproduction is covered by the state.[8] In the USA, 85% of the cost of such procedures accrues to the patient. The price of an in vitro fertilization or a GIFT (gamete introfallopian transfer, in which a mixture of sperm and eggs is inserted into the fallopian tubes) ranges from $7,500 to $15,000; and such cycles are often repeated. The cost of a fertilization cycle for egg donation, on the other hand, ranges between $14,000 and $20,000, depending on the fee the donor receives (usually between $1,500 and $3,000); and there are further separate medical expenses for the surrogate mother and the legal mother, hospital expenses, legal expenses.

Assisted reproduction is part of the vast contemporary market merchandizing the human body. The development of biological technolo-

---

misunderstanding and deliberately ignoring the requests of the sick in their care. As a result, a huge number of people die alone, in severe pain, often attached to mechanical ventilators in intensive-care hospital units. As Doctor Knaus, the director of the Department of Health Evaluation Sciences of the University of Virginia, puts it: "The hospital culture is geared to high-tech treatments. The philosophy is, we have all these machines available and we have to use them."

8 T. Gabriel, *The New York Times*, ibid., January 7, 1996, 18, 19.

gies has led to a boom in the industry of body parts and substances. Thanks to the spread of organ and fetal transplants, the use of fetal materials for medicine and cosmetics, reproductive technologies and genetic engineering, body parts and fluids – from the organs to the blood, from tissue to cells – are now the object of a world-wide industry that draws billions of dollars in profits.[9]

The revenues in the market of reproductive technologies are such that hospitals and clinics compete over their assisted-reproduction services, often publicizing percentages of success that bear no relation to reality. In 1994, for instance, Mount Sinai Medical Center in New York paid four million dollars "to hundreds of childless former infertility patients to settle a suit over false success-rate claims."[10]

The egg, sperm, and surrogate-motherhood marketplace has led to the rise of a set of agencies to match donors with recipients (the Center for Surrogate Parenting and Egg Donation in Beverly Hills is a well-known example). A new type of entrepreneur has emerged, the so-called "donor brokers," most of whom are lawyers, nurses, and psychotherapists, who, for a certain fee, scout for ideal donors and surrogate mothers.

If it is worth relating these facts here it is to draw attention to the real extent of assisted reproduction today. It is necessary to recognize that in reflecting on the consequences of new fertilization techniques, we are not conjecturing over some future reality, but one very much in progress. Not only are we already having to come to terms with the new reality of infertility, derived in part from the effects of birth control on the temporality of conception; not only are we grappling with

---

9 See A. Kimbrell, *The Human Body Shop*, Harper, San Francisco, 1993. It is worth listing some of the body parts that are for sale: cornea, various parts of the ear, heart, valves, lungs, pancreas, liver, kidney, stomach, 206 separate bones, various types of cartilage, 60,000 miles of blood vessels, etc. Kimbrell recalls the case of William Norwood, a 22-year-old, killed in a robbery in 1985; parts of his body were transplanted into 52 different people. Only after was it discovered that Norwood was HIV-positive.
10 T. Gabriel, *The New York Times*, ibid., January 7, 1996, 19.

■   being human:

Some Reflections on Medically Assisted Reproduction

symptoms derived from the alliance between subjective discontents and medical knowledge as it is applied not only to therapeutic but also ideological, political, and mercenary ends; in fact we are already deep in the midst of a social reality in which many new-born children are the product of applied advanced technologies. For many of these children the issue of their birth – an issue that, as ever, raises a host of other questions – enters a new framework. It is well worth dwelling on this new framework for a moment; and it is from the vantage point of these children who've come into the world by means of new applied technologies that we shall try to put forward certain considerations.

All human beings are part of a determined social context that defines the symbolic universe to which they belong. From the moment they're conceived they occupy a precise place in the network of relations that characterize the world of the mother, the father, their families. Beyond a genetic inheritance, a child receives a symbolic, cultural inheritance, marked by the historical and material truth of the generations that have come before it. Whether their conception has been accidental or planned, human beings are marked by the desires of those who have wished to give birth to them, who have wanted them to live and grow. Their identity will be defined not only by a precise genetic map but also by a *symbolic map* that inscribes them in a history – that of their family, their race, their language, their country, their geographical setting. This identity will take shape within the workings of a dialectical process of identification and differentiation between self and other that spans each individual's life.

In the realm of new genetic manipulations or of assisted reproduction, applied medicine, by emphasizing genetic heredity on the one hand and giving absolute priority to reproduction on the other, does not take into account the totality and the complexity of the components that define subjective identity. Content to consider the genetic map and environmental factors (geographic, nutritional, etc.), it pays no heed to the transmission of a symbolic inheritance between generations.

Genetic transmission is in fact *one* of the factors of symbolic inheritance. Human beings must come to terms with all components of their inheritance, articulating themselves – their own self – in a constant dialectical relation with the outer world. The subjective symptoms, be they psychic or somatic, are often one of the ways in which such coming-to-terms is expressed.

What are we to say, then, of the reality the children of the new technologies must come to terms with? Let's state, above all, that these children may find themselves faced with a redefinition of the notion of family. Through the new technologies, five different people can be directly involved in the birth of a child: the woman who supplies the ovules, the man who supplies the sperm, the woman who bears the fetus, the woman and the partner who raised the child. "What does it mean to be a mother, a father, a family member?" some therefore ask.[11] The question is broached above all from a legal standpoint – through the attempt to define these roles in relation to the law. Through special centers, egg or sperm donors act for the most part in anonymity; legally their services end the moment they're paid. Many infertile couples, however, meet these donors. They ask for donors who meet specific requirements, and not only from the standpoint of health or race: they are expected, for instance, to be of a certain height or hair color, to belong to a certain religion, or play certain sports, excel in science or literature. Sometimes the donors are friends.

In the case of surrogate mothers the law ensures above all that the woman has no right over the child she has carried and given birth to. She has a direct contact with the child's legal parents during the pregnancy; sometimes, if they desire it, during its rearing as well.

In a word, then, the law decides who is father and mother. It defines a symbolic and civil function, at least partially independent of biological reality; which is nothing new, given that equivalent decisions have taken place for years in the case of adoption. We should add, more-

---

11 E. Rosenthal, *The New York Times*, ibid., January 10, 1996, 1 and C6.

over, that such a symbolic function does not always correspond to different sexes: in certain instances, the couples in question may be of the same sex.

Yet biological reality counts. And not just genetically. It goes without saying that it counts from a genetic standpoint. To carry a certain genetic baggage rather than another is something that defines a fundamental aspect of individual uniqueness; to know or not know from whom this genetic baggage comes can have both medical and psychological repercussions. Every clinical case demonstrates this; we don't even have to look here to the example of many adopted children raised in happy circumstances, where the need to know about their own origins ends up determining the course of their lives.

What vicissitudes of desire have created a certain child? Faced with this question, we realize that the reality of assisted reproduction calls into question a particular dynamic of desire. If it is true, as noted above, that five people can be directly involved in the baby's conception, then at least as many vicissitudes of desire will be involved in its coming into the world. There are the vicissitudes of desire of its civil parents, who have done so much to have this child, with all the burden of infertility hanging over one or the other, and the meaning of such infertility coloring both their histories. There are those of the surrogate mother, who accepts such a role in order perhaps to feel she's "a good soldier," as Monica Loustlot, a two-time surrogate mother and one-time egg donor, has put it,[12] or to re-experience the ecstasy of gestation, or simply to pay for her own children's schooling. And there are those of the sperm or egg donor, who may feel motivated by friendship, love, a sense of omnipotence, a will to populate the world, or the idea of easy money. Into this web of desires at work in the protagonists of an assisted procreation, we must add the weight of the desire of the doctor who matches and follows them, of that doctor who answers

---

12 J. Hoffman, *The New York Times*, ibid., January 8, 1996, 10.

the question posed by infertility by assuming the role of the creator-father.

A hundred years of analytic practice have shown that the fact that a human being comes into the world through the desire of the other (parents, family, a socio-cultural context, etc.) and that, in the wretchedness of physiological prematurity, he can survive only through such desire, is something that radically determines the orientation of subjective identity. That human desire is the desire of the other is one consequence of this structural condition. The drama of the neurotic must always be defined in relation to a wavering acknowledgment of the desire he/she is the fruit of and whose symbolic debt must be paid. Whether he/she ignores or recognizes this onus doesn't prevent running up against it at every turn. Let us say that the malaise of the human condition will draw comfort from one's recognition of the plot one is woven into; that this recognition will allow a person to interrupt certain transmissions of symptoms between generations.

The circumstances of a birth through assisted reproduction are not without consequences, then. But we must immediately add here that the circumstances of *any* birth are not without consequences. All well and good, but if it is worthwhile reflecting on this question in light of the new reproductive technologies, it is because the current discourse of applied medicine in this field doesn't know, or pretends not to know, the ethical and symbolic implications of its own operations. It is unaware of the scope and import of the vicissitudes of the desire of the parents, the relatives, the milieu, on the newborn child. To shed light on this reality, to take cognizance of it, is one way of gauging the consequences of this new social situation, seriously assessing its ethical implications and lifting one's own ostrich-head out of the sand.

Naively, embarrassedly, people ask what to tell children born by assisted reproduction. According to the *New York Times,* among the parents of children conceived through in vitro fertilization between 1982 and 1992 at Yale University Medical Center, 8% don't want to reveal to their children the circumstances of their birth, 26% are uncertain

Some Reflections on Medically Assisted Reproduction 269

what to do, and 66% plan to discuss it with them. Doctor Dorothy Greenfield comments: "There are a lot of people telling and others who are not comfortable with that. We don't know whether it's important or whether it's like discussing with kids the circumstance of their conception – who was on top, who was on bottom – which we normally don't do."[13] A remarkable statement, all the more so for being made by the director of psychological services of the Yale Center for Reproductive Medicine. She candidly reveals what seems to her an obvious equivalency: that speaking of a test tube birth is the same as speaking of details of a sexual embrace. We learn that, in the judgment of the doctor, there is no difference between the sexual act and the encounter of ovules and sperm in a beaker – same mechanics, after all. And facts are facts. But to smile over what such a vision unveils of the erotic universe of those who uphold it does not eliminate the fact that it generally feeds into the ideology and practice of assisted reproduction. A perspective that fits easily into the puritanical tradition of the U.S. scientific discourse, where, as in some fifties Hollywood film, we are assured from the outset that couples sleep in separate beds.

Advanced technologies in the realm of assisted reproduction exalt the separation between conception and eroticism. Reproduction is reduced to a pure mechanical event, at least in the eyes of technicians who do their utmost to succeed where what they call "Nature" fails. Such a reduction easily fits within the puritanical and capitalistic logic promoted by the body-parts industry: stripped of its erotic nature, the body is reduced to so many components that have above all a commodity and product value. Stellar profits are obtained through the asepticness of an exchange thought to be emptied of any libidinal component – which is obviously sheer illusion. To eroticism falls the task of forming a new enclosure, a wholeness, for a body fragmented by the commercial value of its real parts.

---

13 E. Rosenthal, *The New York Times*, ibid., January 10, 1996, C6.

If in the eyes of technicians, reproduction and libido coincide through a pure useless accident, in the eyes of children, things are a bit different. The technicians are forgetting, or perhaps don't know, Freud's lesson: that children are sexual beings; that rearing, the evolution of subjective individuality, is structurally accompanied by the articulation of a sexual geography that charts a map of the human body. The body is an erogenous site, marked by an exchange with the other that, from the child's very entry into the world, inscribes *jouissance* in it.[14]

Children's longing for knowledge, the same longing that animates their insistent questioning – the same longing that forms the basis for intellectual or scientific research – is born out of sexual curiosity. Such a thirst for knowledge, animated by the child's libidinal development, founders on the mystery of origins, the enigma of sexual difference. In questioning the limits of the knowable, this longing raises the question the human being feels it hangs upon: that of its own origin, its own reason for being.

Let us recall that Freud showed how longing for knowledge emerges in childhood as a consequence "of vital need" ("*der Lebensnot*"): it is when the child feels threatened, for instance, by the Oedipal prohibition or by the arrival of a new-born baby in the family, that the drive for research is kindled. Because of the prohibition that separates the child from the object of desire, the drive-related urge of current eroticism is associated with the sense of threat. As I have stated elsewhere, it is danger that hastens the clinging to causal logic; it is danger that induces production of rational explanations in the face of what takes the guise of the *unthinkable*.[15] And the unthinkable, the unsymbolizable, takes the guise of danger. The sexual theories of children – the phallic theory, the cloacal theory, the sadistic theory of coitus – are exemplary in this regard, in their attempt to offer an answer to the unthinkable aspect of origin, of sexual difference.

14 See translator's note, *Being Human*, 94.
15 P. Mieli, "Dolci melodie," in *Legenda* no. 9, Tarchida Editori, Milan, 1994.

■ BEING HUMAN:

The fact that John Aspinwall, four years old, should explore the dish in which, as he's told, his life originated, and uses this small transparent object as a lens through which to gaze at the camera that captures his image, doesn't alter the fact that his provenance remains as mysterious as the gesture he mimes well represents. More perhaps than the scientists who've brought him into the world, John understands that "seeing" the place of one's own conception, or knowing that one is the offspring of a successful meeting of ovule and sperm, takes away nothing of life's enigma; in no way does it remove the basic question of what meeting of desires one is fruit of.[16]

The empirical data exhaust neither the reality of the fantasy nor the encounter with an unsymbolizable real. To know that the longed-for object is inaccessible in no way prevents us from imagining its love. Knowing how a tumor kills a human being does not release us from the inconceivability of impending death.

What a certain scientific discourse seems to ignore is that the human being is oriented in the world, in a world of objects, moved by desire.

One is therefore made aware here that the question of what to tell or not tell children about the assisted reproduction they are the fruit of, reflects, above all, an ignorance; ignorance of the world of children, which is in reality no different from ignorance of one's own adult world. But let us add that such a question may be the sign of the bad conscience of whoever utters it; the sign, that is, of discomfort aroused by

---

16 John Aspinwall's photo appears in the *New York Times*, January 10, 1996, C6.
    The medical illusion that "seeing" the place of one's own conception answers the human question of the enigma of origins, calls to mind the lesson taught by a noted American porno star, Annie Sprinkle. In an often repeated performance that has been filmed, Ms. Sprinkle, on stage with her legs spread wide, invites the audience to look deep into her vagina with a speculum. One by one the spectators approach, moving through the seduction of this proposal. No need to point out that the gynecological fantasy bursts at once before the dead end of a piece of reddish flesh.
    Which in no way prevents the erotic, fantasmatic implications of sexual curiosity from springing up again like mushrooms, in new woods.

THE TECHNOLOGICAL EXTENSIONS OF THE BODY

the valuing of a practice which, in its very description, suddenly unveils all its symptomatic character.

Anyway physicians, psychologists, specialists, social workers and so on, seem united in paying no heed to what, after a century of psychoanalysis, everyone should know: that what condenses into a family secret is by definition doomed to persist, to return, to weigh upon the life of children and future generations – to produce effects by the very fact of being hushed up.

Let us recall that the truth of their own genealogy does not prevent children from inventing one they prefer. The reality of things, so-called empirical data, far from silencing desire, allows instead for activation of the fantasy. Freud showed how the family romance, that waking, open-eyed fantasy through which children correct the circumstances of their coming into the world – inventing the scenario, say, that they've been adopted by those who claim to be their parents, or imagining that they are the fruit of a relation that substitutes one parent by someone they prefer – is one of the techniques by which children free themselves from parental authority. An emancipation as painful as it is necessary, since "the whole progress of society rests on this opposition between successive generations."[17] Freud distinguishes two phases in the family romance: the one preceding knowledge of the sexual premises of procreation, in which both parents can be replaced by better parents; and a successive phase in which, on the basis of the principle that *mater certissima est*, the father remains the variable. Freud does not fail to point out that this "sexual" phase, as he defines it, of the family romance, revolves around the desire to bring the mother "who is the object of the most intense sexual curiosity, into situations of secret infidelity and into secret love-affairs."[18]

---

17 Sigmund Freud, "The Family Romances of Neurotics," 1909 [1908], *The Standard Edition of the Complete Psychological Works of Sigmund Freud*, vol. 9, The Hogarth Press, London, 237.
18 Ibid.., 239.

■ BEING HUMAN:

Some Reflections on Medically Assisted Reproduction 273

Not only does the fantasy correct one's own genealogy in a way consistent with desire; it also seeks out answers to the question of the mother's *jouissance*. That such *jouissance* is unknown and unknowable, barred by the interdiction that distances her, by the universal prohibition of maternal incest that organizes the social order; that such *jouissance* should by definition be inaccessible – none of this alters the fact that human beings continue to question it. *Che vuoi?* – What do you want? – is, precisely, the question that guides the vicissitudes of one's sexual identity. Taking the form of the horror of castration, the mystery of the *jouissance* of the mother's body represents the impossibility to know as an absence.

We can observe that through interventions that alter the structure of the body, genetic discoveries and assisted reproduction partially change the facts on which the family romance rests. For anyone interested in ascertaining this, the biological father has for some time now ceased to be *semper incertus*, a pure reference of the mother's designation; on the other hand, with surrogate motherhood and egg donation, it is the certainty of the *mater* that is starting to waver. Biological truth gets complicated; we can say that this complication creates new options for the family romance. In a certain sense, it creates variants for it, enriches its pattern.

Perhaps the practice of assisted reproduction also says something about the fantasies, the forgotten family romances, of those who invented it.

Anyway, the questioning of the *jouissance* of the mother, of the one who takes the place of the primordial Other, be it one's real or civil mother, remains central. On the other hand, birth in a test-tube or by a surrogate mother, birth by sperm or egg donation, does not eliminate the question of what *jouissance* has led to one's coming into the world.

For all that assisted reproduction may try to separate conception and eroticism, the eroticism that was chased away from the door reappears at the window – and it does so in the guises of symptom, inhibition, perversion. The children of advanced technology will have to come to terms with such eroticism – whether it's a matter of the libido of infertile parents, potent donors, or busily assisting doctors.

But let us note that assisted reproduction also puts a wrench in aspects of the prohibition against incest. The Reproductive Medicine Society of the United States suggests that donors should not contribute to more than ten births, to reduce the chances – already infinitesimal – that a person could mate with a half-sibling. This suggestion is not an injunction; yet it is likely that in time stringent guidelines will be set. But with or without regulations, the fact remains that society, albeit in a context of improbability and denial, concedes the possibility of a transgression of the order that establishes exogamy between blood relations. The fact that incest with one's father or with siblings often occurs, does not mean that the law doesn't prosecute it. But in the case of the possible consequences of assisted reproduction, society, in not expressing itself on the subject, does undermine its own organization.

A curious phenomenon, which gives rise to the perception to what extent applying new technologies in the realm of the structure of the body prompts a set of transformations that have real, imaginary, and symbolic effects. Might we be witnessing here a new vision of the human, as Serge Leclaire asked some time before his death?

Reproduction by egg donation or surrogate mother even raises the possibility of a transgression of the prohibition that universally upholds the social order, that of maternal incest. Why, after all, shouldn't a child accidently meet up with his/her own biological mother? It doesn't matter whether this actually happens or not; what does matter is that society lets this possibility exist, and at a higher rate of frequency than in the case of adoptions. Between decadence and technological revolution, are we perhaps in an era that augurs a new tragic classicism?

What seems implicitly to oscillate within this framework is the *paternal function,* the function of that signifier, which, intervening to replace the signifier of the mother's desire, guarantees the reproduction of the symbolic order. Ensuring that a vacuum is set up between mother and child, between the human being and its origin, the symbolic function of the *third* ensures the place of a transmission between generations.

The nature of the paternal metaphor shows how it is not necessarily the real father who assumes such a function. Being represented by a signifier, this function can take place in the absence of the real father. This is why, for example, the articulation of a classic Oedipal configuration can be seen at work in cases – increasingly common – where the parents are of the same sex. The vagaries of the subjective structure within this configuration would vary on the basis of the specificity and uniqueness of the individual history, the signifiers, the desires, that have marked it. And such an individual history, in being inscribed, will stretch back at least three generations.

Surreptitiously enough, though, the present-day reality of assisted reproduction is putting the symbolic function to the test; on the one hand, it undermines its very premises, for instance, allowing a fault-line to open up in the incest prohibition; on the other hand, it blithely upholds a separation between the real father and the paternal function, as though such a separation had no importance at all. The fact that the paternal function can take place in the absence of a real father does not mean that the separation between a father and his function is not, both at the individual and collective level, fraught with consequences. We see signs of this in the current crisis of u.s. society; and signs, on the other hand, in the current crisis of masculine identity.

It is odd to observe how a phallic ideology of the sort that inspires certain scientific discourse surrounding the new body-related technologies, when it is animated by the myth of universalism, the elimination of differences – biological, sexual – the myth of the manufacture of an ideal species in which disease, decline, and death will not occur, ends up reducing the male gender to impotence. It is no accident, since such an ideology calls into question the very premises of the symbolic order, promising the abolition of symbolic castration (human beings' uncertainty in the universe, the inaccessibility of the object of desire, deterioration, death).

Reduced to sperm by the practices of artificial insemination, nullified by a collective hysterical position that regards him as a contingency within procreation, a man begins to question a desire to be a father that embarasses him. Confused, he questions his role in transmission. He may dodge the question by going off and playing with new gadgets; he may decide to leave his mark elsewhere, in historical, political, economic, medical or criminal exploits; he may engage in an acting out of carnal violence as a last resort for reuniting fatherhood and eroticism; he may test the weight of his virility in a seductive, brutal comparison with other males or by mounting the slick pedestal of bodybuilding; yet try as he will, a man will find himself ousted from the legation of his function.

Enacting fantasy in reality does not make reality a less real place.

■

Allah Mean Everything!

Amiri Baraka

PART I

Allah mean everything, before the word
Before the slavemasters, before the kings
Everything. So how can there be God, since
Everything is the one, the whole, to understand this is what holiness means. It is the hoarders of the earth, who grasped the earth's flowers like what we had been before climbing the mighty tree, and so could see the sky, could see the sea where we had been, could look down on the ground and declare ourselves, finally, more than it. But they who carried the earth with them, like the dead, were buried under it, took dirt with them, even as the women taught the rest of us, how to stand up straight, and dig the scattered eyes of stars of the other part of our wholeness, where surely we would go and be.
But the soul was from the sun and who did not aspire to be again what it is and we are, unconscious and therefore, small, and without the power and burning and endless self of being and coming and becoming re-being re-seeing, all toward what we were when we knew we were. So those who worshipped what is beneath us, the lowness of this place, which is called terrible and terrifying, because humans have not yet come.
So those who did not dig the above, did not dig the sun, those who created histories of words which dealt with nothing but the transportation for their appetites, who thought there was nothing after the tree, and didn't understand the tree, those who made lies into laws, death into trick endings and mystery. Those who created God because they could not be what they wanted with Good.
Because there is not God, there is Good. Between birth and dimension the wheels of infinity. Death is a choice of ignorance. The Z row the

hole for whole, the nothing when there nothing is nothing and nothing cannot be. Because even in your mind nothing exists as something. A thought.

So the place, the tree, the umbrella of our being, when we first rose, began to suppose we were no longer what we had been, the unknown feelings the biting the search for only food, and the instant death of what we could not change. They became slavemasters and kings and priests, and began to rule the world. Character assassinated women as they threw them from the high place of art, the birth place of what carries a visible soul, the womb. Where the crown of sound, what is. And the motion into breasts under which we come, like the mountains and the sea rolling across the everything when we dig that that's what's going on. To be an animal, instead, these terrorists of the stomach, the bowels, the genitals, the delusion, the submission to themselves, these accumulators, hoarders, who created the whore of themselves and named it God, who died, which is a lie, each time they did......

They built the world into a tower, not the destiny of endless rise, with the deep connection to where we been, so not to got there again.

There is Good and there is Evil. No God and the backwardness of evil worships death, the earth, and calls the sky barren and empty, thinks space is a nothing filled with them and what they know.

To be an animal and forbid humanity from appearing, this is what is going on. The masters of the earth are the creators of death and pain and loss and ignorance and poverty and disease.

Who would want to live in a world full of ignorance violence and murder. The refusal to turn with On and see the sun, and make the soul rest upon this place with one leg, a one-eyed jack, right handed, full of worms. Who wd lie about the world's self. The only Royalty is Reality. The only divinity is the ability to understand the future. There is Good and There is Evil and there is no God and No Devil. But the rulers of the world make themselves rulers measures masters of the terror because they want to eat and have expensive doo doo. The most expensive doo doo is money. And now they are so sophisticated they have

■ BEING HUMAN:

paper doo doo, which makes them think their hineys are clean. The paper is green, so it means to remind you of the flowers and the sun. They created religion to explain why they rule and you are a fool. They overthrew women, so they could be beasts again and eat and bust their nuts, who was their mother now an orgasm and their mother is a mystery they really killed so they could have sex with everything and anything because they were masters so they could do it cheap and not have to give so much money away. Yeh, they created money to fertilize the earth with feces, their symbol, which they tell us means create, if we dump upon the ground, which was the soul, upon with their one leg stood, why they removed one wheel from Good, creating God, and a will the inheritance, which turned their blood to liquid waste.

There is Evil and There is Good. And you will be in pain, the french word for Bread, like their word for civilized is Cop. Which is why the rulers created the POLICE, like mercy is their world for thank you, dig that, and Please is what they make us do to live, cop a plea. So dig the German word for God is Gott, which is what they made of materialism a worship of the earth, of the disease of powerful ease, the word for Good is Gut, the stomach is filled with bodies.

   Don't say these things a head is dropped through the curtain whispering. The spy of the un-is. The dead lie but a lie exists as such only because of ignorance. Without ignorance there could be no lies. Why do the Greeks and Roam mens say Lyre is for music. The strongest material we know is steel, so the grounding of the thieves who make art a thing not creation itself. The belief in electricity perhaps is the beginning of light? The city is democratic supposedly like the lie is ok if it exists as a flag a force a current. When they burned the prophets and changed the meaning to money to surplus, the greed of the owners. What is a slave, a dead life. To rise openly is alive. Why is heaven actually having? Not haven. Why is god a backward dog. Why is Ali Baba the Pope and the forty thieves his cardinals. What is the difference between a sine and sign and sin and without and to connect yes to

THE TECHNOLOGICAL EXTENSIONS OF THE BODY   ■

live El Camino Real. Why is Santa Claus the devil? Why is Halloween the worship of a tribe of killers. What is the skull and cross bones. Why are privateers and privatization slave ship captains. Why is it called the white house? Why was the obelisk painted white? Why is the pyramid cut off near your mother and the eye disconnected and printed in green. Why is the capital where the capitalists rule. Watt does measure mean? Why is Dub Chaint and Dub poetry English? What does it have to do with Dublin. Why are the moors and the moors and the moors the same. The Gypsies and Egyptians and Gyp the same meaning. The black Douglass was a ruler of Scotland. When the anglos killed him they wrote "think on this" under his severed head. Why are the "little people the same as pygmies. Lepracaun means leopard king. Stonehenge is the bridge to the south. Why do the masons say Allah U Akbar when they get their 33.33 from 360 leaves what?

Electrical talk is the same. Resistance, current, ground, watt? Why the Indians say OHMMMM. Krishna is Blue like Monk, and the same word as Christ. Cross...to cross from where, to hang why? Why shd we believe Peter that Christ returned...Paul was not a disciple. They both went to Rome and got wasted hung upside down. Why do they have a candy named Peter Paul Mounds. Why is the rock group called Peter Paul and Mary. Why is the trinity about money. Why is God an adulterer. What is Judeo-Christian? Who invented money? Why is money and slaves called a buck? Is a wooden nickel $3/5^{th}$ of a real one. Why is only a penny brown and got Lincoln on it? Is that why they leave it on the ground.

Nero means black, was he like Stuff Smith. You been to Cambodia, Satchmo whole family up on the side of the mountains.

■ BEING HUMAN:

PART 2

What exists insists and resists, it's
So tragic not to be human. So
Ugly to be ruled by animals.
Who worships God wants good
& is lied to that our pain
is created by a mysterious
Sky Lord. To hide the reality of themselves
The absentee Land lords.
We submit to a lie so the
Devil rules the earth. The Beast
The Bible explains. Who did surely "rise
Smoking from the western sea."
The beast who is a man.

On the money is their I
Disconnected and hovering
Above the abducted pyramid stashed
In the Metropolitan Museum, Too that I
Add their old name, MF, and they are identified as
The ultimate owners of money. The hanged man, given a last meal
Before revealing he is the Limp Peter who lied that Christ
Had risen.

My Eye, they say, stopping the rise the reach
For the Soul in the middle of the
Air, where we have been and will go again
Even now we are the same but that makes no sense
To our tiny brain.

Cash, they say, but if, "what are" get in it, 1929 again, in a car
Can't go far, come out the mountains

Amiri Baraka

Looking for food, we'll say they've made progress
Put caves in the skyscrapers, full of dead bodies
It is their self that is money, for whom they prey
A Predator is Neanderthal, not prehistoric just shy
You should have seen The Mighty, a minute ago, starving in caves
Answer those questions, I say. Out the Caucasus mountains
Across the Black Sea, near Georgia somewhere, eat some Turkeys
Then down, to turn itchy & water roll down they flesh, eat some leaves and lizards, How dreamy to be alone and hunger less, twisting the world into what He need. The answer to the problem is murder theft and greed. Till they skin got tawny, the lips a little puffed, they hair got frizzy, nature changed they talk There was more to see, different kinds of things. They saw a huge fish in the water, made noise, they wd later know as sings. Got down, came down, till way there, they seen some night people wrapped in colors and gold stones who carried long poles and feathers on their head. Even tho terrified they kneeled and begged for bread.
And the world had changed. They said Pain Pain Pain, the other half of The mask. They could have all they need but they had not learned how to ask. Nor to bathe. That's bad they said, We Bad said we. Nature had put on different clothes everywhere new eyes for your nose. It is our home, not a forest but a garden. And the animals the women taught To be with us, give us milk, and honey, and clothes and food, no longer Must we roam the forests everyday, for a month full of food the only pay.
The wise man said, the more time you must spend on seeking food and sustenance the less time you have to practice being human. The less time you have to develop your mind.

So the travelers, the mountain folk, reduced to skin and bones, carried the food everywhere with them, and called our homes The Gotten Of Eaten and were content except for the I's which criss crossed the abundance and seemed to make it disappear. But steal is strong they said to

■   BEING HUMAN:

lay flat and do nothing and be fed by they who have no furs is greatness. It is a pile of bones with knots of red flesh sticking off still.

The Fuhrer had plenty of furs he was a leader. The leader has many furs. So that was paradise, dotted cubes, white cubes black dots, pips, in the east they call it the Tao today it is Dow Jones, the stock market a bingo game. Like religion, a mystery. Everything must remain a mystery in order for evil to rule. The ruler measures.

It must remain a mystery for a few to rule the many. For the hundreds to rule the billions. Everything is a mystery except you got to go to work so you can remain poor and never understand much. Go through the world and never under stand a thing. Except you got to go to work. Go to church all your life and never understand. It's a fraud, a vicious joke. There is Evil and there is good, but not no God and they who say they know God worship the Devil.

They say brain so you will think even thought, Good, is limited and inside, a muscle, not the will of what am you rising, a Black Bird, with burning short stories, history driving you like a jet fire shoots out it's behind as it leaves the earth as space becomes time. Endless yet there is no God only Good, is Evil, reworked as the Devil and who does evil is the only Devil. But this whole endless Amness is Goodness, not a god or invisible dictator with an accent.

So tell the story of the Beast who is a man, came down from the north, snow for eyes, you for food your blood for drink, flatness was a symbol, to lay down was kingly, the body of the being, to be idle and so Duplicate the stone idol, who they word shipped into time, for the God Who had a rep but who did not eat, or take up much space, no agreement was necessary because it was they god in they cave and nobody had to make no noise to oppose it, though each indentation had its own incantation, and the sounds from the ground when the mountain things got down, and wanted to know where they little replacements came from. From the thingless things, with a cave like our place, by why wd the noisy rebe'rs get in there in the beginning. To come out

THE TECHNOLOGICAL EXTENSIONS OF THE BODY ■

from the Cave, like we, which these thingless things make a noise and claim they Have created us again, and we who have some thing, cannot make that thing they pull out of themselves crying and sucking.

For who did not understand got doo doo in their hand, did not re be come and they were struck dumb with filth around them stinking past they regular bad scent. You must understand which hole inside of which is more and which means less.

Then there came the time when they were told they were new and they were nude. Had neither knows nor clothes. But how could you survive in the cold mountains with no clothes or no knows. The furs in strips they had but wore them as mementos of yesterday, they were to keep warm and never for modesty. To wipe the blood and hide the crud there was no word for polite, they were one family, bound by the wizards whose magic brought always the small replacements for themselves. What was left was these two out of thousands who had come down into the desert and on to the great garden where gotten was eaten.

*

But then the day came, when it was no longer the same, they had gotten and they had eaten and the world was limited to what they knew and they knew they didn't know. And what was the twin trees at the center of the hall, which the low 'red of their haven' said they could not take. So even in The Gotten of Eaten, there was to be deprivation of what could not be imagined without imagination.

It was the thing, the dicht that did the harm, so it could exist hanging between their legs and then crawl away and taunt them with what they did not know or understand.

If there was a Good and an Evil, why was it hanging, in Red, on one tree? And if they took the fruit and learned the answer to which they had no question, what would it mean? Except the thing lying on the ground rolled its eyes like citizens of a futuristic state and pretended to

■ BEING HUMAN:

read the newspapers and be just. But it always meant a few, who could survive in the ice, for that was the objective self and its vision.

For Good had done nothing and we changed the Profit's name so it accorded with our having. For Having was the real good. The coming again of surplus, a addition to what they had. The boss is Sir, The animal who first called himself a scientist, somewhere high in the Eastern Steppes when he returned from the under world, where our faces our upside down and the flats and sharps of the thoughts they see are sounds that wont let you lie down and burp the second taste of what we got as steal and the hard art, she held the flesh too close to the flame, what technology gave back from the south, and the steaming meat was easier to eat, for the tiny news that came with cries, like the daily seeking of pain.

What cave, what cave, what cave and where is the what the profit gave. They gave a chronicle of how they returned. Mostly a recipe for ambush and attack. For the taking to rehave the gotten of eaten, so they would eat ever after. And what they would not remember, they lied about, they lie about, like the having gotten, and good became gut, and pain was merely bread, and to be civilized, arranged in a circle of eating was to have a fire arm. For the warmth for the sight for the food for the fight. Still the earth was terra, so why it was named, not to have it or not to have enough, twas best to have it all, to say of it all is gotten and eaten.

The book was to buy the warriors with tales of what did not mean that with rest and snore and fullbellies and a thing wd hold the flag and penetrate, would crawl and lie and create its own world, a scientist of getting and having. We have changed the actual sensuous knowledge of the earth into a slimy crawling animal, spineless with dialectical tongue. And from this beginning it made words that would be history and smite the thingless thing that though we slid from in her womb and she is the namer of all still we are the father and created her from part of the first home she gave us.

We did not understand that what are is was is the difference between

Friend and Fiend and having not the 1st could remember only the latter. Which had to be the claim, not evolution. Jacoub, but a final encore for the pre-human beings you created to be superior to ourselves, by teaching Right without Reason, Logic without Treason.
And when this would not do, nor had they been socialized to love, without that revelation LO!, so we understand, we have to begin, from the beginning and then, SEE! DIG! The expression and the solution, the truth is the vein of our blood, who is the sensuous recreation of ourselves, on that higher ground.
They thought geography, traveling wandering across the earth. Until the thinker caught them in the desert. A flight from a lower self, he sought the thought from on high, was not geography, merely returning north. And the lie and liars turned him white as a flash, having left the south he made an idol out of the law which was to get down, the low and the lowered, who had to flee, for that work of building the sky-scraper for the man god of upper Egypt was a lie, but somebody else's lie, their's, and the prophet came to be a profit and to lead them in a trance across the sand which was water with the flood and being a priest he knew when the tides would lay and led the whitening slaves and fattened wanderers still whiter for God had spake and they were rising, with the geography of the beast north north-east until they reached the palace of stone which they could claim as their own.

And the book of maps and days tells the heroic menu of who they slayed

And of their Lord now high again, and out of sight, to speak with no amendments except in secret by those they hired to rewrite history and condemn those peoples they fired.
But even the tribal gods got left, if they spoke another accent under their breath. The living Gods must be killed. The dead ones rebuilt. And the very sons of their mothers wombs who were born to free they relinquished to their enemy's, claiming only the tomb. And the false Peters became their fathers and the gay anti-Semites their teachers and preachers, the suicided jackasses their bankers, all rushed ahead of them

■   BEING HUMAN:

to Rome, and put their lies in the great tome, they called simply book to hide the fact they knew no other, and all in it was stuff they took. If Dahmer kept a diary it would be a holy book for Cannibals.

There is Good and There is Evil but no God and who worships God worships the Devil, who is Santa Claus.

Judeo Christian is neither Jew nor Christian but A Devil Worshipper and heavy metal user, flag is skull and crossbones, good Friday the day the prophet Jesus slain, they celebrate so he wont come back again. Why else, replace the fish, the mighty wailer, with the cross. How they came down, in the Eastern Kingdom a double cross, because they saw them cross again. The people are nailed spread armed on the cross. Between two thieves helped up the hill by a Negro, a buffalo soldier, off Broadway, he shaved his head and moved to Chicago, wears a wig sometimes and rides a motor sickle cell anemic, the hanged Negro spake aloud from off the cross, he rose his head and cussed out his missing father, and mama mia where was she, below the Negro weeping on her knee. She was told that Peter, the limp betrayer wd return and claim the son had come back, rose again. And so she had the body stole way and where it is it is today.

The liars and thieves who the son had banished thrown out the temple Charging interest for Black studies and various blessings, for what the early profit taking from the prophet who replaced the snake as king, was usury that life itself could be harnessed for fuel, another lie that those who have faith are not fools. Where wailing from the island and staring at the sky, was John, on fire, fuel was desire and the vision of the higher, did cry out about the return of the last people past, Ja je ji Jo ju, away, return, seems a night a day, like the bird, but we was Black with a flaming tale of red, did sing as we rose, brought the light the day sky and everything we knows.

And Jesus was murdered by the state. By the roaming imp highers, of beasts in metal clothes, who worshipped ease and were the creatures of

THE TECHNOLOGICAL EXTENSIONS OF THE BODY ■

the then, The rule of solitary men, Athens who were Heathens, who the Made the women nones. Whose word for heterosexual was prostitute. There fore it was God, the Papal Bull who got under Mary's clothes, so she need not appear in the trinity, it was not adultery but appointm\ent and annointment, that she was taken by the highest of men. Her son was an heir, her husband a carpenter, could never have a share.

But the boy was too dark and named like a Puerto Rican, and he spoke in riddles and parables, like the Black Pip or the Ace of Spades, King Digger Prince Nigger, if he would lie he would not die. But there was no way around his face the darkness in that space, and his words, would not lighten his earthly disposition. He said it is more difficult for a rich man to get in heaven than for a camel to pass through the eye of a needle. And our preaches said yea, and we will get those larger holes and smaller camels by and by. He said I have not come with peace but a sword. He meant history which is mightier than the dick and the dictator. And the sword was the word of what was real, and it was the king, the body of what never stopped. And since they knew they could not cop, they killed him, and made impotence King, Abstraction and Inversion profit and teacher. The lie religion, for the absent and the pleasure seeking, and war the reflection of natural flesh. Which they could still consume from machines and pornography and the stock market.

There is Good but not a God, there is Evil and those who say they represent God worship the Devil for evil created god so they could lie about why good was missing.

There is Good by not know God, there is Evil and the Devil is he who do the Evil. For Good is All of what is every allways eternal. Divine is the seeing of the future, where we begin again, return, go over to the other side as they come round to this.

On the third planet from the Sun, Mercury Venus Earth, and Mars the Red is next, if we pursue our text, for who cannot love must still evolve As animal to yet another animal. And where is the human being we claim to been only, the preachers say, we back slid because of sin, And

■ BEING HUMAN:

so must rise again. It is the place where we lie and steal which much be understood and so reveal us to ourselves, and the world to everything. So that science is the only religion and reality the only royalty and Good the only God!

There is Good and There is Evil and Bad is clean and dirty is the hoarders of the earth. Who will not go anywhere else so attached are they to what remains under us, so they stay as ghosts. Not live but evil, not good but the living dead. Who does the evil is the Devil who does the good is everything the all and this is what is whole and this is what Holy mean.

"A process of insidious
but irreversible metamorphosis."

The Reciprocal Creation of the World and the Subject

Dany-Robert Dufour

"With every tool man is perfecting his own organs, whether motor or sensory, or is removing the limits to their functioning (...). Man has, as it were, become a kind of prosthetic God. When he puts on all his auxiliary organs he is truly magnificent; but those organs have not grown on to him and they still give him much trouble at times. Nevertheless, he is entitled to console himself with the thought that this development will not come to an end precisely with the year 1930 AD. Future ages will bring with them new and probably unimaginably great advances in this field of civilization and will increase man's likeness to God still more. But in the interests of our investigations, we will not forget that present-day man does not feel happy in his Godlike character."[1]

This incurable suffering was none other than the subject of Freud's *Civilization and Its Discontents*.

In 1929, Freud sketched the crossing of air and water authorized by the "gigantic" forces of motors, the overcoming of the narrow limits of vision thanks to microscopes and telescopes, the materialization of the abstract faculty of remembrance made possible by the camera and the gramophone.

Between the thirties and our nineties, in six decades, the "prosthetic god" has perfected its organs and enlarged the limits of its powers in proportions that are at least equivalent to, if not greater than, those of the preceding six centuries: it inhabits space, it gains access to the first formulas of its genetic patrimony, it commands extremely powerful decision-making technologies and means of calculation. It knows how

---

1 Sigmund Freud, *Civilization and Its Discontents*, the *Standard Edition of the Complete Psychological Works of Sigmund Freud*, vol. XXI (London: The Hogarth Press), 91–92.

to master titanic forces....But the old Freudian refrain has lost none of its currency: "Man today does not feel happy" – perhaps even less than in 1930.

So that it might be useful to explore the novel relationship between the technical and scientific states of civilization and the constantly renewed forms of human suffering.

*

Thus, on one side, civilization, the human world, and on the other, the subject. Let us posit the existence of a tangled relationship between the two: the subject creates the world which, in turn, creates the subject. And it is without hesitation that we will leave it to others to address the question of how it all began.

*

Let us examine the first relationship: the human world is the fruit of a continuous creation carried out by a succession of symbolic nominations. "With each scientific revolution, the world changes," says a contemporary epistemologist. This continuous creation of the world is interrupted by profound temporal variations: long pauses followed by brutal accelerations.

The great scansions of the modern era are well known: the Quattrocento, the turn of the century. Not that everything happened in a few days in Florence or Vienna, but in these places everything became suddenly visible. There, visual space submitted to the organization of perspective; this world became disenchanted and was reorganized according to mathematical laws; the vengeful and cruel God became a great architect; the physical world was reconstituted as a continuum. And here, at the turn of our century, the basis of all knowledge, mathematics itself, was called into question before the great wave

■   BEING HUMAN:

The Reciprocal Creation of the World and the Subject

of deconstruction undermined all certainties be they physical, mental, perceptual, musical or visual – everything had to be reinvented, including time itself.

*

Of the second relationship, the creation of the subject by the world, the outlines are known: the Renaissance instituted Man as the measure of all things, and as such circumscribed by boundary figures, those of the primitive, the witch and the madman. Then it was this Man, already suffering from the narcissistic wounds inflicted by Copernicus and Darwin who, in Vienna at the turn of the century was driven from the last preserve of his imaginary mastery: wildness and madness were not external. They were part and parcel of his condition.

*

We do not have much choice: it is upon these two subjects, mutually exclusive and yet concomitantly necessary, the one sure of its reason, the other divided object of its fantasies, the one constantly denying the other, that we must base our inquiry in order to encompass the sense of this adventure.

It is ultimately these two subjects who must be invited to debate and conjecture the new man yet to emerge from the world to come.

*

To think through, today, the new alliance which is being woven between the world and the subject involves sizing up current prosthetic activity. It is intense, proliferating; the current era has fallen under the spell of an uncontrollable burst of prosthetic creation.

We invent sensory prosthesis. The evidence of the senses, the being present to oneself and others, is thereby profoundly altered. Hence forth, man lives in an environment where he is free to immediately disperse himself to the four corners of the universe while every fragment of the world can immediately be projected upon him. The absent can become instantly present and the present can bear the mark of absence. If man can henceforth sample the possibility of consisting in space and subsisting in time in the mode of telepresence, then he also exists in a new temporality, a kind of dilated, polychronic present and a new, videospheric and ubiquitous spatiality. The consequence is that man today inhabits a world in which the illusion of a single reality has notably lost some of its grip in favor of mobile series of "possible worlds" where one may camp, or even jump from one to the other, or rather "surf."

We know how to insert mental calculation technologies in prostheses, which are able to connect in real time enormous chains of consequences right up to the irreversible moment of decision.

We live in the age of assisted procreation which alters familiar generational markers and modifies previously necessary arrangements among men and women.

We create biotechnological prostheses. Because we have recognized the letters which make up the alphabet of life and endeavor to read the writing of which we are the expression, we are beginning to intervene in the course of the evolution of species, to change their destination and even mix them together.

*

But the multiplication of prostheses will not make man happier, anymore than it will encourage him to desist. The race may be exhausting, but it is unlikely that man will meekly return to the shelter of ancient idols to devote himself joyfully to his own unhappiness. Indeed, the more the equipment is developed, the more the essential – what is missing – will appear important and more people will be summoned to work to create the essential prosthesis.

■ BEING HUMAN:

The Reciprocal Creation of the World and the Subject 297

*

How then to define this intense prosthetic creation? Is it related to sublimation as, for example, esthetic creation may be? It does allow death to be usefully adjourned by redirecting to an ostensibly superior cause drives which would otherwise come back to haunt the subject. Here, esthetic creation and prosthetic creation both partake of the phenomenon of sublimation which is, as it were, the crowning achievement of a process wherein the subject is compelled to abandon an imaginary omnipotence in order to come into existence as such; has to accept the horizon of his own death; in a word has to be constituted by his own limits to the extent that this integration takes on the force of law.

Here, we are at the heart of the symbolic function. But esthetic creation is an act destined to remain in the domain of ideas, whereas prosthetic creation returns to the physical and even real places where the limits to omnipotence had been placed. It returns to work materially on the terms of this acceptance.

*

It is sometimes said that the effects of prosthetic creation on the future of civilization must be studied without delay. Perhaps. But not before noting that prosthetic creation being, like esthetic creation, as old as mankind, the alarm should have been sounded long ago.

If there is cause to worry, what should it be about, exactly? The fact that man imagined placing a bit in the horse's mouth, that he invented eyeglasses, the laptop computer or the space shuttle? For one, all the limits are never simultaneously called into question; for the other, those that are, are not, for the most part, the least intangible ones – why indeed put up with a sickness if the cure can be invented, or hunger if more viable edible species can be conceived? So, at what point does prosthetic creation cross the line? We are all the less well placed to know and all the more worried because the current intense era of prosthetic creation places us in a routine of permanent displacement of our points of reference and of permanently shifting landscape apt to be haunted by the transgressive images of literature and esthetic activity: Gilles de Rais, the Sadien hero, Nietszche's Zarathoustra, Frankenstein....[2]

*

The resources of ethics are increasingly called upon to aid the legislator who grapples with the daunting task of deciding what should be authorized and what should be repressed. But, since the field of ethics is still being composed, it risks discovering, under mass democracy, only majority rule as a basis,[3] which corresponds more or less exactly to the dominant fantasies in a period of prosthetic creation, most notably the claim that one day everything will be possible thanks to science. Should we therefore, in order to avoid making a mistake, ring the alarm as soon as the smallest prosthesis changes the subject-world relationship in the least, thereby closing ranks with the modern obscurantist and millinarist batallions who await their time while chomping at the bit?

What to do? Indecision can also be seen among psychoanalysts. They are caught, it seems, between the desire to intervene, faced, as they are, with the increase of certain threats to personal ties secreted by our world, and their constitutional obligation of reserve which is hardened by the idea that the symbolic will always find a way.

---

[2] Monette Vacquin successfully studied and drew out the image of Frankenstein in a fine book *Frankenstein ou les delires de la raison*, ed. F. Bourin, Paris 1989.

[3] See the article by Marcel Czermax and Henry Frignet in *Liberation*, November 17, 1993, "Quel sexe voulez-vous?" The authors analyze the possible effects of the decision rendered December 11, 1992, by the Court of Appeals: by giving the subject the free choice of legal and anatomical sex, this decision can only "incite the subject to abandon his symbolic points of reference in favor of the imaginary ones dictated by majority rule."

■ BEING HUMAN:

Let us accept as a known fact that the fate of our civilization can be grasped through two concomitant phenomena: on the one hand the spectacular and systematic disappearance of any trinitary space (Death of God, reduction of the role of the Father in industrial societies, legal autonomization of the subject, dualization of relationships); on the other, the increasingly blind yielding to the technological benefits flowing from binary logic.[4]

These two types of simultaneous phenomena lead to not insignificant, though different, consequences as far as the position of psychoanalysis in the century is concerned. Faced with the first historical tendency, it seems a given that the disappearance of any viable form of third party can only lead to, among other effects produced on the social bond, the ever increasing demand placed on psychoanalysis to occupy this position and become a substitute for the tertiary function. To yield to this demand would partake of what Serge Leclaire calls, without mincing words, a "major perversion" or "abuse" whereas, he stresses, "it is the psychoanalyst's duty, through the method he employs, and for each person who seeks him out, to firmly keep open the possibility of recognizing and identifying secondary and tertiary voices whose distinction is the necessary precondition for him to be able to truly speak in the first person."[5]

In other words, this opening to a tertiary space, as fundamental as it is to the analytic relationship, nevertheless excludes any straying from the "secular" bounds within which the psychoanalyst generally remains, that is to say any right or qualification to occupy the tertiary position.

---

4 See my essay on the long struggle for influence between trinitary and binary thought, *Les Mystères de la Trinité*, Bib. des Sc. Humaines, Gallimard, Paris, 1990.
5 Serge Leclaire, "Demeures de l'ailleurs" in *Topique*, special issue devoted to François Perrier, Paris 1993.

THE TECHNOLOGICAL EXTENSIONS OF THE BODY

As for participation in the social order, when the psychoanalyst is, for instance, consulted as an expert or consultant to help create new official frameworks for people's care, government or education, he cannot rely directly upon analytic practice without partaking in what might be called, relative to the "major perversion," a "minor perversion" attributable to the psychoanalyst's self-indulgence regarding one of the famous impossible tasks identified by Freud.

What can be said now about the second phenomenon, the increasing yielding of our civilization to the technological benefits stemming from binary logic? Should it involve the same constitutional reserve on the part of psychoanalysts? *A priori,* no, because it seems that psychoanalysts have, at least since 1929, a certain duty to investigate the discontent in civilization,[6] in part because the discovery, in the form of bad news, that *there is discontent because there is civilization* belongs to Freud, and because their position is unmatched when it comes to hearing the multiple forms of the incessant renewal of this discontent and of the death drive's realization in history.

*

Psychoanalysts, beginning with the first among them, proved well situated to testify at the hearing of civilization's discontents – from its origins (lost, and expressed only through myth) to its most recent versions. The reason for this interest is surely linked to its heuristic effects in clinical practice since it is true that every subject is an expression of this discontent. In turn, the capacity to relate any personal history, through the structure of the compound space in which it is inscribed, to a dimension where it appears as a reply to the History of the world,

---

6 The investigation of the discontent in civilization obviously goes back to Freud's first analyses on the nature of the social bond, the founding works of Freudian anthropology among which are *Group Psychology and the Analysis of the Ego* (1909), and *Totem and Taboo* (1912).

is at the origin of the eminent position psychoanalysis occupies in culture.

The question of this position is all the more timely because we are entitled to wonder when these discontents will have reached the point of no return: when does the continuous process of world creation risk becoming a process of subject destruction? Psychoanalysts are certainly busy with the clinical aspects of this question, but it is not impossible to imagine that, in this instance, they could agree to play the role of Capitoline Geese and warn the rest of us about the gang of spectaculars, but also and especially about the insidious progress of a brand new barbarism. The scene of an *encounter* among all those involved in movements of civilization needs to be constructed so that, from their debates, a problematics of prosthetic invention as it relates to the symbolic universe can be worked out. It would be a three-way encounter, such that each would occupy a position irreducible to the other two, occurring among the protagonists of the prosthetic advance of the world (the scientists), those who are in the best position to listen to its effects on personal ties and those who seek to determine its impact on the social bond. This scene, which does not yet exist, cannot be conceived as a simple exchange of information which leaves each one trapped in his or her discourse. It must be open to interruption, to surprise, and to the impromptu, in order to reveal, outside of any immediate legal, ethical or political utilitarianism, a novel and somewhat monstrous object: a thought of places and moments where the great technologies of today (those of assisted procreation, information and artificial intelligence, sensory prostheses, genetic biotechnology) dare to cross the very boundaries which make up the symbolic order. All of which might add up to none other than the latest episode in the relationship between desire and the law.

*

THE TECHNOLOGICAL EXTENSIONS OF THE BODY ∎

New maps must be drawn up that show the new locations of discontent. Places where new forms of "suffering" as well as new forms of *jouissance*[7] are generated, since it is so often said that new paradises await these prostheses.

Some zones are better known than others – it is thus no surprise that psychoanalysts, whose attention to the fate of the Name-of-the-Father is well known, have been particularly interested in procreation[8] and the disruptions created in filiation and the generational cycle by new technologies. But there are other sectors of suffering that are little known, sensory prostheses for example, whose effects on the symbolic function we would be better off knowing.

Sensory prostheses give access, in effect, to new *jouissances* as well as new sufferings in the sense that they use the subject's inclination to play with (or be played with by) the symbolic categories of the here (and hence the elsewhere), the now (and hence the before and the after) which constitute him. By "jouissance" I mean the "vertigo" produced by transporting a visual or auditory here elsewhere or by bringing an elsewhere here.... These technologies give the subject new dimensions which may be ludic but which are certainly dizzying to the extent that they put into play, in the fullest sense of the term, the symbolic points of reference where the subject's self-evidence is constructed ("I" at the intersection of a "here" and a "now").

Where these points of reference have been fixed by the work of language and the use of speech thanks to the venerable personal encounter, a "technology" as archaic and "natural" as the tongue with which language shares a name, constitutive of humanity as such, one can wager that the subject is in a position to use with impunity all the other imaginable technologies, artificial this time, and to use sensory prostheses. These may involve sound over distance (be it drumming, the telephone,

---

7 See translator's note, *Being Human*, 94.
8 A good recent example is Marie-Magdeleine Chatel's book, *Malaise dans la Procréation*, Albin Michel, Paris, 1993.

■ BEING HUMAN:

the intercom, the megaphone, the tape recorder, the Walkman, the net...); image and writing (parchment, books, the Minitel, the fax, the net...); the deferred use of the image which inserts an elsewhere into the subject's here (a narrative elsewhere through the icon, the statue, film, television; a physical elsewhere through the microscope, the telescope...); telepresence which transports the subject's here into the elsewhere of a virtual space....

But whereas, as it may be, the symbolic points of reference (of time, space and persons) are lame (by reason of eventual breakdowns in the archaic technology), the use of these prostheses risks removing the subject even more from these categories and pulling him deeper into a process of psychoticization. The multiplicity of the dimensions offered may indeed become one more obstacle blocking access to the basic symbolic categories, a kind of additional screen, scrambling their perception and adding to symbolic confusion and delirious outbreaks. The contribution of sensory prostheses to social psychoticization will sooner or later have to be evaluated.

In short, while these new prostheses can permit the development of new aptitudes for *jouissance,* they are also well-suited, and for the same reasons, to become the privileged medium for the expression of new suffering. Telephone pathologies are well-known; those of the Minitel[9] and the computer are now appearing; soon we will discover those of virtual space.[10]

*

But these joys and sorrows, most of which are soon to be released, are ultimately of small magnitude compared to the *over-jouissance* as

---

9 Due to undergo a renewal as soon as the visiophone, which allows you to see your interlocutor goes, on the market in the very near future.
10 I have attempted to treat this topic in an article for the journal *Mscope*, "Prothèses sensorielles et fonctions symboliques," No. 5, Sept. 1993.

well as the total misery linked to a crossing of the boundaries that constitute the symbolic order and within which desire is contained. What would happen to a prosthetic activity that had arrived at its point of inversion, able to transform man himself into a super prosthesis – temptation of a *jouissance* all the more intense since this paroxysmic calamity would definitively put an end to all other forms of misfortune? Crossing the symbolic boundaries means giving life solo, escaping filiation, reducing the other to the same.... What should we think, for example, of a process for transmitting life which could escape filiation and engender itself laterally? Would this not be tantamount to obtaining something like an ancestor/brother where there would be a realization, through a novel path, of what the incest taboo, the foundation so far as I know of all social bonds, proscribed?

*

It appears that different types of prosthetic activity are deeply congruent. At the center there is something like a digitization of the world. Of the world and the subject: the deciphering of the number of which man is the expression is probably one of the key points of this harmony.

After this rendezvous between man and his destiny, only one more step would be needed for the deciphering of the genome to lead to the eternal *young man*....

If the dawn of such a day arose, it is probable that humanity, in an ultimate mirror stage would wake up in the skin of a civilizational Dorian Gray, lost at the very moment of finding itself.

■ BEING HUMAN:

When Science Remakes the Body

Jean-Pierre Lebrun

Today, science remakes the body. At first glance, why complain? Surely, we witness the benefits every day. Nevertheless, we cannot be content to sit back smugly when current scientific progress presents us with important new questions that beg our consideration.

The advances of science, its accomplishments, and, most of all, the social ideal it promotes, profoundly, often unwittingly, and in novel ways, subvert our ethical points of reference. In this regard, it is worth noting that in the French language the noun "inhuman" did not appear until after the Second World War, as if we had to wait for the development of techno-science to produce the horrors we have since termed "crimes against humanity."

My aim here is not to disparage science in any way. My hypotheses are not concerned with science as such. Rather, I shall examine the implications that science carries with it and which seem to have been blindly espoused by our society.

To give an example of the subversive effect produced by the discourse of science, I often tell a story concerning the problem of time. Before the Copernican revolution, the Earth was considered a flat surface with Jerusalem at its center. The Sun moved according to a circular trajectory, perpendicular to the Earth's surface. It rose in the East and set in the West continuing its tour into "the waters of the underworld" until it reappeared in the East the following morning. In this earlier view of the Earth, time was the same for everyone; the Sun rose and set at the same moment for all. The discoveries of the 16th century, however, would overturn this view of the world. Science would completely upset this earlier theory, which nevertheless had the advantage of a unique point of reference, because from the moment we realized that it was, in fact, the Earth that turned around the Sun, and not the other way around – the question would have to be posed; how were we to determine the time that would serve as the point of reference?

In 1600, Rudolph II of Hapsburg called a scientific colloquium in Prague to attempt to respond to this question. The leading figures of the time were invited to participate. Ultimately, the question was left unanswered because there was no external reference that could universally authorize the scientific determination of a meridian that would serve as a reference point for all. We would have to wait three more centuries until Greenwich for a purely arbitrary decision in 1884. One can only imagine the chaos in which those who organize intercontinental communication would have found themselves. Today, the consequences of science concern us in our everyday lives. Only since the systematic use of telephones and jet travel do problems arise because of time differences. Fortunately, the consensus of Greenwich, despite the fact that it was utterly arbitrary, protected us from what could have been a very disorderly world.

This little story gives us an appreciation of the sense of how much our bearings were changed by the substitution of a scientific world view for a religious world view. Moreover, this was precisely what was at stake in the Galileo affair, which could very well be interpreted a posteriori as the stigmata of the new situation in which two readings of the world confronted one another for the first time. Beyond a particular conflict between a man of religion and a man of science, this was the first time that the authority of the church was countered by that of science. And that was how "Galileo and his struggle against Rome were brought about by the event that the possibility of affirming, 'this is scientific' constitutes."[1] Beyond simply a personal conflict, this was the conflict of two different conceptions insofar as what was at stake

---

1 Isabelle Stengers, *L'invention des sciences modernes* (Paris: La Découverte, 1993), 87.

■ BEING HUMAN:

was the very basis for the legitimation of authority itself. Galileo's trial signaled the twilight of the legitimacy authorized by God's omnipotence, in favor of a new legitimacy science permitted. It signaled the beginning of the end of a legitimacy founded on the authority of the author and its replacement by one which would be founded on an authority derived from the internal coherence of the statements themselves.

Thus, the development of modern science upset the place of religious authority, and in so doing, produced a new modality of the social tie. What would henceforth be the motor, what would now rule would no longer be the master's enunciation, his pronouncements, but rather a knowledge of statements themselves, of an anonymous array of statements. Scientific knowledge would henceforth serve as our compass – and this has not been without consequences.

Under a religious world view, the social tie could be read in the well-known adage; "The church must be kept in the center of the village!" There was no reason to invoke God to do so. The effects of the marking of the social by religion were apparent in the mono-centric and vertical social organization of society. On the other hand, the social tie induced by the development of science spontaneously promotes a pluralistic and horizontal, non-transcendent organization of the social dimension. The church is no longer alone in marking the center of the village. The new boutiques of knowledge are multiple, each one just as good as the next. We need to shed some light on the impact of this development.

The argument I shall put forth is based on my work as a psychoanalyst, with reference to the work of Freud and Lacan insofar as what the first inaugurated by the discovery of the unconscious and the transference, the second fulfilled by referring these to the function of speech and language.

The human being's inscription in language carries with it a set of rules that function like inescapable laws in order to ensure the perpetuation of the social tie. When these laws are not respected, probably because they are not properly identified, the social tie may be imperiled as it submits completely to techno-scientific determination.

The fact that we inhabit language radically subverts the natural order of the living being since it breaks with an earlier adaptation to the world, regulated by signs. And while we agree with what common sense tells us, namely, that what is important between subjects is that they communicate, if we look a little more closely it is actually impossible as well as naïve to limit language to this sole function of communication.

"What is language good for," asks Lacan, "if it is not made to signify things expressly, and I want to say that this is not at all its primary purpose, and if it is not for communication either? Well, then, it's simple. It's simple and capital; language makes the subject. And that is damn well sufficient. Because otherwise, I ask you, how would you be able to justify to the world the existence of what we call the subject?"[2] Thus, what is peculiar to language is that it makes the subject. This supposes that in the act of speaking itself we recognize not only what speech transmits as information, but also what it establishes through the very fact of enunciation.

We can now claim that there are two sides to speech, and that while the speaking being uses speech to exchange information, to communicate, in this very speech he also establishes himself as a subject. In this case, the passage from the continuity of his presumed being to the discontinuity of his speech will involve the irremediable loss of a part of himself. The thinking and speaking individual is thus condemned to only being able to say a part of himself; he will never be able to say it all. Thereafter, the human subject, far from being a full, continuous subject can only be glimpsed, his being is condemned to having itself "represented."

The specificity of the subject is created by the fact that he partici-

---

2 Jacques Lacan, *Petit discours aux psychiatres*, (unpublished).

pates in the economy of language. We can, therefore, no longer speak of "instinct" with regard to the human world since instinct presupposes an adequate object of satisfaction. For humans who inhabit language, we must speak instead of the drives, that is to say of an impetus which necessarily means that the object of satisfaction is not attainable as such, since such an object only exists through words, and thus it will already have been amputated by what the detour of symbolization implies.[3]

Ineluctably then, human desire is constituted by the material of language, that is to say, by the material of the Other. Even if we agree that the mother is the child's first Other, we should point out that what she actually does is to make present the Other to the child as the place of language. This lets us appreciate how the Other resides deep within, and that it is in the relation to this Other that we are constituted as subjects.

However, while the fact of consenting to language constitutes the subject, it further implies that something comes between myself and my alter-ego. Language functions like the presence of a wall, a "wall of language." As an irreducible third party, language is as much the boundary I come up against, as it is the medium that permits exchange.

Symbolization introduces a way of taking pleasure specific to human beings, meaning that the object of satisfaction is worth more when it is absent than when it is present. Or, we could say that every object of satisfaction is available only against the backdrop of the totally satisfying object's unavailability. The object of human desire, insofar as it is subservient to language, is thus structurally of a signifying nature. It is this fundamental unsatisfaction that characterizes an economy of the signifier. Whereas an economy of the sign supposes, on the contrary the satisfaction of possessing the object. We can now further understand the double movement that language supposes; language both imposes a form of satisfaction and a boundary to satisfaction. Language is the

---

[3] Serge Leclaire, "Détour" in *Etudes freudiennes* 25 (April 1985).

THE TECHNOLOGICAL EXTENSIONS OF THE BODY ■

cause of *jouissance*[4] because it introduces a specific, permissible, language-based *jouissance*. But language is also what limits or stops *jouissance,* insofar as it is by way of language that an absolute and instinctual *jouissance* is kept at bay, a *jouissance* we usually call forbidden – incestuous *jouissance*.

Such is the specificity of Lacan's reading of Freud. While Freud put sexuality at the heart of the subject's unconscious, Lacan taught that in that sexual realm, the question of language is already at stake. Freud's pansexualism is the epic version of the structural fact language constitutes in the human being. And while Freud is the one who introduced us to the question of lack in the register of having – since the first way that we encounter lack is, to refer to the title of a Hemingway novel, by confronting our own having or not having – Lacan showed that even before the encounter with lack in the register of having, this question is already situated in the register of being, precisely to the extent that a speaking being is defined by the possibility of making present what is absent and making absent what is present by way of language.

The incest prohibition now takes on a much broader meaning than what is understood by its common definition, since the incest prohibition anchors itself in the impossible congruence between the world of words and the world of things. To consent to the incest prohibition is to accept the renunciation of the certainty of things in order to consent to the uncertainty of words. This also means a renunciation of narcissism and a move toward the other. And finally, the incest interdiction puts an end to all dreams of totality because no totality is possible. The very fact of speaking makes such totality lost forever.

This is why we should understand the implications of what it means to speak. Far from being a human faculty among others, language doesn't simply separate us from animals, it definitively constitutes our membership in the human community. This necessitates a renunciation of im-

---

4 See translator's note, *Being Human*, 94.

■ BEING HUMAN:

mediacy, the need to be able to defer and differentiate in time and space, and the separation from the universe of things in order to enter the universe of words. If, according to Hegel's formula, "the word is the death of the thing," we can say further that the word is the implementation of the prohibition of incest with the thing. The world of words requires that we free ourselves from the world of things.

In sum, language itself is the implementation of the incest prohibition and thus it gives rise to an irreducible unavailability: that the impossible be an integral part of the human condition. It is this renunciation of the *all is possible* that psychoanalysts call "castration." The refusal to consent to this renunciation corresponds to the persistence of the incestuous wish or of the fantasy of infantile omnipotence.

It is this same renunciation, by virtue of the incest prohibition, that is the origin of the social tie, above and beyond all cultural differences. We will have to continue to refer to the incest prohibition if we want to give a backbone to the ethical examination at the heart of our techno-scientific culture.

Indeed, today we witness the appearance of a host of new problems that are posed for the first time and are directly related to the development of techno-science. Today, we can extend our natural life span. We can choose our anatomic sex, or the sex of our children. We can consider the possibility of cloning. We can "establish paternity" on the basis of genetic testing with a heretofore unimagined certainty; we can even "establish paternity" after death.

I contend, in this regard, that the appearance of these new questions as well as our difficulty finding adequate answers to them originate in the profound change that the development of techno-science has engendered in discourse itself.

Through its own functioning and by way of the predominant and idealized place it occupies in our society, contemporary science has induced the disappearance of the distance between words and things at the same time as it unwittingly promotes the return of the immediate.

From the moment the scientific model becomes the model for the social tie, instead of facing irreducible non-adequation and the ineluctable encounter with lack, instead of confronting the category of the impossible, the contemporary subject is invited to cast off these structural vices. He is encouraged to regard sexual difference as though it were a matter of instrumentality, of male or female connectors. Whereas the religious context modestly covered up sexual non-adequation, the scientific context would have us believe that we can actually arrive at a possible adequation, so long as we trust scientific progress. Medicine, for example, dispels the real of anxiety by treating it as an avatar susceptible of being erased by the appropriate sedative.[5]

Contaminated by the implicit structures of scientific processes, our society fosters a belief in the resolution of sexual non-adequation. Now, as we have just seen, even before the question of sex is posed for a subject, even before he encounters lack, in the register of having, he had already encountered it in the register of being, insofar as he is a speaking being, and thus already in a state of radical non-adequation, not just to the opposite sex, but to the world of things in general. For the laws of language to which he has consented have marked him with the irreducible impossibility of adequation; and these are the laws that structure sexual difference for speaking beings. The consequences of the subversion introduced by the signifier are, on the one hand, that anatomy is insufficient to account for sexual identification in speaking beings, and, on the other, that it is by way of their inscription in language that men and women can be called such, at the same time as this inscription renders their relationship impossible. Thus we can take our argument a step further and say that sexual difference is first and foremost a difference of position.

The implicit models put forth by the discourse of science, however, let individuals believe that they are able to finesse this difference of position.

---

[5] Jean-Pierre Lebrun, *De la maladie médicale* (Louvain-La-Neuve: Ed. De Boeck-Wesmael, 1993).

■ BEING HUMAN:

How does science come to nourish such an illusion? In this regard, we shall consider the evolution of the concept of authority. Where does authority come from? There are only two answers to this question.

Charles Melman has said: "There are two ways of validating a statement. The first is to say it. From the moment that I say it, this saying becomes a referent. And whoever I may happen to be, it is the saying that is authoritative, that creates both the authority of the statement, and also, strangely, an authority that reflects back upon the statement's author. Notice that if this speech finds sympathy in a listener, among a public, its authority will be reinforced, established by way of it. We can see how hysterical sympathy contributes to the authority of the statement. The other way to validate a statement pertains to its logical consistency, the manner in which it is constructed, of which mathematics are the culmination. Certainly, there is no reply to a correct mathematical formulation. Its consistency is authoritative. Notice also, that in this logically constructed statement, the enunciator is excluded. He is – why not put it like this – foreclosed. It doesn't matter who makes this logical statement. It is self-evident to all. At the same time, all the authority is now based on the statement whereas the enunciator of this logical statement becomes only its parasite, something that disturbs the statement, and risks introducing some error that, in general, undermines the logical statement."[6]

Thus, authority is derived either from the place granted to the speaker, or from the coherence of what is said. In other words, either it is derived from the fact of saying – which is what creates the authority of the leader, the father, the boss, or anyone who is not on the same footing or level as others – but also that of everyone who speaks "on their own authority." In the second instance, authority is derived from the coherence of what is said. This is what establishes the authority of the

---

6 Charles Melman, *Colloque de Cordoue,* Conclusion (Paris: Association Freudienne Internationale, 1994), 483.

expert, of the scientist, when he is in his area of competence, but it is also a position that everyone takes up when he refers to a body of knowledge in order to make a decision.

This double legitimacy of authority coexists in each one of us inasmuch as it is a consequence of the fact that we are speaking beings. To bolster what we say, we sometimes refer to the simple fact that we said it, but it also happens that we refer to a body of knowledge.

Nevertheless, what is usually called *authority* denotes the first type of legitimation. Authority comes from the Latin, *auctoritas,* which comes from *auctor,* which designates the author, the founder and the one who is responsible for a work. Authority implies the notion of beginning and it is the fact of "being the first" that can give authority. We see this in the law of primogeniture. This is also the modality of authority evoked in what we generally call the *argument of authority.* In other words, traditionally, authority comes from speaking from the position of head, from the enunciation itself. Now, we could say that it is precisely this "argument of authority" that was upset, if not impeached by scientific modernity. We have seen a shift to authority based on the logical coherence of the statements themselves. Authority is granted to the person who has been enabled to occupy the position of authority, not on the basis of his speech, but, rather, on the basis of his acquired competence.

With the development of modem science, the *argument of authority* was beaten back and replaced by an authority stemming from logical coherence. We have seen that this was at stake, for example, in the Galileo affair.

Similarly, during the Valladolid controversy the papal prelate – at the request of Charles v who had asked whether the Indians were human beings – had to decide between the arguments of Sepulveda and Las Casas. Las Casas pleaded in favor of recognizing the humanity of the Indians. Sepulveda left the field open – with the help of prayer – to the final judgment of the prelate who would ultimately be the authority in the matter. Here we find a simple way of deciding between two contradictory positions, further simplified by the fact that the authority could not be challenged. We get a sense of the extent to which the place of authority has been shaken up if we imagine what would happen should a scientific controversy be decided in such a fashion today. No one would accept that an argument of authority decide a problem of this type. But does this mean that there are no conflicts and controversies in other domains that could not be decided in this fashion?

Our difficulty exercising authority nowadays is situated here. The delegitimation of the argument of authority, so salutary for scientific controversy, has gradually led us to believe that this argument can never be used. The more knowledge prevails, the more the argument of authority becomes obsolete and is decried as an abuse. In other words, the development of science has had the effect on social organization of legitimatizing an enunciation on the basis of its statements to the detriment of the place traditionally occupied by the "voice of authority" itself.

With this displacement in place, and thanks to the evolution of technique, we are witnessing the loss of our traditional bearings, for these two modes of legitimation do not have the same characteristics, and privileging the second is not without consequences.

While each of these two modalities of the legitimation of authority ultimately aim at introducing a third party, and thus authorize an end to the controversy, their functioning is far from being identical. What introduces the third party in the scientific controversy, what allows for arbitration, is the confronting of the model with the real. The scientist initially elaborates a model that will then give him the tools to examine the real and by way of which he will finally obtain (or not) the confirmation of the validity of his model. In the case of science, what sets a limit to what might otherwise remain only an imaginary confrontation, is the fact that the symbolic is confirmed by the real. The third party is produced by the model's adequation to the real. This adequation, however, implies the forgetting of the symbolization that had been necessary for it to succeed in the first place. Legitimation by the adequation of the model to the real erases, indeed it almost renders invisible, the work of enunciation that was nevertheless indispensable to it. In the Valladolid controversy, on the other hand, the question was settled and a decision was made solely on the basis of the prelate's authority. The argument of authority could be used by the one in authority. In this case the third party was established by the enunciation of the one in authority.

The major consequence of the difference between these types of legitimation is that the argument of authority does not hide the source of its legitimacy. It does not hide the dimension of appearance in which the enunciation takes place. This leads us to situate it in the register of a *certainty* that we will qualify as *uncertain,* whereas the authority of scientific enunciations, on the other hand, lays claim to a *certain certainty.* Thus, for example, the recognized authority of the word established paternity – but always with the well known caveat – *mater certissima, pater semper incertus.* Whereas, the authority that we give to genetics with regard to the same question allows us to produce, and therefore induces us to demand, guarantees concerning the paternity of a subject under the pretext that it is possible to scientifically identify the biological father.

How great are the effects of such a subversion? Today, most knowledge is scientific and has effects in the real. The authority of science gives the general public the impression of an uncontested legitimacy and casts itself as all the more consistent and better of an authority since it refers to the register of *certain certainty.* In avoiding the dimension of enunciation, it lets us believe that we can escape the arbitrary nature of the argument of authority. It presents itself as an impartial third party, outside of all subjective involvement. In other words, science lends itself to letting us think that we can do away with castration.

But what could be better than not having to doubt? Why not benefit from the advantages of science? Must we be so picky, now that we finally dispose of rigorous methods for solving thorny problems? Do we have to return to religion, alone in being able to guarantee the dimension of enunciation?

■ BEING HUMAN:

Of course, I am not proposing a nostalgic return to the enunciation of religion, or of a traditional father, much less to that of the ancient master. First of all, because it is not within our power to stop progress, but most of all because psychoanalysis, itself aligned with the development of science, allows us to realize that the change of legitimacy does not, *ipso facto,* result in the abolition of the argument of authority. These two legitimacies are linked and just because one is more prevalent today, doesn't mean the other one disappears. To be able to inflect the position of father or master does not mean to stop using it. For to do so would be to actively and unwittingly participate in a virtualization of the symbolic. In fact, if what marks the move from an imaginary to a symbolic economy is the difference in place, or the fact of being caught in the system of language, as we stated earlier, the pure and simple erasure of the difference of place results in a regression from the symbolic register towards the imaginary register, the eclipse of the signifier by the sign, or, to put it more simply, in what we could call dehumanization.

And this for the simple reason that the symbol is only constituted through the loss of the thing, contrary to the sign that remains tied to the thing. Let us return, for example, to the question of the determination of paternity. As Lacan indicates, "The paternal function is incomprehensible in human experience without the category of the signifier... the sum of these facts – of copulating with a woman, that she then carries something for some time in her belly, that this product is ultimately ejected – will never lead one to constitute the notion of what it is to be a father. It is necessary that the elaboration of the notion of the father, by a process produced by the whole interplay of cultural exchanges, have been raised to the level of primary signifier and that this signifier have its consistency and its status."[7] Paternity, then, is the erasure of the biological, in just the same way that the signifier requires the

---

7 Jacques Lacan, *Le Séminaire, Livre III, (1955–56), Les psychoses,* (Paris: Ed. du Seuil, 1988), 329.

THE TECHNOLOGICAL EXTENSIONS OF THE BODY

erasure of the sign as a trace of the thing. To put it in more radical terms, paternity is the register of language, and as such, it even supposes the erasure of the reference to the progenitor.

By turning to genetics in order to solve the problem of paternity once and for all, we are lead to believe that it is the real of the progenitor that is the foundation of paternity and that absolute transparency can be attained. But this is to forget that the consent to paternity was always supposed to keep the question of the biological father in limbo. Thus, the movement currently advocated, away from the signifier and towards the sign, in letting us believe that genetic analysis will be able to resolve the question of paternity, ratifies that the *certain certainty* authorized by science can substitute for the *uncertain certainty* of the father's word. When a judge has recourse to genetics in resolving a case of paternity, he is not actually concerned with paternity, but rather the legitimacy on the basis of which he will be able to make and sustain a judgement, in other words, an enunciation. The recourse to science is a long way from having done away with language. It only displaces the word, but in so doing science creates the illusion of having left language behind.

The category of *uncertain certainty* is restricted by the generalized movement modeling what is implicit in the discourse of science. By dint of believing that there could be a way of referring only to *certain certainty*, without risk, without error, the dimension of enunciation has been gutted. At the same time, we can refuse all arguments of authority and even discredit all authority under the pretext of a possible abuse. In so doing, today's society – without even knowing it – advocates the evacuation of the difference of place. And we will have to make an effort in order not to believe that we can definitively go beyond this difference of place, in a process of generalized "same making," a "final solution" to alterity.

Traditionally, the subject was confronted by the law and it was in relation to the law that we could define neurosis: confrontation with a too-powerful father in the case of obsessional neurosis and with an impotent father in the case of hysteric neurosis. At present, the problem of the subject seems to be displaced; he no longer knows how to situate himself. He calls on paternal authority, but at the same time, he seizes upon the discredit attached to the exercise of *uncertain certainty,* as an alibi, so that he does not have to abide by its dictates. The contemporary subject – strengthened by a science that allows him to remake his body – feels constantly authorized to choose, at the very site where, in the past, his body drew a boundary. Among the examples of "choice" we encounter today is the option of choosing one's sex. But this is a paradoxical privilege, since in order to choose, one must consent to a loss of certainty. And when today's subject can choose everything, he finds himself often incapable of choosing anything at all, since choice implies a renunciation. He who can do everything is the least apt to do anything since he is not ready to lose what he does not choose. And what looks like a privilege given to us by science turns into just the opposite. Being able to choose everything makes one that much less apt to give up the losses implied in the choosing.

THE TECHNOLOGICAL EXTENSIONS OF THE BODY ■

Medical Discourse, Science, and the "Talking Cure"

Annick Galbiati

> Docteurs, prenez garde... Souvent les maux dont on vous parle prennent racine dans la tourmente morale. Ce sont les brimades subies et les perpétuelles contradictions qui s'accumulent quelque part dans le corps et l'étouffent.[1]
> — MARIAMA BÂ

Since 1929, the discontents of civilization have multiplied. André Green does not hesitate to use the term "disease."[2] The diagnosis for medicine is scarcely better; witness, among others, J.P. Lebrun's excellent book, *De la maladie médicale*. In order to give an idea of the effects of "medical discourse" understood as a sum of determinations of the physician/patient relationship, the Belgian psychiatrist and psychoanalyst J.P. Lebrun relates the following clinical anecdote: A patient complains to his doctor that for about a week, since the last village fair, he has felt "like he was pissing blood." Listening to him, the physician writes a single word: *hematuria – presence of blood in the urine.*[3]

From the patient's utterance to the physician's inscription, which is to say its inscription in medical discourse, what process was at work? The term *hematuria* says nothing about what happened at the fair. While this single word *hematuria* is perfectly adequate for medical reception, it must be noted that a reduction has taken place, eliminating the patient's way of speaking, his chronology and his history. From the

---

1 "Doctors beware... Often the complaints you hear are rooted in moral torment: insults endured and perpetual contradictions accumulate somewhere in the body and stifle it," from *Une si longue lettre* (Dakar: Nouvelles Editions Africaines, 1996), 67.
2 See: André Green, "Culture et civilisation: malaise ou maladie?" in *Revue Française de Psychanalyse*, 4 (1993): 1029–1056.
3 Jean-Pierre Lebrun, *De la maladie médicale* (Ed. De Boeck-Wesmael: Brussels, 1993), 43.

physician's point of view, the inscription *hematuria* is exhaustive, but it excludes everything that was particular to that patient, and tends to annul everything about the subject that could be signified. The physician transformed a singular complaint into an objective symptom and treated it as a sign; in other words, as "something for someone." This "someone" is not "anyone" but someone specific, the physician, here being caught up in a certain idea of "medical discourse."

This process of reduction – excluding the subjective dimension – ensures medicine's generalization and efficacy. It appears to be a structural necessity, but it is probably not without consequences for the patient as well as the physician.

Freud, after and before many others, noted very early on that there are psychological and transferential aspects to the practice of medicine. While it is nearly certain that medical consultation can also entail interpretation and truth effects for the patient, it is undeniable that these effects occur and act in an almost occult manner on the boundaries of medicine's current knowledge and established domain. What medicine relegates to the category of the "physician/patient relationship" and chalks up to "personal factors" – is this not the very thing upon which is based the transferential relationship that led Freud to establish the field of psychoanalysis? "The marginalization of the patient in favor of the disease," writes J. P. Lebrun, "is combined with the eviction of the physician as subject in favor of his function as agent of scientific medicine. For this exclusion of the subject does not only apply to the patient, it also affects the physician. And it is this double exclusion which must be recognized if one is to grasp what causes the dehumanized character of today's medicine."[4]

Also located outside the scope of scientific medicine is the psychoanalytic distinction between need, demand and desire.[5] The reduction

---

4 Ibid., 48, 49.
5 Jacques Lacan, *Ecrits* (Paris: Ed. du Seuil, 1966), 627.

of the subject's desire to a demand for help seems to induce a kind of inflation and exacerbation of the latter, which could easily turn into a requirement, in the name of a democratic right to health and even to children. The "treatment of the demand," to use current terminology, requires the response to the demand for care to be increasingly fitting and increasingly effective. This being the case, does medical discourse care about the subject's suffering, inasmuch as it is stamped regardless of circumstances with the seal of the human, caught up, in other words, in a singular expression and a particular story? Its essential aim is to isolate, target and combat disorders and dysfunctions raised to a kind of absolute value, separated and disconnected from the "rest," that is to say from the subject and what the subject confronts. Hence the success of new therapies which, by conforming to the evidence of symptom and demand, subscribe to this ideal; hence also the success of so-called alternative medical treatments, which are reputed to lend a more holistic or more sensitive ear.

The "effectiveness" label participates in the supply status of contemporary medicine, which constitutes a language in which the question of the difficulty of existence can be articulated. "The general practitioner is one of those today who knows best that the complaint addressed to him is very often, inevitably, non-medical, but that it is formulated in medical terms, given, on the one hand its limited possibility of being able to be heard otherwise and, on the other, the physician's need to provide an answer, be it an inadequate one... Think of the incalculable number of marital or work-related difficulties which nowadays are seen as the province of medical language – which does not mean that any disease is involved. It seems that we see the physician less often for a real consultation and more often for what might be called medicalized circumstances... We must observe that, were medical hyperconsumption to be seen in this light, it would literally result in a health care financial scandal. It is our hope that a serious study will one day specify the sums that physicians "unknowingly" devote to reassuring their patients, for lack of being able to stand or to

■ BEING HUMAN:

help stand anxiety, discontent, uncertainty or the absence of a guarantee."[6]

As far as "supply" is concerned, there is a great difference between 17th century medicine, which had very few remedies to offer the sick – such as bleeding and a few pharmaceuticals – and the current physician who commands an impressive array of therapeutic tools. The difference is not only quantitative, it is also qualitative. Whereas in the 17th century the diffuse belief predominated that the cure for disease depended on fate rather than on art, the contemporary discoveries of medical science have given rise to great hopes. They have also mobilized immense pharmaceutical business interests, legitimizing the idea that there could be a "remedy," a therapeutic object to compensate for "ill being." Statistics show that France holds the world record for the consumption of psychotropic drugs. The United States is probably comparable. Statistics also show that what is called the "placebo effect" is effective in at least 40% of cases regardless of disease type. This, when the effectiveness of standard medications rarely surpasses 60–70%.

In *Science and Truth*, Lacan links the very existence of psychoanalysis to the emergence of scientific discourse. "It is unthinkable," he writes, "that psychoanalysis as a practice, that Freud's unconscious as a discovery, could have occurred before the birth of science in the 16th century, which is called the century of 'genius'... Contrary to what is repeated about a supposed break of Freud's with the scientism of his time, it was this very scientism which led Freud, as his writings show, to blaze the path that will forever carry his name. This path never deviated from the ideals of said scientism, since this is what it is called... and its mark is not contingent [to the path blazed by Freud] but an essential part of it."[7]

We are thus invited to rethink the relationship of medicine to psy-

---

6 Lebrun, *op. cit.*, 51.
7 Lacan, *op. cit.*, 857.

THE TECHNOLOGICAL EXTENSIONS OF THE BODY ■

choanalysis, in terms of their ideals and their inscription in the discourse of science. What, for instance, in the relationship to these ideals, constitutes the specificity of analytic practice? How is it different from medicine and science while still related to both? I shall address these questions by first attempting to put them into perspective.

If we refer to Henri F. Ellenberger's major work, we learn of or rediscover, among a multitude of other things, the existence since earliest antiquity of two types of medicine: one profane, more effective in treating physical than mental or emotional illness; the other sacerdotal run by priest-healers, "medicine for the soul."

As Ellenberger points out, the true ancestors of the modern physician were the profane healers whereas the official healer was more like the priest's ancestor. For many centuries, the physician and the priest-healer lived side by side: Cos, the birthplace of Hippocrates and his school, was also famous for its Asclepeion (the Asclepeion was a temple devoted to Asclepeios – Esculape – one of the gods of medicine which inherited his emblem – the caduceus, Esculape's serpent, is the symbol of power over life and death.) Galien, the most famous physician of the 2nd century, did not hesitate, in certain cases, to turn to the Asclepeion of Pergame.

In the context of Western thought, under the sign of monotheism, the two modes of healing continued to be practiced. Physicians cared for the sick, convinced all the while that it was God who did the actual curing. God had assumed a tertiary role, which current ideology now tends to ascribe to science. This leads Pierre Benoit to write, "God's share in medicine, healing and death is closely bound to the mysterious facts of the patient's life...this share of the cure left to God, has been replaced by scientific knowledge. In short, it is now 'I bandaged him. Science healed him.'"[8]

The advent of science in the modern sense of the term has indeed

---

8 P. Benoit, *Chronique médicale d'un psychanalyste* (Paris: Ed. Rivages, 1988), 144.

■ BEING HUMAN:

profoundly changed these two tendencies. Begun in the 19th century by François Magendie and Claude Bernard, the transformation of Hippocratic medicine into scientific medicine is today complete. Science is said to have been born twice, once in Greece and a second time in the 17th century. The Master of Cos was already involved in making medicine into a rational practice, at a time when everything that upset the natural order was explained by supernatural forces. Yet it was in a vigorous movement of opposition to Hippocratic medicine that Bernard and his teacher Magendie founded modern medicine.

"Hippocratic medicine is a medicine of observation," notes Canguilhem.[9] "It is passive, contemplative and descriptive like a natural science. Experimental medicine, from the outset, views itself as a conquering science." Its goal is discovery and the possibility of action based on a hold on disease's real. Current medical discourse was born in a rationalist coup, affirming its intention to break away from Hippocratic tradition. Bearing witness to this is what Lichtentaeler in his *Histoire de la Medecine* calls Magendie's "revolutionary program" which, although dating from the early 19th century, is absolutely modern. In it Magendie enumerates the principles which must be respected in order to integrate medicine into science.

François Magendie, French physician and neurologist, shared an enlightenment education with the physicist Laplace and the chemist Lavoisier. As early as 1809, he opposed the vitalism of Bichat who, although wanting medicine to share in the dignity of the experimental sciences, did not want the language of medicine to be based on that of physics.

The six master ideas Magendie then elaborated in order to give medicine new foundations – those of medical rationalism – were not organized into a "plan," but were so often repeated and so closely

---

9 G. Canguilhem, "L'idée de médecine expérimentale selon Claude Bernard," in *Etudes d'Histoire et de Philosophie des Sciences* (Paris: Ed. Vrin, 1983), 131.

related to each other that one is free to speak of a "revolutionary program."[10]

Magendie begins by noting that physiology and medicine are not yet sciences. Medicine is still dominated by empiricism; man is always trying to impose his conception of things instead of grasping the laws of nature.

We should note, concerning "these laws of nature," that those which relate to the nature of language are never taken into account in the scientific methodology, the way other instruments of investigation are, be they conceptual or pertaining to measurement. Does this mean that language is considered a perfect tool, about which there is nothing to say?

Freud always wished to include psychoanalysis among the sciences of nature, and not among the human sciences which, according to him, were lacking in rigor. In the process he showed that the goal defined by Magendie – to set aside personal convictions – was much more difficult to attain than had been thought. If convictions originate in unconscious and indestructible desires, how can they be "set aside"? Won't the desires in question necessarily reappear in the real, unseen but effective, all the more effective because unseen?

According to Magendie's program, physicians lack the scientific objectivity of physicists and chemists who rely on experimental data. We cannot fail to note, along with the spectacular growth of medical science, the fulfillment if not the triumph of Magendie's wish.

Physics and chemistry are not only models for physiology but also its major supports. The laws that govern the inanimate world must be applied to the human organism. It should be noted that this will to "return to the inanimate" is exactly how Freud defined the death instinct drive.[11] In line with Magendie's wish, and encouraged by new techniques – such

---

10 C. Lichtenthaeler, *Histoire de la médecine* (Paris: Fayard, 1978), 402–412.
11 S. Freud, "Au-delà du principe de plaisir," in *Essais de la psychanalyse*, (Paris: Ed. Payot, 1981), 82.

■ BEING HUMAN:

as organ transplants – our contemporaries are surpassing each other; the fragmentation of bodies, the sale and trade of organs, don't these lead to a breaking apart of human material into more or less interchangeable parts like those of a machine?

The ensuing articles of Magendie's program confirm at the same time his modernity and his systematic exclusion of a subjective part of reality, which ought to be called, strictly speaking, "human"; since the human subject has to contend not only with the real of the body and his image, as does the animal, but also, with the reality of language in its symbolic, subjective and polysemantic dimensions.

An experimental method which bases its deductions on the data only, is supposed to replace the systematic approach of the nosologists who follow Hippocrates. It should be noted that, starting in 1980, the DSM (Diagnostic and Statistical Manual of Mental Disorders) seems to be directly based on Magendie's program: by defining disorders solely on the basis of facts and behaviors, excluding all verbal expression. As for content, Magendie wishes to apply the principles of physical and chemical science to the living.

Magendie reasserts that physiology can only be scientifically explored through physics and chemistry: to the extent, however, that there are particularities specific to life that are not found in inorganic bodies, physiology is an autonomous science.

Pathology is pathological physiology. Magendie states elsewhere: "Medicine is the physiology of the sick man." Clarity was not the least of his merits. This program meant that for the first time – in 1861 – a physics test replaced the philosophy test in the entrance exam for medical school.

If we now consider the "medicine of the soul" side of things, to which we referred earlier, the "will to science,"[12] did not take long to manifest itself. In the 18th century Europe of the Enlightenment, Mesmer and his theory of "magnetic fluid," directly inspired by Newton, finally tri-

---

12 See I. Stengers, *La volonté de faire science. A propos de la psychanalyse.*, coll. "Les Empêcheurs de penser en rond," (Paris: Ed. Laboratoires Delagrange-Synthe Labo, 1992).

THE TECHNOLOGICAL EXTENSIONS OF THE BODY

umphed over Gassner, the great exorcist. Franz Anton Mesmer, an Austrian physician and philosopher who had come to France with his "tub," had his moment of glory, until the affair had created enough fuss, in 1784, to be referred to two royal Commissions. The scientific authorities of the time, such as Franklin, the astronomer Bailly, and Lavoisier, refused to recognize the physical existence of "magnetic fluid" without, however, refuting the authenticity of the phenomenon or the possibility of therapeutic effects attributed to imagination.

The phenomenon in question was followed by many others during the 19th century, successively referred to as "magnetic," "somnambulistic," then "hypnotic." The names Puységur in France, Braid in Britain and, later, those of Charcot, Liebault and Bernheim, remind us of the amount of this kind of activity during this period.

A follower of Mesmer, the Marquis of Puységur, Lord of Buzancy, was able to plunge his peasants into a state of "magnetic sleep" from which they emerged after a number of sessions, cured. Mesmer uses what we would now call "abreaction," whereas Puységur is said to "invent the cure." He invites his peasants into his home, or "sees them at regular intervals, for indefinite periods of time, until they no longer need him and return to their own lives. All he has to do is provoke sleep, the somnambulistic state, and listen to them, then awaken them."[13]

With Braid, a Scottish surgeon, "magnetic sleep" became the object of scientific study. In the middle of the 19th century it acquired the name "hypnosis." Charcot did not neglect this method in order to better explore the workings of "major hysteria." He even secured the re-examination of magnetism by the French Academy of Medicine in the 1830s, in the wake of a major condemnation which had prohibited further consideration of the matter. As for Bernheim, in Nancy, in his "Village University," fascinated by the cures brought about by "kindly

---

13 See: D. Lévy, "D'où la psychanalyse est-elle sortie?," (December 8, 1993), in the Seminar "Bords actuels de la psychanalyse," A. Galbiati, *Cahiers du Cercle Freudien*, II (1993–94) (unpublished).

■ BEING HUMAN:

Dr. Liébault," he distinguished between "suggestion" and "hypnosis." He showed that suggestion can be totally effective outside of the hypnotic state, even that the hypnotic state is a result of suggestion. Freud, who went to Nancy several years after his stay at la Salpêtrière in 1885, would benefit from this lesson.

Indeed, when Freud started as a neurologist in the 1880s "it was the era when, in Vienna, neurasthenia was commonly diagnosed as a cerebral tremor."[14] Thanks to Charcot he discovered a method able to create the reproduction of the spontaneous triggering of symptoms and their suppression. This led him to use suggestion therapeutically, with or without hypnosis. Freud became increasingly committed to the idea of the existence of "powerful psychic processes which nevertheless evade human consciousness."[15]

With Breuer, Freud's method became "cathartic." It prompted him to explore the genesis of the symptom: hypnosis allowed him to broaden his patients' consciousness and to "put at their disposal knowledge they do not possess in the waking state."[16]

When, ten years after his stay at La Salpêtrière, he published *Studies on Hysteria* with Breuer, it should be noted that his interest had shifted away from the words of his teacher Charcot: he had started to pay greater attention to the hysterics own words, and he worked to decipher their language. A language which probably more than any other says something about the relationship of the speaking subject to his body and his sexuality, a language which, through all the ills which populate doctors' offices, tells to what extent the imaginary, the symbolic and *jouissance*[17] can imperil health, can subvert the good working order of the body in the silence of the organs.

---

14 S. Freud, *Selbstdarstellung, Sigmund Freud présenté par lui-même* (Paris: Gallimard, 1984), 21.
15 *ibid.*, 30.
16 *ibid.*, 47.
17 See translator's note, *Being Human*, 94.

THE TECHNOLOGICAL EXTENSIONS OF THE BODY

The Interpretation of Dreams, Wit and the Unconscious, The Psychopathology of Everyday Life, bear witness to the discovery of the unconscious; not the unconscious as the irrational, as that which eludes the conscious (a theme that fascinated the romantic 19th century) but the unconscious as an unusual, decipherable knowledge which is structured according to laws which turn out to be homologous to the laws of language.

Under the heading of dream, of wit, of lapses – those singular manifestations, those hiccups of daily life – what Freud and Lacan reveal is not only the effect of language upon human subjectivity but, more radically, the manner in which language constitutes, in its symbolic dimension, the very existence of the human subject.

One day Frederic II (1194–1250) got the idea to find an answer to the question: "Is there a natural language?" Forty babies were separated from their families, who spoke different languages, and put into the care of nurses who received the following double order: They were to make sure the babies were well cared for and lacked nothing, but under no circumstances were they to be spoken to. Not one word. The "experiment" ended tragically: deprived of the bath of language, not one baby survived. This might serve as an apologue – taking them to the extreme – of the ravages of a certain scientism.

If the Cartesian Cogito can thus be considered the birth certificate of the subject of science, it must immediately be added that this subject has been "forgotten" if not forclosed by scientific discourse, and, with it, the relationship to truth that Descartes, in the 17th century, leaves in the care of God. Following the Cartesian break, we can say that Freud was able to tie knowledge to truth by "secularizing" it, by making it fall away from divine aspiration, and find it back in its place, where it speaks, in language, in the language of the body. By paying attention to the comprehension and treatment of hysterical phenomena, by seeking to study a field traditionally set aside as belonging to the domain of the irrational or the sacred, Freud invented a new rationality: he showed

■ BEING HUMAN:

that symptoms stem from a meaning, that they can be deciphered and reduced, and thereby an unknown knowledge can be restored.

Lévy notes that psychoanalysis is an offshoot of scientific medicine:

> On the one hand the idea of neurosis, of a disease originating in the "nerves" which is not a form of madness, and on the other, the idea of a "process," borrowed from Jackson, a great American neurologist, namely the concept of an apparatus performing a function and being able to adapt to this function without it being necessary to suppose a conductor, a subject, an intention, at the beginning or at the end. Take away the nerve and you have a psychic apparatus which includes the unconscious.[18]

Psychoanalysis stems from the application of the scientific method to medicine, and one is tempted to accuse current medicine of "too much" or "not enough" scientificity when it fails to recognize the existence of another logic, of a "measure" or "ratio" different from calibrations, scales of evaluation and statistical curves.

Psychoanalytic practice shows how a symptom has a specific logic where the real of the body is bound to its imaginary and symbolic dimensions. Current medical discourse hardly cares about the imaginary and symbolic implications of the subjective experience of the body. Yet these implications exist nonetheless. They are always present when the body is afflicted with disease. Psychoanalysis could be located on the "edges" of the field of medicine, but these edges are yet to be constituted. I shall explain. Doesn't psychoanalysis deal with what medicine relegates to a kind of occult zone? Couldn't psychoanalysis "humanize" medical discourse by taking into account the patient's singular reality, the effects, on the human subject and his body, of his belonging to the domain of speech and language?

---

18 D. Lévy, *op. cit.*

Towards an Epistemology of the Unconscious[1]

Antonello Sciacchitano

> Yes, what is it that forces us to admit that there exists a strong antithesis between "true" and "false"?[2]
>
> – F. NIETZSCHE

KNOWLEDGE VERSUS TRUTH

I am going to present certain premises on the logic of the Unconscious, from which you can draw certain conclusions. What does a psychoanalyst speak about? You would expect an analyst to speak the truth about the subject, to speak of the subject's desire. You would expect a psychoanalyst to speak of the mythic forms in which desire is involved. Indeed, myths are important because they help us to speak the truth.

But I am not going to speak about the truth. I am going to speak of the subject's knowledge as an effect of the Other, of the Unconscious. And the Other, as Lacan put it, is another name for language.

The Other forces the subject to say what he or she does not know. Lacan posits the distinction between knowledge and truth. Alain Badiou, the philosopher, one of Lacan's students, proposed the following dichotomy: Truth on the side of the infinite, knowledge on the side of the finite. I'll speak on the side of knowledge, of the finite, and address the logic of the knowledge that the subject does not know he or she knows.

---

[1] This text represents the transcription of a talk given at Après-Coup Psychoanalytic Association on June 6th, 1996.
[2] *Jenseits von Gut und Böse*, 34.

Toward an Epistemology of the Unconscious

MIND, NOT MIND, NEVER MIND

What do I mean by logic? If you open up any standard, well established book of logic such as the one by John Dewey, you will read that logic is the art of reasoning well. This is not what I mean by the word logic. I know that there is good reasoning as a matter of fact. But I do not believe there is any such art (or algorithm) of producing all and only good reasoning. This is a narcissistic illusion that arises from the idea that somewhere in my Ego there is such a thing as a mind, which applies its well-experienced logic to the analysis of the world where I live. I am speaking of the kind of logic used by English scientists and philosophers to produce many developments, for instance artificial intelligence and cognitivism. I am well aware that cognitive science originated in America. It is awkward for me to say here that the mind does not exist and its logic is a problem for the psychoanalyst. In the USA this is probably heresy. I am saying it quietly, but I am saying it.

The illusion behind the belief that there is a mind lies in the supposition that to be and to think are the same thing. This old illusion goes back to the time of Parmenides; and by way of Descartes, who said "I think therefore I am," if I say "I think then I am," we come to today's cognitivism. Psychoanalysis can only smile at the affirmation that there is a mind. This notion arises from the impotence that the speaker feels in the face of the real. In front of what escapes our awareness or in the face of the unthinkable, the "reasoning" is as follows: momentarily my reason has failed me, but a moment before I had it. This is not true. It is not true that I was at any moment able to speak the truth about the real. As a matter of fact, to be and to think do not overlap.

The mind as a self-adjusting mechanism to the real does not exist; it exists as a narcissistic illusion. I want to underscore that this is not pessimism on my part. Even if it is true that the mind is nonexistent, even if you do not have in your head some sort of universally functioning computer that you call your mind, you can still do a lot. One can still do a lot on a local level, on a partial level, connecting up tiny particular facts. Analytical experience is clear on this. The work the mind can do is to pursue the signifiers, the elements of its own language. This is a high intellectual performance, a linguistic one.

If it is not a matter of reasoning clearly, logic is a way of speaking clearly. I shall be speaking of logic in this sense, as a way of speaking well. I shall speak about it as an amateur mathematician.

The Freudian unconscious is a container that does not contain itself, like Russell's sets. Bion says this as well. When he speaks of thoughts without a thinker, he distinguishes true thought and false thought. The true thought has a thinker and false thought has no thinker. Unconscious thoughts exist, without a receptacle containing them. In a sense they are scattered in the colloquial space among speakers. I prefer to say that they are independent of any predefined social code.

This poses a problem. It is difficult to speak of logic without there being a container for these thoughts. Can you speak of logic without presupposing a mind that operates according to a logic of identity and difference? This is not an easy matter. In fact it is an impossible one. You have to try anyway. You know that Lacan speaks of the real as the impossible. There is a quip he made: he said that logic is the science of the real, and, therefore, it is a science of the impossible. And this is what we're going to be doing: a little bit of the science of the impossible.

We are encouraged by famous examples. Freud spoke of logic: for instance, when he said that in the unconscious there is no such thing as denial; when he proposed that in the unconscious there is no such thing as contradiction or temporal representation.

THE COLUMNS OF BINARISM

The most important aspect of this discussion is the attempt to go beyond binarism. My purpose is to approach three, to get over the binary operation of true and false, of good and bad, right and wrong, which are theoretical notions, however necessary they are to actual practice

■ BEING HUMAN:

and, naturally, to politics as well. They become matters of opposition like "us/them," with the same paranoical structure as the opposition "me/others."

I can cite one instance of Freudian reasoning in this. How can you conceive of the continuity between sadism and masochism within a binary arrangement? You cannot deal with it within any operation that declares one side good and the other bad, that sets up such oppositions. Drives are neither good nor bad, neither conscious nor unconscious, claims Freud.

True and false are truth values that get assigned to various human discourses but do not exhaust the possibilities of speech. Language is a trinary structure. Writers, artists, poets know this better than scientists. Stories, myths, fables, reflect such a structure. The *third* can also be approached in a formal way. You can approach it with mathemes, not only with mythemes.

Mythemes stand on the side of truth and mathemes on the side of knowledge. Nowadays, almost all mathemes are binaristic in nature. But binarism does not account for the trinary dimension of language. Therefore, we need first to weaken binarism. The job is not a light one, because binarism is found within two fundamental subjective registers. At the core of the imaginary register, we find the binary operation of symmetry: my peers and me, my double and me, my neighbor and me, my mirror image and me. It takes two for narcissism to be. Another core of binarism is within the symbolic register; and this may be a slightly harder nut to crack. For the signifiers that form such a register, the binary system has validity. The fact that a signifier always refers back to another signifier, is a binary operation. This means that one signifier is never self-referential. It cannot speak of itself. It is not the same even with respect to itself.

Our battle is against binarism in order to defend the subject of desire. I am speaking in soldierly terms: binary thinking is based upon four columns. And there is a fifth column, a secret one; I have to leave some surprise for the end.

THE TECHNOLOGICAL EXTENSIONS OF THE BODY ■

I've numbered the four columns, upon which the logical binarism of true and false lies, in order of hardness. The hardest is the first. I call it univocity. Univocity is calling a spade a spade. Its law is the principle of identity: *if A, then A*. It allows only one level of truth, only one truth-value, the true, for every speech. Univocity strongly implies one.

Bivalence is the second column. Here there are two truth-values. There is the true and the false and nothing else. Not paradoxically, bivalence is a generalization of univocity. It presents univocity in a black and white movie.

The third column is the principle of non-contradiction. I am going back to Aristotle here. Non-contradiction means that you cannot have at the same time both *A and not A*. Black cannot be white and white cannot be black.

And last but not least, the fourth column is the excluded middle. This means that *A or not A* is a logical thesis, a tautology of binarism. In the binaristic context, colors others than black and white are not allowed.

THE WEAKENING OF BINARISM

To weaken binarism, we have at our disposal as many strategies as columns. First of all, we could attack univocity. We shall not follow this approach. Why? Because we know our limits. We should build a theory of ambiguity. This logic is in our everyday practices of transference (a right love for wrong people) and of ambivalence (love and hate for the same people). But we do not know how to formalize it, while still keeping the essential ambiguity of the practices within the theory. We have to formalize a logical system without formalizing it – a task with which our intellectual tools cannot yet cope.

The weakening of bivalence may be achieved introducing polyvalence. That is a set of truth-values which contains properly the binary set of true and false. Polyvalence is a latent form of binarism. It naively dares to tackle binarism, introducing directly a third element or more. But it fails to introduce a third dimension. It is easy to verify that polyvalence reduces itself to bivalence through the division of the original set of truth-values into two sets: the designated truth-values and the undesignated.

The weakening of the principle of non-contradiction generates a minimal logic. We can call it the dictionary logic, because as in a dictionary false coexists with true and black with white, without forcing the system to collapse. We feel that this weakening is too strong and we do not adopt it.

The weakening that interests us occurs through the suspension of the excluded middle. This move produces Brouwer's intuitionistic logic, which we prefer to call epistemic or effective logic, following a suggestion of Lorenzen. The reason for choosing such a name lies in the empirical observation that this logic allows to treat the function of effective knowledge and not only the function of the truth, like classical logic. Actually, within the context of Brouwer's work, it is possible to define some epistemic operators that transform a formula into a sentence with a content of knowledge. Theorems about these operators mimic the behaviour of the epistemic construction that Freud called the Unconscious. This point seems to be interesting for an epistemology of psychoanalysis. I shall discuss it without going too technical.

The suspension of the excluded middle (EM) has some simple consequences. The most immediate is that the EM itself and all the theses deriving from it in binary logic are no longer valid theses in this new logical context. For instance, the law of double negation (DN), which makes it possible to delete the negative of the negative: if *not (not A)*, *then A*, is no longer valid in epistemic logic, because it is a consequence of the EM. Our idea is to utilize the EM and DN to define operators ε and δ that transform every formula X in a new one: εX or δX. While presenting their properties, we shall show that the new formulae share some features that are epistemic in nature. Two examples will make the procedure clear.

THE EM EPISTEMIC OPERATOR ε

The ε operator (epistemic operator) transforms every formula X into the assertion εX, which now simply means: X or not X. The reason why we call ε an epistemic operator lies in two quite immediate theorems concerning ε. For the psychoanalyst they are interesting, because they give the structural and ethical basis of unconscious knowledge.

The structural theorem is easily written as a conditional: *if X, then* εX. It means that everything can be known if it can be expressed in any suitable language. This theorem excludes every mystical approach to the Unconscious as something ineffable. It sets language as a precondition of the Unconscious and suggests a linguistic ontology: *to be* means *to be articulated in words*. It is not only metaphorically that we can claim: our knowledge may go only as far as our words can reach. It is just a matter of time. Time in this logic becomes epistemic time: the time to know. Freud proposed the same principle with his aphorism: *Wo Es war, soll Ich werden*. We can translate it here as: *where words are, there is my knowledge*.

Obviously, it is impossible to invert the structural theorem. But we know a quasi-inversion: *if not* ε *notX, then notX*. This is the principle of *in dubio pro reo* until one has proof to the contrary. As a consequence, the subject is born in the Unconscious as guilty. His primary sense of guilt, which not paradoxically at all come before and not after committing a crime, tells us the very truth about him.

The ethical theorem makes a categorical statement: *not* εεX. It means that the subject of the Unconscious cannot not know, even if it is ignorant. The Unconscious is a knowledge we do not know we have. In our formalism: *if not* εX, *then* εX. If you do not know, then you know. The paradox is only apparent. The subject may learn a lot of his own unconscious desire…if like a detective he agrees to spend time and money to follow the elements of the unconscious significant chain. *Scilicet, tu peux savoir*, "you may know," translated Lacan. We add an ethical but crucial premise: *si tu veux*, "if you want."

■ BEING HUMAN:

Among the many other features of ε we stress idempotence and a negative peculiarity, which goes back to the ontological proof of God's existence.

ε is idempotent, i.e. *εX if and only if εεX*. In terms of unconscious knowledge, we can interpret this theorem in two different ways, the first negative and the second positive. The first says: "I regret, there is no meta-unconscious." The Unconscious of the Unconscious is again simply unconscious. Lacan's formula is similar: "There is no Other of the Other." The positive interpretation is a clinical one and resorts to epistemic time. "What you learned about your knowledge, you already knew." This is a self-evident fact in the analysis of the slip of the tongue, which anticipates a truth the subject will know later.

In such a way we prove that the Unconscious is a knowledge that is less at the subject's disposal than the not conscious. As a rule it becomes conscious but in a negated form. For the sake of completeness, we have to observe that the negation of idempotence is also a form of knowledge: *If not εεX, then εX*. Here we find the core of the unconscious knowledge as not knowing one knows (*das Unbewußte,* following Freud).

We conclude our discussion of the ε operator by pointing out a not-theorem involving the relationship between ε and the existential operator. We meet here a counter-ontological (more precisely counter-existential) argument: *if there exists an x which we know satisfies X, then it does not necessarily follow that we know that such an x exists*. Within this logic it is impossible to prove the existence of God as *id quo maius nequit cogitari*. The epistemic logic is prudent. By suspending the EM, it puts out of its field any form of omniscience, that of God first. Knowledge is always partial with respect to truth, existence, truth of existence and existence of truth, as we see every day in analytical practice, where it may happen that the obsessional subject knows (or better, assumes) what kind of object causes his desire, without being able to recognize it when he meets it.

THE DN EPISTEMIC OPERATOR δ

Now I have come to a parallel discussion of the δ operator. Since the DN is a logical consequence of the EM (that is, when the EM is true, the DN is always true too), the δ operator is weaker as ε in the sense that *if εX, then δX*. This means that, if δ simulates unconscious desire, in any event of desire we may find epistemic markers. In our opinion this trait makes the Freudian concept of desire (*Wunsch*) quite different from any other: biological, phenomenological and so on. As a consequence of the weakness of δ with respect to ε, all not-theorems about ε are still not-theorems about δ. But some positive analogy is also preserved.

First of all the structural theorem: *if X, then δX*, is still valid. This means that where there is a linguistic structure, there is desire. In Lacanian terms: Unconscious desire is the speech of the Other. Unconscious desire is not a matter of the Ego, because the Ego is immersed in language and language is not the exclusive domain of the Ego, nor one organ of it, as in Chomsky's psycho-linguistic interpretation. Our theory puts psychoanalysis on a not-anthropomorphic plane, which in our opinion is a more solid foundation than any other that anthropology may offer.

And the differences now. Idempotence has only half validity. *If δX, then $\delta\delta X$*. It is clear that when I desire, I desire to desire. In Lacan's theory this aspect is formulated with more subtlety: when I desire, I desire the desire of the Other. But the contrary is false: *If $\delta\delta X$, then δX* is no longer a theorem in this logic. The desire of desire is nearer to the subject's (demand of) love than to pure desire (of the Other). The not-equivalence between love and desire is a commonplace in analysis of transference. From a theoretical point of view the non-equivalence between δ and $\delta\delta$ opens to an infinite series of operators: δ, $\delta\delta$, $\delta\delta\delta$, $\delta\delta\delta\delta$ ...all different but similarly built. We recall that Matte Blanco had pointed out the relationship between Unconscious and infinite sets.[3]

---

3 I. Matte Blanco, *Unconscious as Infinite Sets: An Essay In Bilogic* (G. Duckworth Press, London 1975).

■ BEING HUMAN:

In the place of the ethical theorem we have a different version with respect to that of ε, which it would not be wrong to call the Oedipus version, that is, δ *not* δX. Oedipus's *mè phunai*, to never have been born, is in epistemic logic translated into *to desire not to desire*. Beyond the screen of tragedy our epistemic logic interprets this desire as the desire not to know. The theorem which combines ε and δ, knowledge and desire, makes possible the interpretation: *If εX, then εδX*. This theorem justifies stating that the epistemic scope of knowledge is the field of desire. Therefore, people prefer ignorance (and hate any practice of knowledge like mathematics, philosophy and psychoanalysis) because knowing means to know (something, not all, about) desire. And desire is something, following Freud, that may lie beyond the principle of pleasure.

It is noteworthy that the last theorem is not true in general for any formula X and Y. It is quite easy to verify that, given *If X, then Y*, it does not necessarily follow that *if εX, then εY*. Here we are confronted with the intransitiveness of unconscious knowledge, which makes up the main difference with respect to conscious knowledge. The latter "downloads" itself on the object, the former restrains itself within itself. In a sense, unconscious knowledge is inhibited, also with respect to self-knowing. It is this epistemic inhibition which hinders analytical work. When such an inhibition gets itself, at least a bit, out of the subject's way, he or she can be regarded as analytically "healed."

The interaction we pointed out between the ε and δ operators, which restores transitivity between unconscious knowledge and unconscious desire, is worth studying in depth, not only as a typical trait of this epistemic logic, but also for practical reasons. The healing that psychoanalysis produces concerns more the reform of the human intellect (*Spinoza dixit*) than magical manipulations of emotions.

COMPLETENESS, INCOMPLETENESS AND PARTIALITY

We have to recognize that effective logic is consistent. Therefore, our epistemic logic, which is based upon it, is also consistent. But what is the meaning of consistency? Logicians have identified various definitions of consistency, syntactical and semantical. The following are the most important.

From the syntactical point of view, absolute consistence comes first. A logical system is absolutely consistent if not all formulae are theorems. Obviously, intuitionistic logic is absolutely consistent, because at least one formula, that of the EM, is not a theorem.

Then comes relative consistency. A logical system is relatively consistent if there is no formula X such that both X and *not* X are theorems. Intuitionistic logic is relatively consistent. (Let us suppose that X and *not* X are theorems. Then, by the Duns Scoto's law [*if X then* (*if not X, then Y*)], Y is a theorem for every Y. At that moment the EM becomes a theorem. A contradiction.)

The semantic definition of consistency introduces the notion of interpretation. A logical system is semantically consistent if its theorems are true in any interpretation. In this sense consistency is the same as correctness. The demonstration that effective logic is correct is long enough but not difficult and we need not go into it here.

The definition of syntactic completeness depends on that of consistency. A logical system is absolutely complete if, for every formula Y, either Y is a theorem or, upon the addition of Y, the system becomes absolutely inconsistent. What would you guess about effective logic? Is it absolutely complete or not? Strangely enough it is not. You may add the non-theorem of the EM and the system does not collapse to absolute inconsistency. (When you work with propositional intuitionistic calculus, the addition of the EM transforms it in its classical propositional calculus, which in its turn is absolutely complete.) The same has to be said about relative completeness. A logical system is relatively complete if, for every Y, either Y is a theorem or, upon the addition of

■ BEING HUMAN:

Y, the system becomes relatively inconsistent. Effective logic is relatively (and infinitely) incomplete. You may add upon it however many terms you want from the series $\delta X, \delta\delta X, \delta\delta\delta X$...and you'll always fail in completing the system in relative sense.

The semantic notion of completeness is weaker than the syntactic one. A logical system is semantically complete when its valid formulae (that is, formulae which are true in every interpretation) are also theorems. Thanks to Kripke, since 1963 we have known the right semantics with respect to which the intuitionistic system is semantically complete and the effective proof of the fact. I shall not go into details about this. I want only to point out the main difference between classical and effective semantics, which may turn out relevant to our epistemology.

Classic semantics is about the one. In it there is only one world or one epistemic state. The statement X establishes a correspondence between descriptions of the world and the real state of the world X. The statement is true when it corresponds truly to the state of the world, and false otherwise. The binarism of true and false, which we have already discussed at length, is the syntactical reflex of this semantic monism. Effective semantics breaks away from the uniqueness of the epistemic state (and therefore from binarism). Effective epistemic states form a set, even an infinite one. (This result was anticipated by Gödel in 1932.) Kripke introduced in this set a reflexive and transitive relation. This makes the truth as well as the falsity of a statement depend on the state of all the epistemic states, which are *downward* with respect to that in which the statement is actually uttered. As a consequence the evaluation of its truth requires time: an epistemic time.

Incompleteness, infinity and time are the ingredients of the psychoanalytical epistemology, which we have assembled in our epistemic logic. Can we sum up these three words just in one? I propose partiality. Why? Because partiality is at the very core of both the practice and the theory of psychoanalysis. Partial is the overcoming of the inhibition to knowing. Partial is the overthrow of repression. With respect to the Thing a person tries to speak about, he/she always lacks the words to tell the whole business.

Where does all this partiality come from?

ε and δ correctness

It is only fair to report the correctness of the epistemic operators.

The syntactic correctness of epistemic operators is a consequence of the syntactic correctness of effective logic. In this context we are interested in semantic correctness. We show that for every formula X, εX and δX are correct in that they have a model, that is an interpretation which is true. The theorem is easily predictable, since semantic correctness is a weaker property than syntactic correctness. Nevertheless, the argument is for us so interesting that we supply two independent demonstrations of it.

A) ε*X has a model.*
First demonstration. Suppose the contrary. This means that within every epistemic state the interpretation of εX is false, that is *X or not X* is always false. It follows that *not X* is false. Therefore, there is a world in which X is true. This provides a model of εX.

Second demonstration. Suppose an isolated epistemic state Γ in which the formula Y is true. *First case.* X = Y. X has a model and therefore *X or not X* is true in Γ, that is εX, has a model. *Second case.* X ≠ Y. Not X has a model and therefore *X or not X* is true in Γ, that is εX, has a model.

There is no other case since every formula is a finite expression and in finite domain EM is a valid principle.

B) δ*X has a model*
First demonstration. Suppose the contrary. This means that within every epistemic state the interpretation of δX is false, that is *if not not X, then X* is always false. It follows that *not not X* is true, according to Frege. Therefore there is a world in which X and *not not X* are true. This provides a model of δX.

Second demonstration. Suppose an isolated epistemic state Γ in which the formula *Y* is true. *First case.* *X* = *Y*. *X* and therefore *not not X* have a model in Γ. It follows that *if not not X, then X*, is true in Γ and δ*X* has a model. *Second case.* *X* ≠ *Y*. *X* is false in Γ. *Not X* has a model in Γ, therefore *not not X* is false in Γ. It follows that *if not not X, then X* in true in Γ and δ*X* has a model.

We conclude as before.

It is noteworthy that our epistemic operators give a model, that is a representation, to every formula, even if contradictory. In a sense Freud's thesis that the Unconscious, as epistemic construction, does not know contradiction is justified in our logic.

A SCIENTIFIC FOUNDATION OF PSYCHOANALYSIS?

Science and psychoanalysis have much in common, at least their origin in the *Cogito*. Both are epistemic processes. Both are Cartesian processes. But science and psychoanalysis are also quite different, on which everybody can agree. Science seeks certainty as objectivity; the price it pays to reach this goal is the annulment of any scientific consideration concerning the subjects that science is made of.

On the contrary, psychoanalysis is concerned with subjects. The certainty she seeks is subjective and concerns the single subject. That is why psychoanalysis can never be a science, even if it is presented in a highly formalized way as here. Formalism is a necessary condition to build a scientific model, but it is not sufficient. Indeed, in mathematics there are a lot of formalisms that find no application in science (for instance the pure theory of number) and there are formalisms that are not scientific (for instance in astrology). Typical of the current pseudo-scientific ideology, is the false implication: *if it is mathematics, then it is scientific.*

Psychoanalysts can find their own niche, which is not scientific, within mathematics. Our effective formalism has been introduced here in order to cope with the not scientific, epistemic exigencies that the trinary structure of language introduces in the subject; the logic exigencies and the ethic responsibilities of the Unconscious. By weakening binarism, effective logic makes room for a third element between those, for instance, of the mother/child couple: a third dimension, that of language, of the symbolic function of the father. This is an operation that gets out of scientific range. In effect, I do not know if up to now intuitionism has ever found any applications in the hard sciences, I mean physics or chemistry. I have never heard of Nobel prize winners speaking in intuitionistic terms. In *The L.E.J. Brouwer Centenary Symposium*,[4] I do not find any scientific or technological application of Brouwer's thought. Is effective logic useless? Today's technology (above all, information technology) and modern pragmatism prefer hard-line binarism. This is self-evident. Binarism is more reliable, even in coping with uncertainty. The same modern calculus of uncertainty, the theory of probability, relies on the binary axiom: *the probability of A + the probability of not A = 1.*

As we have shown in our presentation of the four columns of binarism, there are many possible ways to weaken binarism with escalating effectiveness. We have lingered over Brouwer's form of weakening. We shall present in a moment another and even more effective way to weaken binarism, developed by von Neumann. This multiplicity of theoretical approaches is a good sign that we are out of the domain of science. From Aristotle through to the present big project to unify the physical forces, science works ideally within the context of one: one world, one theory. We have just seen that scientific binarism is a consequence of science's monistic ontology. At the same time, breaking down this monism, for instance in Kripke's intuitionistic semantics, weakens binarism and leads the way to the infinite. The situation is similar to that with which we are confronted in psychoanalysis, where

---

[4] Edited by A.S. Troelstra and D. Van Dalen, North Holland, Amsterdam 1982.

we experience the effects of a structure, that of the subject in relation to the Other, which is too complex to be entirely represented within a single model. Theoreticians can only present different partial models of it and content themselves with comparing and discussing them.

We are in position to explain this phenomenon. What cannot be forced into a unit is language. The disciplines, like science, that succeed in keeping language and its effects on the subject out of their own field, have their unity guaranteed by binaristic logic. The others, like psychoanalysis, are "condemned" to a not-unitary ontology and to a weakened binarism in logic. Accordingly, this means that any project of transforming psychoanalysis into a science, in order to restore its unity and to give it solid foundations, represents a "logical" attempt on its life.

NOT ALL CAN BE TOLD

The battle against binarism is not definitively won until we tackle the problem of the fifth column. Up to now the difference between hard and soft binarism is only nominal. In a sense, lacking the EM and its consequences, intuitionism seems to have fewer theorems than binarism. In 1968, Schütte stated precisely what here "fewer" means, proving Gödel's conjecture (1932) that a formula is effectively valid if and only if when transcribing negation as necessity of negation and implication as necessity of implication, it becomes a valid formula within the Lewis's modal system s4. But on the other side, the set of intuitionistic theorems is infinite exactly as the set of classical theorems. This suggests that the two sets may differ from one another in that their elements are merely different transcriptions of the same basic elements. What if they are not? If not, are we faced with a paradox? If yes, have we done a useless exercise, since the two logics, that of truth and that of knowledge, are basically the same, except for a transcription?

When toying with the infinite one must be prudent. That is the very lesson of intuitionism. The analyst puts it into his practice by reckoning with a lack in the structure producing the subject, that is language. But what kind of lack may language suffer?

Let's start from the inscription we read on one of the fragments of the fifth column of binarism now knocked down, which lies on our epistemic ground: *You can tell all*. The column of the classical speech meant: *About all you can tell, when it is true, that it is true and, when it is false, that it is false.*

We just debunked one half of this principle when we quoted the structural theorem of knowledge: *if X, then εX*. Then we stressed that it is impossible to invert it. We affirmed that everything can be known, if it can be expressed in any suitable language. But it may happen, as in the case of unconscious knowledge, that the subject knows something, but nonetheless he is not able to find the right words to express it.

Does this mean that it is *in general* impossible to built an epistemic logic? I am an analyst. I do not seek a general formalism valid for all cases. I restrict myself to the particular ones that are suited for particular situations. The route of the analyst goes from particular to particular.

Brouwer's logic, upon which my epistemic logic is based, originated at the start of the century as an attempt to contain all mathematics within itself. My desire is actually analogous to the logician's in wanting to keep discourse open, also in clinical practice.

To the subject of a discourse you can always apply the predicate "true or false": the first to doubt this principle was Wittgenstein. The doubt was in the air. The doubt started with Russell and Cantor. With the antinomy of Russell and Cantor it was borne in mind that you cannot assign true or false to everything; you cannot say whether everything is true or false. I do not want to go into this business, which is very complicated. But at the end of his *Tractatus*, Wittgenstein writes a single sentence, number 7, which states explicitly: what you cannot speak about you must be silent about.

The first true blow to the fifth column was delivered in 1931 by Gödel, with the theorem of incompleteness, which says: "If a logical system is coherent, then there exists within it a truth that cannot be proven." It is worth lingering over this for a moment. Much of the work we have done so far has been on the syntactic level. Broader than the syntactic is the semantic one. I have alluded briefly to Kripke's system of connections between states of knowledge. Generally the semantic field is broader than the syntactic. This is a constant in mathematics. When you set up a writing system, one field closes off, and you lose something outside. That is precisely what Gödel's theorem is saying. Try to build an axiomatic system of arithmetic. There exist several: the Peano system is one of the first. Your system would work, but be careful. If it is not contradictory, eventually you will find sentences within it that will be true but that you cannot prove. Gödel gave an example of one of these unprovable sentences that is about the coherence of mathematics. You cannot demonstrate the coherence of arithmetic within arithmetic. The mathematician must accept the coherence of arithmetic as an ordinary empirical datum without being able to ascertain certainty as meta-arithmetical truth.

Gödel's theory is a syntactic one. The same incompleteness may be proved at semantic level. Two years later Tarski proved it is impossible to define a truth predicate, which states that a sentence is true when it is true and false when it is false. Since this is precisely the definition of truth according to Aristotle, Aristotle's truth predicate cannot be built. Binarism becomes basically ineffective with respect to its own premises, when it is forced to tell the truth of the truth.

"NOT ALL" CAN BE TOLD

Between Wittgenstein and Gödel and Tarski, there is an author, a genius in his ability to simplify, about whom I would like to speak briefly. He is one who said something truly definitive about the matter of all and not all.

Johann von Neumann is the name of that author. A man who invented game theory for economists, quantum mechanics for physicists and who built the first computer. A lesser known fact is that he started, at the beginning in 1925, to develop an axiomatic theory of sets, without any contradictions, which was not perfected until 1965 by Gödel and by his Swiss friend, Bernays. Let me give you briefly the concept, insofar as it can be useful for psychoanalytic practice.

It is a matter of abandoning the principle of logical closure according to which everything can become a matter of logic and logic can make a judgment by establishing true and false about everything. You have perhaps already heard this. As Lacan said, truth can only be half spoken. What Lacan meant in a metaphorical and aphoristic sense corresponds to what von Neumann in a more scientific way affirmed about functions that cannot become the argument of other functions or classes that cannot be elements of other classes. Simply put, there exist certain discourses that do not have to become subjects of other discourses.

Here you have to be careful. There is latent a temptation to mysticism. If truth cannot be said, then I shall not say it. And I am going to take shelter in the night of God or in the ineffable. The fact that you cannot say everything in no way reduces the need to say everything you can say. The structural impossibility of telling all does not relieve your responsibility to say everything you can. And Freud's fundamental rule of course is this: say everything, knowing that you cannot tell all. This is a way of respecting certain linkages, certain articulations, chains, ties, connections, restrictions at work in every language.

Respecting these linguistic restrictions is a precondition for listening to the unconscious. The Freudian unconscious is an effect of natural language, not a biological fact. So to listen to the unconscious you have to listen to language. But in listening to language you have to respect the laws upon which it is based. The general characteristic of natural languages is that they are incomplete. The term *general* is itself incomplete. It means that each natural language, enumerated one by one, is incomplete. It does not mean that all are incomplete. The use of plural and singular points out the difference, which is structural. Various authors have said this in various ways. Freud, for instance, did not say that the speaker cannot say everything about everything. He affirmed that a death drive is silently working in language. Such a drive "cannot say what it wants."[5]

Lacan developed a different idea. He spoke of the signifier of the lack of the Other. I want to pause briefly on this signifier, to give one model, among many others, of what it means. I think my model concretizes very well what would otherwise be the very abstract notion. What at first must be humbly admitted is that this notion is too big to be conceived. It is certainly too big to reduce it to a subject of a discourse (or of any predicate), so that you can assign it a true or false value.

The very notion of the bigness of a totality immediately appeared as problematic, at the outset of Cantorian set theory. The attempt to make set theory rigorous has always been an attempt to respond to the problem that the "too big" posed. Von Neumann, for instance, distinguishes between very big classes and slightly smaller sets. What does "big" and "small" mean in this context?

The difference, which seems to be a quantitative one, is topological. The "too-big" is too big to find a place in any predicate, so it cannot become a subject of discourse. The less "too-big" can become a subject of discourse. It finds a place without contradiction in a predicate that affirms something about it as a whole. The first are totalities bigger than the second ones in the sense that they are not a whole; the others are smaller in the sense that they can be conceived as a whole. Likewise, von Neumann distinguishes between two kinds of bigness, the "too-big" and the slightly less "too-big." The first kind of bigness, the really big, is called precisely a class, a proper class. The other bignesses, the smaller, are sets. Von Neumann's set theory is the theory of classes.

---

5 "*Es kann nicht sagen, was es will.*" Sigmund Freud, "Das Ich und das Es," in *Gesammelte Werke* , vol. XIII.S. Fisher Verlag, Frankfurt am Main, 1940, 289.

All the bignesses are classes, but he distinguishes two different kinds of classes. The sets are classes in that they belong to classes. To the contrary, proper classes, the big ones, do not belong to classes. This solution avoids antinomies.

Next step. Sets belong to classes, but what does this mean, this belonging? It means that a single set can be reduced to a unit and this unit can be planted within a class. It is the same discourse as before. The set can be a subject of discourse through any predicate. In this case the discourse takes the set as a whole and places it within a class. But this discourse does not hold true for proper classes. Proper classes are too big to be reduced to a unit. They are too big to become an element for another class. As such these proper classes lack something. They lack unity (and I should point out to you that *unità* means both unity and unit in Italian). Which does not eliminate the fact that you can still write something that says precisely that something is lacking. Lacan wrote it this way: Signifier of the lack. And he added that this is a strange signifier, it cannot be uttered but it can be written.

I consider proper classes models for excessive bignesses, such as "language," "the feminine," and also "the paternal." "The Other," "the feminine," "the paternal" are instances of *not all*, according to Lacan's terminology. This terminology can be criticized because it is negative. I prefer using positive terminology such as proper classes.

There comes an interesting corollary in relation to proper classes. I said before that proper classes cannot be reduced to being elements of another class. If this were possible, they would no longer be proper classes, but they would be sets. What would happen if I took two proper classes and I tried to couple them?

I take two different classes, for instance, to make the reasoning concrete, the *two* sexes. (I underscored *two* because I do not know yet if they are effectively *two*. For the human baby there is only *one* sex: the phallophorus.) At least one sex, the female, is a proper class, warrants Lacan. About the other, the male, we do not know if it is proper or not but we can hardly endure this kind of ignorance. The result that follows is a strange one. Now we must admit that a proper class belongs to a couple. A couple is a set. But a proper class cannot belong to any set. A contradiction? Not at all. It suffices to recognize that the couple we have built is the empty set. Is this a good idea? It seems to be. Indeed, nothing belongs to the empty set. Not even our proper class, the feminine gender. We are forced to admit that the two sexes are less than two, and that the sexual relationship cannot be written at all.

"Why do people say artificial mind
and not artificial soul?"

What Do Cyborgs Eat?
Oral Logic in an Information Society[1]

Margaret Morse

> "Well, I'll eat it," said Alice, "and if it makes me grow larger, I can reach the key; and if makes me grow smaller, I can creep under the door: so either way I'll get into the garden and I don't care which happens!"[2]
>
> – LEWIS CARROLL

INTRODUCTION: BODY LOATHING

For coach potatoes, video game addicts and surrogate travelers of cyberspace alike, an organic body just gets in the way.[3] The culinary discourses of a culture undergoing transformation to an information society will have to confront not only the problems of much-depleted earth, but a growing desire to disengage from the human condition. Travelers on the virtual highways of an information society have, in fact, at least one body too many: the one now largely sedentary carbon-based body at the control console suffers hunger, corpulency, illness, old age, and, ultimately, death. The other body, a silicon-based

---

[1] Condensed by the author from *Virtualities: Television, Media Art and Cyberculture* (Indiana UP 1998), ch. 4.
[2] Epigraph on a 19th century human growth hormone from Chapter One of Lewis Carroll's *Alice in Wonderland*. The outcome of the pill is revealed in Chapter Two: "'Curiouser and curiouser!' cried Alice. 'Now I'm opening out like the largest telescope that ever was! Good-bye feet!'"
[3] Allucquére Roseanne Stone has nominated this attitude "cyborg envy" in her article "Virtual Systems," in: *Incorporations*, eds. Jonathan Crary and Sanford Kwinter (New York: Zone, 1992), 609–621. Manuel De Landa's *War in the Age of Intelligent Machines* (New York: Zone, 1991), a history of technology from the machine point of view, implies not only a fundamental problem of incompatability, but of ultimately opposed interests of humans and machines.

surrogate jacked into immaterial realms of data, has superpowers, albeit virtually, and is immortal; or, rather, the chosen body, an electronic avatar "decoupled" from the physical body, is a program capable of enduring endless deaths. Given these physical handicaps, how can organically embodied beings enter an electronic future? Like Alice, this requires asking ourselves if and what to eat.

Some theorists in future-oriented subcultures who have wholeheartedly embraced technology, (or who, as critics, at least speak from its belly) have posed the union of machine and organism as the hybrid meld, the *cyborg*, a "human individual who has some of its bodily processes controlled by cybernetically operated devices."[4] However satisyfing such an imaginary blend might be, the actual status of the cyborg is murky as to whether it is a metaphor, a dream-like fantasy and/or a literal being; and, its mode of fabrication and maintenance is, practically at least, problematic.

Consider such a mundane and practical problem as, what do cyborgs eat? After all, the different nutritional requirements of silicon- versus carbon-based intelligence of the mammalian persuasion are not negotiable in material reality. The alimentary process and its beginning and end products, food and waste, tie us inextricably to the organic world. Both the need to eat and the pleasure of eating are part and parcel of the condition of mortality which electronics are spared. It is unlikely that the very notion of "breakfast" (or lubrication cycle? power feed?) would have much meaning for the relatively immortal and virtual parts of a creature that might suffer obsolescence, silica fatigue, and sudden crashes, but not hunger or death. Willing the cyborg into

---

4 Definition from *The American Heritage Dictionary*, cited in Gabriele Scwab's "Cyborgs, Postmodern Phantasms of Body and Mind," *Discourse* 9 (Spring-Summer 1987), 64–84. Donna Haraway writes (1985) "By the late twentieth century, our time, a mythic time, we are all chimeras, theorized and fabricated hybrids of machine and organism; in short, we are cyborgs. The cyborg is our ontology; it gives us our politics." In "A Cyborg Manifesto: Science, Technology and Socialist Feminism in the Late Twentieth Century," revised in her *Simians, Cyborgs and Women: The Reinvention of Nature* (New York: Routledge, 1991), 149–181, quotation, 150.

being appears to be the equivalent of wishing the problems of organic life away – yet, unless the human is erased entirely, food and waste will enter the cyborg condition.

The more immediate question then is, what do humans who want to become electronic eat? For we are no longer talking about metaphors or electronic prostheses that extend organic body functions (the way McLuhan understood the media, for instance), or even Frankensteinian reassemblage or Tin Man-like displacements of the organic body part by part. In this more *mechanical* sense, cyborgs with heart monitors, organ implants and artificial limbs already walk the earth. The contemporary fantasy is rather, how, if the organic body cannot be abandoned, it might be fused with electronic culture, in what amounts to an *oral* logic of *incorporation*.

I. ORAL LOGIC: THE DIALECTICS OF INCORPORATION

> As she withered, sucked of energy, he became more alive and animated. When she brought our tea, her face was clouded and dark, her shoulders bunched and turned in. He had eaten her alive. I sat amazed watching this psychic cannibalism.
> – WM. PATRICK PATTERSON,
> *Eating the "I": In Search of the Self* [5]

While the late stage in the evolution of the process of identification paradoxically depends on distance, its earliest manifestion, fusion of

---

5   Published in San Anselmo, CA; Arete, 1992, 284. The excerpt continues: "I had always thought that one class of beings eats another; that all forms of life, gross and subtle were engaged in a kind of perpetual eating or, as Gurdjieff called it, 'reciprocal maintenance.' The different classes (the vertebrates, invertebrates, man and angels) are separated by what they eat, the air they breathe and in what medium they live. It had never occurred to me that within classes of beings the strong psychically feed on the weak. The waitress was 'food' for her boss."

oral incorporation, is a more-than-closeness that involves introjecting or surrounding the other, (or being introjected or surrounded), and ultimately mixing in a dialectic of inside and outside that also can involve a massive difference in scale. Bodies in oral logic can be very small (usually but not always the eaten) or immense (often, but not always the eater). The body of the other can be a intrauterine-stomachic-intestinal interiority or virtual void within which one is "immersed," (consider, for instance, electronic encapsulation of the body in virtual reality) or as small as a smart pill that one ingests. There also appears to be a dialectic between eating/being eaten and sucking out, piercing, and fragmenting the body (as if into food) versus resurrecting it into wholeness or preserving in a incorruptible state. Note that unlike identification, incorporation does not depend on likeness or similarity or mirrors in order to mistake the other as the self; in an "oral-sadistic" or "cannibalistic" fantasy, the introjected object (electronic machine or human body as other, depending on who eats whom) is occluded and destroyed, only in order to be assimilated and to transform its host.

Eating

One method of cyborg construction is by means of introjection and absorption. That is, "the subject, more or less on the level of phantasy, has an object penetrate his body and keeps it 'inside' his body.... [Incorporation] means to obtain pleasure by making an object penetrate oneself; it means to destroy this object; and it means, by keeping it within oneself, to appropriate the object's qualities. It is this last aspect that makes incorporation into the matrix of introjection and identification."[6]

---

6 J. Laplanche and J.-B. Pontalis, *The Language of Psychoanalysis,* transl. D. Nicholson-Smith (New York: Norton, 1975), 212. An oral topology of covering, engulfing /being

That is, *pace* Brillat-Savarin, you are what you eat or introject. Therefore, to become a cyborg, a partly human, partly electronic entity, a human must eat the stuff of cyborgs. Indeed, cannibalistic fantasy plays a great part in this oral logic "marked by the meanings of eating and being eaten."[7] As J. G. Ballard's aphorism on "Food" proposes, "Our delight in food is rooted in our immense relish at the thought that, prospectively, we are eating ourselves."[8]

Currently, when we want to introject cyborgs, "smart drinks and drugs" will have to do. Built along the analogy of smart appliances, houses and bombs, the adjective *smart* attributes some degree of agency and, at times, even of human subjectivity to the object world. "Smart" pill and powder cuisine consists of vitamins and/or drugs, laced at times with psychotropics and aimed directly for the brain. To the cyberpunk culinary imaginary, these chemicals are decidedly Utopian, a kind of lubricant or "tune-up"[9] for wet-ware that breaks the blood-brain barrier, makes neurons fire faster and encourages dendrite growth, *not unlike* the networks linking the electronic channels along which information flows.

But the more fundamental, albeit speculative answer for humans who want to transcend the organic body and its limits, i.e. who want to be cyborgs, is to eat *nonfood,* that is, food that negates the very idea of the organic or "natural" value of food. Vitamin gels and chemical soups qualify precisely *because* they blur the categories of food and drugs, anticipating the advent of what futurologist Faith Popcorn calls "food-ceuticals."[10] Capsules of what are tantamount to brain chemi-

---

covered or engulfed also plays a role in the logic of some video games and computer displays and interaction, for instance, with the Pacman or the Graphic User Interface of Macintosh and Windows.
7   Excerpt from the "Oral Stage (or Phase)," *Ibid..,* 287.
8   In: "Project for a Glossary of the Twentieth Century," *Incorporations,* eds. Jonathan Crary and Sanford Kwinter (New York: Zone, 1992), 277.
9   Metaphor courtesy of St. Jude, *Mondo 2000.*
10 Faith Popcorn, *The Popcorn Report* (New York: Bantam, 1991), 66f. Durk Pearson & Sandy Shaw® of Designer Foods™, however, emphasize that, "All ingredients [in their

cals, condense "intelligence" into a magical essence or fetish for transforming the human brain into a high-performance electr(on)ic machine.

Smart drugs are chemically targeted at the brain – but they are considered efficacious at bringing the flesh at the console along for the ride: for instance, despite the fact that he reports spending "all my time lying flat on my back on my waterbed with my computer," Durk Pearson of Durk Pearson and Sandy Shaw® claims to have "good muscles," thanks to smart nutrition.[11] The strategy is not only to feed the mind but in the process to *purify* the body of organic deterioration. For would the ideal cyborg, an electronic *kouros* or imaginary of machine/human perfection, have any need of flesh? To become cyborg, one does not eat the apple of the knowledge of good and evil, but something more like the body of the deity, the host of disembodied information. For most of all, *nonfood* is a transitional substance that incorporates the desire for change as abandonment, or at least a neutralization of the organic body, (or "meat") and its contemporary identity.

Being Eaten

On the other hand, the fusion of organic and electronic must also logically include the possibility of being eaten by electronic machines.[12] Some scientists of artificial intelligence anticipate such an event as ecstasy. For instance, Hans Moravec, author of *Mind Children,* foresees

---

SMART Products] are recognized and utilized by the human body as a food." (*IntelliScope: The Newsletter of DesignerFoods™ Network* 2, n.2 (February 1993), 4. Despite its name, the newsletter also disclaims that "Smart Drinks" will improve IQ, though they will increase well-being, energy and clarity, especially in someone who is "feeling constantly tired, drained or stressed out."

11 "Durk and Sandy Explain It All for You," *Mondo 2000* 3, Winter 1991, 32.
12 Karl Abraham set the oral-sadistic stage concurrent with teething and the phantasy of being eaten or destroyed by the mother, whereas Melanie Klein placed infantile sadism throughout the oral stage.

leaving the organic body behind like an empty shell, after what amounts to having the brain scooped out, emptied, exhausted bit by bit, by one's own advanced robot mind-child. Brain cells or wetware would be displaced with silicon, byte by byte.[13] (This fantasy is elaborated at length in Harry Harrison and Marvin Minsky's 1992 science fiction novel, *The Turing Option*.) "Downloading consciousness" into a computer, achievable according to Moravec by the mid-twenty-first century, would simulate brain functions but at an incomparably faster speed. Gerald Jay Sussman, a professor at MIT, once reportedly expressed a similar desire for machine fusion as the wish for immortality:

> If you can make a machine that contains the contents of your mind, the machine is you. To hell with the rest of your physical body, it's not very interesting. Now, the machine can last forever. Even if it doesn't last forever, you [notice this logical lapse or, perhaps, metalepsis] can always dump it onto tape and make backups, then load it up on some other machine if the first one breaks....Everyone would like to be immortal....I'm afraid, unfortunately, that I am the last generation to die.[14]

Far more recently, Larry Yeager's confession of "why I fell for artificial life" expressed a similar desire to "live on inside the chips."[15] Yet, in O. B. Hardison's image of humanity's immersion in the machine in *Disappearing Through the Skylight*,[16] remnants of the organic body are nonetheless retained within the greater body of technology and

---

13 Hans Moravec, "The Universal Robot" in: *Out of Control: Ars Elecronica 1991*, eds. Gottfried Hattinger, Peter Weibel (Linz: Landesverlag, 1991), 13–28, citation 25.
14 Cited in Jeremy Rifkin's *Biosphere Politics: A New Consciousness for a New Century* (New York: Crown, 1991), 245, from Grant Fjermedal, *The Tomorrow Makers* (New York: MacMillan, 1986), 8. Moravec and Sussman appear to entertain the questionable idea there will still be an "I" and a "mind" capable of "thought" in this future union of brain and hardware.
15 Cited in the publicity for *Ars Electronica 93: Genetic Art, Artificial Life*.
16 Subtitle, "Culture and Technology in the Twentieth Century," (New York: Penguin, 1990).

silicon-based intelligence, much as the mitochondria within human cells remind us of our origin in the sea and in asexual reproduction.

Compare these images of incorporation *within* machines, be it as enthusiastic vision or warning, with the image of disembodied, artificial intelligence in the romantic imagination: a miniature artificial man, the homonculus, was a created product of mind kept in a bell-jar. His greatest desire was to dissolve himself in the ocean, perceived as a female realm of pure body, in a kind of death wish of erotic fusion with undifferentiated nature itself.[17] In contemporary future-discourse with its various degrees of hostility to organic life, intelligence which breaks its corporeal container is seen as simply joining its like in a great digital sea of data. So, the virtual realm is tied symbolically to *immersion* and all its attendant hopes for transcendence and, in this case, inorganic rebirth. However, death-wishing and repudiation of the organic body – insofar as they apply to this and not some afterlife or spiritual plane – adopt a kind of psychotic and fatal reasoning, only to be haunted by the very parts of the organic world they fail to register. For, of course, the scientist only apparently usurps motherhood with the extra-uterine development of the robot-child; his subsequent immersion in a sea of data is implicitly a symbolic return to the first inner space, the womb, much as the fantasy of being eaten by machines evokes phantasies of being eaten or destroyed by the mother.

This phantasy of being eaten by machines which are, in some confused or unspecified way, part of the natural world is graphically visualized in the controversial (and hence widely censored) industrial music video, "Happiness in Slavery" by Nine Inch Nails, (directed by Jon Reiss, 1992.) A man (and, by implication, another man, *ad infinitum*) is shown "submitting to ritualized sad-masochistic relationships with devouring machines" (publicity release). The man is played, significantly enough, by Bob Flanagan, a recently deceased performance art-

---

17 See the adventures of the homunculus in Goethe's *Faust II*.

ist with cystic fibrosis whose larger subject was illness, pain and sado-masochistic pleasure.[18] In the video fantasy, the man's nude body is pierced by pinchers and grinders, put into "some kind of disposal system" (PR) and ultimately "ground into meat." (*Hollywood Reporter*, Nov. 25, 1992.) Oddly enough, this strange kind of bachelor machine is "servicing the MAN's desires," (evidently for castration, penetration, death and complete fragmentation.) The result appears to not only mix machine and human fluid, but to nourish the natural world which appears to consist largely of writhing worms (or, at another level, geometrically multiplying castration symbols). In spite of most waking experience to the contrary, the ecstatic affect and the symbiotic relation of nature and machine is not articulated – it is just there. Here is not a proposition about reality, but a fantasy which reaches back into experiences in infancy. The posture of the devouring machine, leaning over the restrained Man in "Happiness in Slavery," in order to take a bite, is a reminder of the archaic mother and the wish for self-annhiliation to which she is ultimately linked. In this case, the oceanic feeling and ecstatic transcendence of the body occurs by means of pain – in a way that Flanagan's installation work suggests is specific to this culture and its denial of illness and death.

Second Skins

In this highly reversible logic in which subject and object are not clearly differentiated, rather than being eaten, one can try to *become the other* by "getting into someone else's skin." That is, the cannibalistic phantasy of introjection has a counterpart in the reverse gesture, that of covering oneself with the other as a means of self-transformation. En-

---

18 This writer is not unaware of the metaphor of oral sex as "eating" – but here eating is figured directly and sexuality reappears in terms of sado-masochistic affect – as pain and ecstasy.

tering this skin envelope also suggests that where the space is not already void, one is scooping out or evacuating the other either from without or within. That is, the issue is not one of replication or cloning of bodies, but a struggle for sovereignty over the same body as host of identity and subjectivity. In this instance, what was once "other" (machine) becomes the "self" and vice versa. In the process, identification also shifts toward the electronic.

In some historical societies, where ritual involved human sacrifice, the "skin ego," or envelope of identity and self, could literally be transferred by flaying a human victim and wearing his or her skin. In one record describing Aztec ritual, for instance, prisoners who were made to impersonate gods were then flayed. "Other men donned the skins immediately and then took the names of the gods who had been impersonated. Over the skins they wore the garments and insignia of the same divinities, each man bearing the name of the god and considering himself divine."[19] Tzvetan Todorov contrasts such Aztec sacrifice of victims socially very much like themselves with the societies of massacre associated with the conquistadores, whose cruelty grew in proportion to the distance from the observation and control of their own culture. Literalized skin symbolism has a sad counterpart today in the flaying and facial mutilations which have been reported in the states of the former Yugoslavia, inflicted on people who are only partly other, belonging to essentially the same language and ethnicity, but to a different religion and culture. The desire to literally cut away or obliterate the outer identity or "skin ego" of an other – not to mention destroying that close-other from within via hunger, rape and torture – owes something to both sacrificial rites and a society of massacre.

When political boundaries fall apart, ego and identity are also threatened with fragmentation, and they must be radically fortified. Even for an individual in the relative calm of the post-Cold War United States,

---

19 Tzvetan Todorov, *The Conquest of America*, transl. Richard Howard (New York: Harper and Row, 1985), 158.

an ordinary skin may no longer be enough to contain the ego or protect bodily fluids from escaping or pollution and irritants from the outside world from entering. Consider the announcement of a new product, a transparent SmartSkin™ or "ultra-high-molecular-weight cellulose polymer" which is permanently electrically charged so it firmly binds to the skin surface, now made slick, smooth and, as a result, youthful. When used with BETAMAX CAROTENE+™, a melanin layer (or "suntan" without the sun) is also produced, in effect marrying body- and techno-chemicals into what is literally a second, fortified skin.[20]

However, skin egos are usually less literal, though there is always some sensuous element or other that envelopes the body, for instance, the muscular skin earned through weight training (also accompanied by a preferably sunless tan) or a symbolic skin applied via tatooing or writing. The ultimate in second skin is electronic, as in the data suit, helmet and gloves of virtual reality. Under an electronic skin one can adopt virtually any persona and experience a written world of images and symbols *as if* it were immediate experience. Indeed, it is as if the body were immersed in unframed symbols themselves, without need for distance or reference.

Immediacy and Ubiquity

Such introjecting and enveloping responses do depend on a sense of presence and participation in the *"imaginary,"* but in a way in which "the distance necessary to symbolic functioning seems to be lacking."[21]

---

20 SmartSkin™ is a new product introduced in the *Intelli-Scope* above. For a psychoanalytic elaboration of the concept of the skin ego, especially the second skin, with case histories, see Didier Anzieu's, *The Skin Ego: A Psychoanalytic Approach to the Self* (New Haven: Yale UP, 1989), especially the chapters on "The Second Muscular Skin," "The Envelope of Suffering," and "The Film of the Dream," for a discussion of the concept of the skin ego.
21 Tzvetan Todorov, *op. cit.,* 158.

Oral logic can be as archaic and *immediate* as an infant at the breast or as immersive as the fetus in the womb. However, the expressed belief that virtual reality provides an unmediated or "post-symbolic" experience of externalized mind is an illusion fostered and supported by an oral logic of incorporation. In the state of immersion, it seems that one doesn't symbolize flying, one *does* it, just as virtual objects *are,* albeit only via an electronic skin. However, this illusion is possible only because the second skin (or "interface") that mediates the virtual world can also conveniently mask the very apparatus of that mediation. (Walter Benjamin noted similarly that the film image is the only place in which the technological apparatus is invisible – because it is carefully organized to be out of frame.) In both cases, mechanical and electronic, this absence of the apparatus from awareness furthers psychic regression.

While "cannibalistic fantasy" may have its prototype in infancy, it is far from restricted to the past or to infantile or regressive aspects of life. One could call "eating"/"being eaten" and "enveloping"/"being enveloped" deep metaphors that pervade even the most advanced cultures and the highest art forms. Perhaps not surprisingly, the imagery of piercing and engulfment is ubiquitous in the technological realms of laboratory-created immersive virtual worlds, as well as in high-tech war. But even certain philosophies could be lambasted by Jean-Paul Sartre for introjecting the world into the "rancid marinade of Mind." Sartre's disgust with "knowing as a kind of eating" is a lucid polemic against oral logic which has much in common with Brecht's attack on the "culinary" aspects of illusionistic theater.[22]

Although oral logic is conceived as a stage of development in the infant and child before the development of language, it evidently coexists with the logics of other stages of development into adulthood and throughout culture in general. Furthermore, oral logic is hardly

---

22 As in "His eyes devoured her." Sartre's essay on "Intentionality," (1939) was reprinted in: *Incorporations,* 387–391, citations 387 and 390.

■ BEING HUMAN:

restricted to the thematics of food, just as food is not exclusively "oral," but rather participates in the constitution of a full range of oral, imaginary, and symbolic subjectivity. Considering that food itself is the liminal organic substance at the boundary between life and death, need and pleasure, it is also the symbolic medium par excellence: a particular cooking process not only transforms nature into culture (as elaborated in Claude Levi-Strauss' work in structural anthropology), it also offers the means of exchange or communion between the body, the world and other human beings and defines a culture *per se* in its specificity. There can be wide differences in the perceived immediacy or degree of mediation of the body in relation to food, from the fully enculturated eating of an organized meal with utensils to feeding at the breast (or, inversely, the imaginary of being devoured by the mother) to imagining oneself inside the mother's body or as immersed in oceanic oneness with the all. This range is comparable to the difference, say, between symbolizing the other, "interfacing" with it, or wearing brain probes and "jacking in" or being "immersed" in a digital sea. Thus, subjectivity includes processes of incorporation, identification, and symbolization, and oral logic is a constant part of that range of subjectivization.

However, when fragmentation and fortification of the ego become strongly thematized, it suggests a situation of cultural distress. The contemporary prevalence of the imagery of horror and disintegration – fragmented, dismembered, or, for that matter, mismatched, multiple or decaying bodies and lost parts – namely, the cannibalistic fantasy of the body treated as food suggests that something fundamental is "eating" our culture. Perhaps because we live in a situation of epochal cultural change, the envelopes of cultural identity – the body image and skin ego – seem to have been torn beyond repair. But how can the body be resurrected when it is so loathed? We are in a strange situation when the desire for fusion and wholeness presupposes, at least in representation, the repudiation or disavowal of the body and the negation of food itself.

The answer of electronic culture to a vision of confusion and waste is largely one of "purification" or disembodiment. However, this option may be of limited value, for how can the response of culinary and corporeal negation afford to be more than a minor and transitional phase when food confronts cultural change? Food is at once a symbol system and organic fuel; thus there are limits to *nonfood* – we humans have to eat or perish. However, a tacit assumption of the contemporary technological fantasies is that we perish *because* we eat. The desire for an evolutionary transformation of the human has shifted focus from the preparation for the journey into "outer space" from a dying planet to the virtual "inner space" of the computer.

II. FAILED FOOD IDEOLOGIES AND THE REAL

The failure of a food discourse can be read in terms of survival of a population; the body itself is the surface written and sculpted from within in terms of well-being or eating disorders, disease and death. Which is to say, food ideologies and the symbolic order have an ineluctable organic limit that might be thought of as the intervention of the *real*. The consequences of the socio-political and ideological dimension of food can be seen in malnourishment for some and overabundance for others that cannot be blamed on nature. To an outside observer, the most striking American culinary metaphors might be too much and too little: Lack of food for significant populations of children, the working poor, and the homeless for at least part of each month, and poor quality food for the rest of it.[23] On the other hand, the significant numbers of overweight Americans suggests a perverse situation of unwanted abundance. In Berkeley, there was a good-hearted but absurd sounding

---

23 Many different kinds of theorists, from anthropologists like Mary Douglas to political scientists and activists suggest that hunger is most often a result of lack of legal entitlement and social inequities rather than scarcity. "Standard Social Uses of Food: Introduction," in: *Food in the Social Order: Studies of Food and Festivities in Three American Communities*, ed. Mary Douglas (New York: Russell Sage Foundation, 1984),1–39.

■ BEING HUMAN:

organization for passing along the calories of dieters to the homeless – yet it would take such an actual transfer of caloric capital on a massive scale to have an effect on national health. Considering this caloric imbalance, the existing American fast techno-food system has discredited its democratic and Utopian promise.

Fat: The All-Too-Visible Flesh

However ubiquitous cannibalistic fantasies in our culture, it did not prepare the American public to regard the satellite images of taut skin over fleshless skeletons (with eyes that could return our gaze) in Somalia with anything but horror. The images on American television at dinner time conveyed not just human suffering in a situation of social and moral collapse, it confronted Americans with what amounts to our own negative image. (The story of a reporter drinking diet soda amidst the starving is already legendary.) The famished bodies of Somalians were as if sucked dry, excavated by an immense malignant force in a grotesque and tragic exaggeration of a fleshlessness so often idealized (but so seldom achieved) in America. When we find emaciation in America, it can well be malnutrition, but it can also be a fashionable result of will power *and/or* the effects of an eating disorder (such as bulimic purging). The American body may often be hungry, but it achieves most cultural visibility in the United States not in terms of fleshlessness, but as fat.

While some critics have labelled the press coverage in Somalia "disaster pornography," the gaze of American viewers is certainly not one of guilty pleasure, but of guilty repulsion. Advisors are reported to have counseled American soldiers embarking with foreboding for Somalia, that it was normal to feel sick on encountering victims of famine.[24]

---

24 PBS radio, *All Things Considered,* December 10, 1992.

Considering the relative lack of discourse on hunger in the United States, it is ironic how much public support there was for what amounts to a militant response to the devouring land of Somalia. But while we are involved as a nation in the resurrection of flesh and the restoration of bodies elsewhere, our own American flesh – when not covered by hypertrophied musculature or second skin – can be an object of ambivalence and even moral revulsion.

In America, "fat" is a stigma and the sign of the self-indulgent behavior of someone who has let themselves go. Yet, the fat body has an intimate and causal link not only with a poor quality, fast food diet, (shared even by the soon labelled "fast-food" President Clinton, counseled to freshness by celebrity chefs and schoolchildren) but with the life-styles of an information society. Consider, for example, the link between a massive increase in pizza deliveries in Washington and the high-tech planning for virtual war in the Gulf. But the link is even more direct: recent research shows that just by watching the tube we can put on weight.[25] All too visible flesh puts what some consider a crisis of (in)visibility in the age of computers in another light. It is true that without visualizations for the monitor screen or phantom journeys "inside" the computer via virtual reality and science fiction, computer events offer little for human eyes or ears to perceive. But to see the problem of information as invisibility alone is to disavow the intimate connections between the virtual electronic story world (what happens inside computer systems and networks) and the very visible and increasingly repulsive and/or wasted world of the organic body outside the screen.

---

25 Dr. Robert Klesges put on a tape of the television dramedy *The Wonder Years* and measured the metabolic rate of 32 half-normal, half-obese girls from 7 to 12 years old. All showed a drop in metabolic rate, but that of the obese children was especially striking, suggested a reason why heavy-use of television and obesity go together. The study was cited by Jane E. Brody in "Literally Entranced by Television, Children Metabolize More Slowly," *The New York Times* (April 1, 1992), B7.

■ BEING HUMAN:

Fresh Failures: Tourisma and Frankenfood

Perhaps those very symbolic analysts who eat fast food when working can best afford to be "foodies" during leisure hours. For while the once Utopian food counter-discourse of *freshness* in America still retains many populist and progressive aspects it had in the 1960s, such as engagement in practical issues of healthful ingredients, pesticide use, and small farm production, in the 1980s, it evolved into an elite restaurant culture of "foodies" grazing on tiny portions. While the values of *freshness* may harken back to a pre-industrial food system, (indeed, what French farmers with pitchforks are now fighting to preserve from McDonald's), restaurant food art has actually come to rely on expensive ingredients delivered by air and van from small or boutique farms. There are many signals in current food discourse that the "nexus of haute cuisine and counterculture" which opened in 1972 is over.

The social failure of the "fresh" is a sad one that includes food critics as well, who rarely addressed what a homeless person eats for breakfast or – until the recent programs for school gardens and organic lunches – what is being offered in school cafeterias today. (Certainly, one of the fundamentals of human intelligence, not to mention social justice, is having enough of sufficient quantity and quality to eat.) In what appears an oblique response to a more general economic decline, high-art restaurant culture seems to have backed away from eclectic postmodern concoctions in favor of comfort food that doesn't draw attention to itself, a zero-degree of culinary writing. The results are simulations of food like Mom would have made had she cooked a simple ethnic cuisine without the schmalz. There is also a fundamental threat to the ethos of freshness in its very nearness to the *raw* – the growing mistrust of nature itself. Now technological advances have produced "fresh" food that virtually can't decay, via irradiation and genetic engineering. Nuking microbes to preserve food suggests a state of un-death, rather than freshness. Meanwhile, the genetic material which allows vegetables to decay can be inverted, producing a kind of "zipper" effect against rotting. Such "frankenfood" (referring, of course, to cuisine by Mary Shelley) has been perceived as more than mere chemical adulteration by advocates of freshness: "Don't put trout genes in my tomatoes!" exclaimed a San Francisco celebrity chef in disgust, in a talk-radio interview. And no wonder, since by simulating freshness, genetically-engineered and frankenfood blurs the most fundamental distinction between industrial fast food and its culinary postindustrial counterideology; it may look unspoiled, but it is symbolically as well as genetically contaminated. In retrospect, it appears that what is at stake in fresh cuisine is not freshness *per se,* but a continuum of ripe- to rottenness; the fresh is a vote for the organic itself and the mortality with which it is inevitably at one. Unfortunately, such an ideology appeals largely to those with the means to hold decay at bay.

Yet even the discourse of health appears to have in part abandoned freshness. Now there are ex-"foodies" concerned enough with health- and intelligence-maintenance to have stopped eating unhealthy, and, in the extreme, dangerous substances (namely food!) from depleted and polluted soils lacking in the trace minerals and electrolytes needed by the brain. [26]

Since food is a lived metaphor of culture itself, it should not be surprising that a culinary system could emerge in the United States that has many analogs with the computer. After all, the television/microwave, and food/word processing have been image-supports of the fast and fresh, the two great food-ideologies of the post-World War II era. But the computer and cyberspace have come to be reference points for what amounts to a post-culinary discourse. Perhaps, as opposed to Alice's "very small cake, on which the words 'EAT ME' were beautifully marked in currants," an electronic culture virtually confronts us

---

26 One San Francisco ex-restaurateur and coowner of the mail-order firm Smart Products, says "I don't eat breakfast anymore. I don't eat lunch; I drink smart drinks," that take one minute to stir and one to consume. Mark Rennie in "Smart Drugs' True Believers: Highly Developed Thoughts on These Additives for the Psyche," *San Francisco Chronicle* (March 4, 1992), D3., and in personal telephone interview, April 2, 1992.

▪   BEING HUMAN:

with the directive, "DO NOT EAT." For what cyborgs eat (and what evidently incorporates people into cyborgs) negates the very idea of food as mediation between the organic body and the natural world.

III. POST-CULINARY DEFENSE MECHANISMS

The *negation* of the organic body, its nourishment and all that it stands for can occur in many different cultural fields and adopt many different means, for instance, forms of psychic defense such as: *repudiation, denial,* or *disavowal,* (in a discussion based here on the useful distinctions of Elizabeth Grosz.) Furthermore, the body-machine relationship can be inverted, much as the body may (at least in fantasy) be turned inside-out, in surrender to the collapsing boundaries between the symbolic and meaninglessness, as in Julia Kristeva's concept of *abjection*.

Purification Strategies: Repudiation and Cyberpunk

Repudiation is the "rejection of an idea which emanates from external reality rather than from the id. It is a failure to register an impression, involving a rejection of or detachment from a piece of reality."[27] What certain cyberpunk fantasies namely fail to register is the organic body itself. If, according to Allucquére Stone, *cyberspace* is "a physically inhabitable, electronically generated alternate reality, entered by means of direct links to the brain – that is, it is inhabited by refigured human

---

27 Grosz, "The psychotic's hallucination is not the return of the repressed, i.e. the return of a signifier, but the return of the Real that has never been signified—a foreclosed or scotomized perception, something falling on the subject's psychical blind-spot. The subject's perception is not projected outward onto the external world. Rather, what is internally obliterated reappears for the subject as if it emanates from the Real, in hallucinatory rather than projective form. It confronts the subject from an independent, outside position."45f. Grosz.

'persons' separated form their physical bodies, which are parked in 'normal' space,"[28] the next question might be, how does an organic body "park"? The answer of William Gibson's sci-fi novel *Neuromancer*, the bible of cyberpunk, is to submit the body to pseudo-death in the coffins and loft-niches in the desolate landscape of Chiba. While the surrogate Case travels cyberspace, ("Unthinkable complexity. Lines of light ranged in the nonspace of the mind, clusters and constellations of data. Like city lights receding." p.67) his organic body is evidently in a state of suspended animation, neurally sustained by a fantastic pharmacopia.[29] The junction between the human brain and the computer (or "a graphic representation of data abstracted from the bank of every computer in the human system"– note slippage between computer/human) seems to consist of electronic impulses between implanted chips and brain chemicals enhanced by drugs. In terms of oral logic, machine penetration of the brain allows the fusion of electronic-human chemicals which in turn allow the virtual traveler to be enveloped in the electronic skin of cyberspace – at the cost of leaving the meat behind.

But the flesh is left gladly. The desire to repudiate the body that pervades the novel – a desire which, of course, can't succeed in reality without falling into a completely psychotic state – is rooted in a fundamental and pervasive revulsion with "meat." Even the sexual description reads like a SCSI port docking. As for food, while there *is* a restaurant in *Neuromancer*, the Vingtième Siècle, where steak is served, Case isn't hungry (despite Molly's cry, "They gotta raise a whole animal for years and then they kill it. This isn't vat stuff.") because his brain was

28 See Stone, note 2, cited from 609.
29 "Case sat in the loft with the dermatrodes strapped across his forehead. [....] Cowboys didn't get into simstim, he thought, because it was basically a meat toy. He knew that the trodes he used and the little plastic tiara dangling from a simstim deck were basically the same, and that the cyberspace matrix was actually a drastic simplication of the human sensorium, at least in terms of presentation, but simstim itself struck him as a gratious multiplicaton of flesh input." Willam Gibson, *Neuromancer* (New York: Berkley, ©1984), 137 f.

"deep-fried," and the "aftermath of the betaphenethylamine made [the steak] taste like iodine." (Corpulency will never be a problem for this hero.) Ultimately, (im)mortality and the (in)capacity to love become the issue in the novel, though not in the way it might in Chiba, but rather in the ethers of cyberspace, suggesting that for virtual bodies (and, by extension, for cyborgs) the issues of organic life are not evaded, they are merely displaced.

Beyond Operation Margarine: Simulation and Olestra

Denial is a way of negating the corpulency-prone, mortal body (i.e. "the piece of reality") at the *symbolic* level. Grosz describes it thusly: "By simply adding a 'no,' to the affirmation, negation allows a conscious registration of the repressed content and avoids censorship. It is a very economical mode of psychical defense, accepting unconscious contents on the condition that they are denied." (45) Of course, *denial* as the process Roland Barthes identified in his well-known "Operation Margarine," can also work at a far more conscious level of willful public and self-delusion. *Mythologies* describes the process of denial, when confronting unpleasant cuisine like artificial fat, at a time when the "natural" was preferred. (Today, of course, a preference for the artificial prevails among many initiates of an information society.) First, the fact is affirmed (i.e. *yes,* it is margarine), then denied, (*but* it is, in effect, butter) in an act which innoculates the discourse against the artificial, after which the fact may be ignored without embarassment.

On the other hand, *Olestra* does Barthes' margarine one better: "It [sucrose polyester] retains the culinary and textural qualities of the fat, but in a form that the body is unable to digest. Result: a fat that passes straight through the body." As a result, one needn't even try to negate or repudiate the body – its cravings have no consequences for health or mortality. One can "let oneself go" or "carry on" as before and nonetheless be pure, because *Olestra* is not merely artificial (like margarine

– which remains a fat, with all its consequences for the body.) It *is* the simulacrum of food – a solution to the fat accumulating on this side of the television screen and computer monitor would require no change in life-style to purge the flesh. It is rather a process which makes junk food – itself already a concoction of artificial food ingredients plus sugar and fat – subjunctive or contrary to fact food, that is, "the realization of an impossible dream."

Interestingly enough, this "fat-free fat," another guise of nonfood, is the opening metaphor for artificiality *per se* in Benjamin Woolley's introduction to *Virtual Worlds*.[30] For Woolley, this culinary by-pass operation raises the issue of what remains real in an increasingly artificial world. Unfortunately, his argument resorts to denial in another mode; for Woolley eventually finds reality itself to reside not in a visible material and physical world, but "in the formal, abstract domain revealed by mathematics and computation." (p.254) The capacity to mathematically model or *simulate* what is otherwise beyond human perception is thus more real than whatever passes for reality itself.[31] So, on one hand, there is a "food" which treats the body like cyborgian steel conduit, but allows us all the incomparable pleasures of junk food; on the other, there is Wooley's implicit rejection of the visible and manifest (here the "flesh" and its problematic nourishment) in favor of truth in pure mathematics – an act of denial. Neither Olestra nor math ultimately evades the effects of the real.

---

30 Benjamin Woolley, *Virtual Worlds: A Journey in Hype and Hyperreality* (Oxford: Blackwell, 1992), Introduction, 1–10.
31 In a contrasting view, Paul Virilio in *The Machine of Vision* castigates the very will to mathematical power as the occasion for voluntarily blinding the horizon of sight and hearing. My reference is the German translation of *La machine de vision*, *Die Sehmaschine*, (Berlin: Merve Verlag, 1989), 171.

■ BEING HUMAN:

Virtual Reality and Telepresencing: The Body Disavowed

In virtual reality, the "meat" body is not "parked" but rather mapped onto one or more (or even shared) virtual bodies; at the same time, the organic body is purified by being out of the frame, hidden from view for the virtual traveler, in favor of a cartoon-like graphic world. It is as if one were able to shrink down to whatever shape desired and enter one's TV, transformed into fully-rendered, life-sized 3-D space filled with virtual objects (and possibly other virtual personas) with which one could interact. To "enter" or "stick your head in" such an immersive artificial world with a technological second skin – or helmet and gloves – simultaneously blinds the virtual traveler to the world and obscures his/her organic body *and* the machine apparatus which sustains the virtual world from view. Meanwhile, an organic finger merely points inside the data-glove and the surrogate body flies at great speed through an artifical world, promoting an impression of disembodied superpower and almost omnipotent thought in a persona that is freely chosen, not contingent and limited by flesh. According to Michael Naimark's "Nutrition" segment of the tongue-in-cheek video documentation, *Virtuality, Inc.,* virtual reality even has application as a diet tool – in a way not unlike Olestra: While a helmet with television eye-screens blinds the immersee to what we see are actually virtuous crackers, she munches with moans of pleasure at virtual cherry pie.[32]

The seduction and playfulness of virtual reality is based on this very disparity between organic and virtual bodies – and the power to erase the organic one from awareness, if only partly and just for a while. To the degree that the duality of worlds (a reversal of the everyday situation of mental invisibility and physical visibility) remains conscious to

---

32 With students at the San Francisco Art Institute, 1990. A prior piece, *Eat,* from 1989, explores the absurdity of a virtual restaurant, with video projection meals like "Jackson Pollock" and "eat" and a button that can lead to surprising effects – like the Heimlich manoeuvre.

THE TECHNOLOGICAL EXTENSIONS OF THE BODY ■

the one immersed in a world in which s/he possesses superpowers, the situation is one of *disavowal* or split-belief. ("I know it's just a computer-generated display, but nevertheless....)

However, once one switches point of view from internal to external and from the virtual to the organic world, virtual reality takes on a wholly different guise. To the outside observer, the virtual cherry pie eater's improbable gustatory moaning resembles a regression to infancy; meanwhile the flailing motions of "flying," or other virtual locomotion suggest an actual situation of helplessness and vulnerability in physical space.[33] Indeed, immersion in the artificial realms of information presupposes not only an electronic skin, but also a womb-like fortress of safety from the physical world before the user can enjoy apparent invulnerability.

This link and disparity between the two worlds can serve virtual play; but they can also be electronically linked to cause real effects on other distant (or microscopic) parts of the physical world via *telepresence*, as in, for instance, long-distance, robotic brain surgery or precision bombing. Organic bodies in the referent world (i.e. those without access to the virtual system, who are to be operated on, bombed, etc.) need to be prone, anesthetized, or otherwise disempowered in order to be vulnerable to the actual remote operator of any robotic agent. It is as if persons with material bodies were confronted with phantoms; or, as if they were put into a story as characters within which other characters are surrogates of the author and enjoy his or her special privileges and ultimate invulnerability. The danger in the electronic divide between "symbolic analysts" encapsulated in global cyberspace, and those outside is a kind of willful blindness that supports that maldistribution of power – not to mention calories and culinary capital – that is, ultimately, a negation of the social contract.

33 It is true that virtual reality interfaces of glove, helmet or suit map the physical body onto the other virtual body – but not 1:1. The immersive illusion depends on kineasethetic sensations in the physical body moving, but the machine tracking body co-ordinates allows for locomotion in a very circumscribed area in physical space.

■ BEING HUMAN:

Furthermore, responsibility for the organic consequences of remote action is all the easier to deny: consider the virtual conduct of the Gulf War and the many opportunities it offered for denying the connection between war and human suffering. Yet, even for relatively invulnerable warriors beyond the phantoms and under the technological superskin, the temporarily invisible flesh that suffers hunger and that needs to go to the bathroom is still there, its demands merely deferred.

Smart Fetishism: Do Not Eat

Smart drinks and drugs are the ultimate fetishes for initiation into the cyborgian condition. Like tiny introjected phalluses (i.e. undecidably organic/electronic and thus both), they offer a kind of magical thinking, the promise of human transcendence, with the alibi of science. This imaginary ideal nourishment is a techno-food reduced to its bytesized chemical constituents like decontextualized data for intake into a brain conceptualized very much like a computer.

At present, the *smart* actually consists largely of vitamins and drugs developed for the treatment of Alzheimer's, Parkinson's, AIDS, cancer and other diseases – suggesting its deepest rationale is fear and its modus operandi is the preemptive strike. Some people who once preferred baby vegetables now supposedly eat a chemical litany of what isn't (but then again it is) really food.[34]

---

34 In terms of description of effects, Ward Dean, M.D., and John Morgenthaler's *Smart Drugs & Nutrients: How to Improve Your Memory and Increase Your Intelligence Using the Latest Discoveries in Neuroscience* (Santa Cruz, CA: B&J Publications, 1990), for instance, reports that piracetam promotes the flow of information between the right and left hemispheres and "might increase the number of cholinergic receptors in the brain. Older mice were given piracetam for two weeks and then the density of the muscarinic cholinergic receptors in their frontal cortexes was measured. The researchers found that these older mice had 30–40% higher density of these receptors than before. Piracetam, unlike many other drugs, appears to have a regenerative effect on the nervous system."(44.)

THE TECHNOLOGICAL EXTENSIONS OF THE BODY

In addition to ex-foodies, cyberpunk circles and techno-music clubs have made the smart (and psychedelic) the beverage of choice, in lieu of snacks, alcohol and coke. The odd result is a mix of the discourses of health, space exploration and junk food. No wonder smart drugs enjoy an at best quasi-legality, (but then so should barbecue potato chips.) While the psychotropics are on FDA schedule one (illegal), the nootropics or smart drugs have had a hazy legality, though FDA raids on vitamin dealers has put even largely vitamin "smart" drinks on notice. Perhaps, as with rock culture, the semi-outlaw status serves the ethos of "smartness" as a counter-discourse with a program for, if not social, then evolutionary change out of the human condition.

There are certain recurring features in the very limited literature on smart drinks and drugs in how-to-books, manifestos and ads in *Mondo 2000*: a) smart nonfood tastes bad, medicinal, in fact; b) smart drugs are better than nature, once one achieves the right "fit" between brain and chemicals; and, c) they result in better performance of mental tasks. To at least one counter-cultural theorist, insofar as they are pyschotropic, smart drugs are in fact, *Food of the Gods,* at once archaic and posthistorical tools toward the next phase of human evolution toward colonizing the stars. The following surmises part of the rationales behind these features.

Bad Tasting Medicine

Once we have entered a realm of negation and nonfood, we have left the voluptuous effects of food on the "gastronome's body" of a Brillat-Savarin, or the sense of "well-being" Roland Barthes describes as *cenesthesia* – the total sensation of the inner body or bowels – as a cultural value far behind.[35] For if food is the manna of fullness and pleasure,

---

35 Roland Barthes, *The Empire of Signs*.

■ BEING HUMAN:

nonfood is bad tasting medicine, that can – precisely because it is disgusting – be eaten with pleasure, much like the ecstatic response to devouring machines. Prophets of food-ceuticals as longevity enhancers, Durk Pearson and Sandy Shaw, described with particular relish one of their concoctions (arginine and co-factors) that not only didn't taste good, but smelled like "dog vomit."[36] Pills as electronic metaphors for firing synapses may not taste very good. Evidently, savor and taste are not the primary issue when "smartness" or health are at stake.

The underlying image may be a mixture of medicine and the future-food of aerospace and astronauts; *but* it is not the legendary unpalatability of K-rations or MRE's. Rather, ingesting this space age oriention is part of the magic of preparing for a future in which more and more demands will be made on mental performance. While or even because they are not so very good to eat, smart drugs may be literally good to think. Smart-drug discourse vacillates between medicine that is "good for you," an ascetic or virtuously masochistic negative desire to transcend the body, and a cerebral high that invokes a body image of wholeness and perfection at one with the future.

Better than Nature

Smart discourse how-to's are largely about the right dose and the proper *fit* between chemicals and the brain. Of course, "fitting" the various entities, reality statuses and modes of electronic culture together is the

---

36 "Durk and Sandy Explain It All for You," *Mondo 2000* 3, Winter 1991, 32. Like Alice, the discourse of smart drugs and drinks is all about "too much" and "too little." Terence McKenna in *Food of the Gods: The Search for the Original Tree of Knowledge, a Radical History of Plants, Drugs, and Human Evolution* (New York: Bantam, 1992 – to be distinguished from the H.G. Wells novel with the same main title,) also posits a distinction between bad drugs (which include television) and good drugs (largely hallucinogenics) as one of "fit" fostered by co-evolution of plants with shamanistic practices from archaic times.

general practical problem a global information society must solve in order to come into being; composing cyborgs is but an especially difficult instance of it. Note that the cyborgian direction of fit between organic and electronic is heavily weighted toward the latter.

Smart drug "fit" is not based on existing "natural" quantities – neurochemicals are too costly for the body to make in beneficial amounts.[37] However, according to Terence McKenna, nature has offered psychoactive drugs, which are not merely smart but which he claims have spurred human mental evolution, in abundance. McKenna views the fifteen thousand years of cultural history in between the archaic period and the present as "Paradise Lost," a dark age of ego-imbalance, to be abandoned, along with "the monkey body and tribal group," in favor of "star flight, virtual reality technologies, and a revivified shamanism."[38] Again, the archaic and the electronic are united.

Smart Performances

It is informative to consider what "smart" means in the context of this drug discourse, where "learning" is defined as "a change in neural function as a consequence of experience."[39] In the drug descriptions, smartness implies more effective neurotransmissions; however, in drug testimonials by users, "smart" is not described in terms of higher cognitive processes but as the ability to retrieve trivial or obscure information in the context of school or work. This information recall is prized largely for its exchange value or as evidence of performative ability and instrumental reasoning capacity: A secretary given a raise by her boss to buy smart drugs becomes "more alert, and intelligent-acting

---

37 Pearson and Shaw, 168.
38 "Durk and Sandy Explain It All for You," *Mondo 2000* 3, Winter 1991, 32.
39 Ward and Morgenthaler, 206.

and she smiles more. She is overall a much better employee." A student is enabled to become a math major and get a job in Silicon Valley. A graphic artist is able to work all night and present her work the next day with a smile. A father in his 40s is given Hydergine by his son, and to the son's amazement, the father recalls "family vacations, picnics and holidays" that happened in his 20s (!).[40] (Note how often "smartness" is something desired of someone else.)

Smart drugs also reportedly rejuvenate sexual performance a good twenty years.[41] Yet, often, however facetiously, descriptions of sexual and keyboard activity are mutually substitutable or are metaphorically intertwined: "It started so innocently: just a snort of vasopressin before sex, or before getting down with your keyboard..."[42] So, smart drugs may enhance cognition and sex, (or they may not) but many of the motivations the discourse implicitly suggests for taking them suggest a situation of stress and fear for the future, plus wishful thinking. My personal reading of the connotations of smart drugs extends beyond the personal Utopian quest for health and longevity to include loss of faith in our ability to survive a toxic natural and social world without medicinal help, as well as guilt and despair with the arrangement of our social-communal world. The half-secret and intensely shared present tense of a rave is a substitution for such communality. Capsules of "information" are at once a kind of sympathetic magic which allows the body to converge with computers and apotropaic magic which holds all sort of plagues now loose in the world at bay.

---

40 Testimonials from Ward and Morgenthaler, 179–184 and Morgenthaler "Update."
41 See especially Pearson and Shaw, but his theme is universal.
42 St. Jude, *Mondo 2000, 5, 38*.

THE TECHNOLOGICAL EXTENSIONS OF THE BODY

Margaret Morse

Contamination Strategies: Excremental Art and Cyborg Initiation

If nonfood is a cuisine of lack or waste, nothing or excess to be received with loathing, then rather than being taken in, nonfood may be something that is spit out. And rather than purifying the body into electronic wholeness, the negation of the organic may wreak a transformation via waste and *defilement,* in which bodies are violently implicated, torn open, their inner linings exposed to the world, allowing bodily fluids and food waste to smear the boundary between inside and outside, and between self and other.

Such deliberate violation of the surface of the body or skin affects the symbolic as well, by undermining the boundary which produces recognition, identity and meaning and by separating it from a ground of meaninglessness: namely, a deliberate evocation of *abjection.* Many historical and contemporary cultural rites throughout the world propitiate the spirits of the dead with food and libations. (Thereupon, for example in Chinese rites, humans may eat the delicious offerings to ancestors with gusto – in effect, incorporating them; other cultures prefer to leave the food and wine of the dead or of sacrifice to deity to evaporate and decay.) Considering widespread practices of making offerings to inanimate spirits, it is not so strange to subject immaterial projections or the ghosts in machines to symbolic exchange with death. What are these devouring ghosts but alienated human agency, otherwise locked out of ripeness and development in time? Implicit in the very question of what cyborgs eat is an accomodation of the human to the machine. The better question could be: how can cyborgs incarnate the human condition? that is, how can cyborgs become meat?

∎

"Your Wish Is My Command": Human Communication with Magical and Mechanical Agencies in Norbert Wiener's Cybernetics

Salvatore Guido

PORTRAIT OF A CONCERNED SCIENTIST

> *Gray.* Would you not say, Dr. Oppenheimer, that such an attitude might imply something like divided loyalty?
> *Oppenheimer.* Divided between whom?
> *Gray.* Loyalty to a government – loyalty to mankind.
> *Oppenheimer.* Let me think.... I would like to put it this way: if governments show themselves unequal to, or not sufficiently equal to, the new scientific discoveries – then the scientist is faced with these conflicting loyalties.
> — HEINAR KIPPARDT,
> *In the Matter of J. Robert Oppenheimer*, II.2

Norbert Wiener (1894–1964) was never called before anything like a Personnel Security Board for questioning. Unlike Oppenheimer, the atomic physicist and so-called father of the atomic bomb, Wiener, the father of cybernetics, had dissociated himself from the wartime Manhattan Project, concentrating instead on how to shoot German aircraft from the sky over London. But he, too, was appalled by the decision to deploy the A-Bomb against human life. The legacy of Oppenheimer's divided loyalty and his treatment at the hands of his government as a security risk informed Wiener's mature life. Throughout his career, he remained hypersensitive to the use and abuse of scientific knowledge by governments concerned only with their own survival and expansion. "We have contributed to the initiation of a new science," he wrote, referring to cybernetics, "which...embraces technical developments with great possibilities for good and for evil. We can only hand it over into the world that exists about us, and this is the world of Belsen and Hiroshima. We do not even have the choice of suppressing these new technical developments. They belong to the age, and the most any of us can do by suppression is to put the development of the subject into the hands of the most irresponsibile and most venal of our engineers."[1] At one point, Wiener considered leaving his post at MIT rather than participate in government sponsored military research. He reconsidered, however, after realizing that his refusal to participate did not prevent others from doing so. "The best we can do is to see that a large public understands the trend and the bearing of the present work... there are those who hope that the good of a better understanding of man and society which is offered by this new field of work may anticipate and outweigh the incidental contribution we are making to the concentration of power (which is always concentrated, by its very conditions of existence, in the hands of the most unscrupulous). I write in 1947, and I am compelled to say that it is a very slight hope."[2] Wiener opted to address his concerns to the scientific community as well as to take his crusade to the public sphere. If knowledge is power, he thought it best that it not be kept a government secret.

Meanwhile, in the post-war world of Big Science, research was no longer the exclusive property of the academic scientist, but was funded and administered by large government bureaucracies. There was a division between those who produced scientific knowledge and those who managed it. Accordingly, Wiener advocated for a scientific education in which an intense specialization in free research was complemented by a broad interdisciplinary and humanistic education. He hoped

---

1 Norbert Wiener, *Cybernetics or Control and Communication in the Animal and the Machine* (Cambridge, Mass.: The MIT Press, 1961 [second edition]), 28.
2 Ibid., 28–9.

■ BEING HUMAN:

thereby to encourage a vigilant scientific culture capable of evaluating the production and circulation of scientific knowledge in the light of socially determined human values. But Wiener was also aware that the cultivation of intellectual integrity was threatened by good old-fashioned human weakness. He viewed his colleagues – scientists, engineers, computer programmers – and intellectuals generally, as bearing a heavy burden of responsibility. Such people were in the position of the shaman of other cultures, part of an elite with access to power. As such, the new intellectual entrepreneur, a trafficker in the precious commodity of powerful knowledge, was susceptible to the temptation to substitute personal gain for responsibile social action. After all, it is precisely those in positions of power who are in position to abuse it. Drawing upon Christian theology, Wiener moralized the predicament of the scientist: "So long as we retain one trace of ethical discrimination," he wrote, "the use of great powers for base purposes will constitute the full moral equivalent of sorcery and Simony."[3] A distant relative of the figure of Faust, the legend of Simon, fully developed during the first four centuries and widely circulated throughout the fourteenth to sixteenth centuries, is best known for the conflict it portrays between the Apostle Peter and Simon Magus in Asia Minor. A professional magician, Simon desired glory and was not above deceiving others to achieve the end of being taken for what he was not: an exalted power. As the story has it, he asked to purchase the magical powers of the Apostles for his own benefit after converting to Christianity. In the apocryphal *Acts of Peter*, he challenges St. Peter by flying on the strength of his gods. Peter, invoking the name of God against the demons who supported Simon, sends him crashing to the ground in defeat. Posterity will always remember him for the connection of his name to what Dante, among others, classified among the greatest of sins, *simony*.

---

3 Norbert Wiener, *God & Golem, Inc., A Comment on Certain Points where Cybernetics Impinges on Religion* (Cambridge, Mass.: The MIT Press, 1964), 52.

Undoubtedly because he lived in the paranoid post-war world dominated by the logic of the balance of terror, Wiener often framed the problem of social homeostasis in terms of the destabilizing potentiality represented by the rigid ideological machinations of hostile governmentalities. "Permanent homeostasis of society cannot be made on a rigid assumption of a complete permanence of Marxianism, nor can it be made on a similar assumption concerning a standardized concept of free enterprise and the profit motive. It is not the form of rigidity that is particularly deadly so much as rigidity itself, whatever the form."[4] Rigidity is not homeostasis. In fact, rigidity of any kind spelled the death of homeostasis.

As a mathematician, Wiener could certainly appreciate the elegance of von Neumann's theory of games, but he could not deny that the model described "a welter of betrayal, turncoatism and deception, which is only too true a picture of the higher business life, or the closely related lives of politics, diplomacy and war... There is no homeostasis whatever."[5] The ontology of the other portrayed by the game theoretic model, especially in the version that resulted from its appropriation by military strategists, was typical, according to Wiener, of power hungry organizations like governments and their armies, determined to defeat the other side at any cost. In such a scenario, each side defines the other as one who will stop at nothing in its effort to win. The enemy will not hesitate to deceive, so *we* must be prepared "to prevent the other side from winning on the basis of a certainty that we will not bluff."[6] Wiener refused to believe that he must be complicit with such a game of deception in the quest for scientifically valid knowledge or in matters of ethics. He followed Augustine's demonology by calling this manner of demonizing the other the Manichaean demon. He found this demon

---

4   *Ibid.*, 83; emphasis added.
5   Wiener, *op. cit.* (1961), 159.
6   Norbert Wiener, *The Human Use of Human Beings: Cybernetics and Society,* Da Capo series in science (New York: Plenum Publishing Corporation, 1954 [second edition]), 35.

■ BEING HUMAN:

palpably incarnated as the enemy other in the military appropriation of the game-theoretic model of rationality as well as in all forms of ideological dogmatism generally, whether political or religious. By contrast, the Augustinian demon represented the kind of disorder associated with the quest for knowledge of nature and was characteristic of open systems like those of living phenomena generally. As a matter of fact, the body of Wiener's work is marked throughout by a fundamental division of the demons confronting human existence into two kinds, according to the strategy required to deal with each. "I have already pointed out," he writes in one of his many references to demonology, "that the devil whom the scientist is fighting is the devil of confusion, not of willful malice. The view that nature reveals an entropic tendency is Augustinian, not Manichaean. Its inability to undertake an aggressive policy, deliberately to defeat the scientist, means that its evil doing is the result of a weakness in his nature rather than of a specifically evil power that it may have...."[7] Such is the demonology that animates Wiener's vision of Cybernetics. On the one hand, there is the Manichaean demon that is determined to win and will resort to disimulation to do so. It would not hesitate to bluff or deceive in order to outwit its opponent. On the other hand, the Augustinian demon, characterized by the "evil" of chance and disorder, is incapable of willfull deception. The Manichaean or Gnostic demon might be compared to antagonistic forces at war, while the Augustinian demon is best understood in terms of the forces of nature.

Thanks primarily to his analysis of the predicament of the physicist in pursuit of scientific knowledge, Wiener formulated his Augustinian alternative to the Manichaean demon. He recognized that in the game with nature, the quest for scientifically reliable knowledge does not depend for its success on a strategic game aimed at overcoming an active resistance offered by nature. Science is not at war with nature.

---

7 *Ibid.*, 190.

THE TECHNOLOGICAL EXTENSIONS OF THE BODY

Resistance to the knowledge of the workings of nature is not an active force emanating from nature that seeks to remain undetected. It is only the passive resistance and "evil" of disorder that must be transformed into order. More to the point, knowledge of the working order of nature, once gained, is secure, and the quest for further knowledge can safely be built upon it without fear that nature might secretly change the laws under which it has been revealed to operate. Nature does not deceive or surreptitiously change its mind: we can count on the eventuality that it will not secretly change the rules of the game. It is only our knowledge that is limited. In the quest for scientific knowledge, the resistance to be overcome is on the side of the *scientist*. From this perspective, the demonic is not taken to be an independent, evil force as Gnostic dualism conceived of it. The Augustinian demons do not threaten the integrity of being human from without. Incomplete expressions of unfulfilled desire, these hungry ghosts haunt our very thoughts and strive to influence the will. Unchecked, they may even take possession of the soul.

MECHANICAL SLAVES?

> "The dream of scientific technics promises a boon that was long thought impossible: inexhaustible riches combined with a liberation from toil. Ellul compares this dream to the legend of Faust and concludes that in his contract with la technique, man did not read the fine print...."[8]

In many respects, Wiener was under the influence of nineteenth century debates regarding the social impact of the machine. Ubiquitous

---

8 Langdon Winner, *Autonomous Technology: Technics-out-of-Control as a Theme in Political Thought* (Cambridge, Mass. and London, England: The MIT Press, 1977), 187.

■ BEING HUMAN:

references to Samuel Butler's *Erewhon* throughout Wiener's work testify to his concerns regarding the intrusion of machine technology into a society that is not yet ready to defend itself against its destabilizing potential. The difference is that whereas the first industrial revolution witnessed "the devaluation of the human arm by the competition of machinery... The modern industrial revolution is similarly bound to devalue the human brain at least in its simpler and more routine decisions."[9] No longer limited to processing routines that merely perform pre-established functions, the learning machines that interested Wiener entailed the performance of programs designed to directly assist the human thinking process. Their high-speed computational ability, coupled with their capacity to store information, endowed these mechanical agencies with memory and learning capabilities, allowing for the application of mechanism to higher human cognitive faculties. But while some researchers in artificial intelligence make ambitious claims to the effect that human cognitive abilities are quite literally a species of computer-like operation and that the human brain and the computer are sub-species of a single information-processing system, Wiener cautioned that such a view encourages the confusion of human judgement with computer power, thereby further exasperating the all-too-human wish to divest ourselves of the burden of making difficult decisions and judgements.

In the book published in the year of his death, *God and Golem, Inc., A Comment on Certain Points where Cybernetics Impinges on Religion*, Wiener returned to his abiding interest in a theory of technological politics. "What we now need," he wrote, "is an independent study of systems involving both human and mechanical elements. This system should not be prejudiced either by a mechanical or antimechanical bias. I think that such a study is already under way."[10]

---

9 Wiener, *op. cit.* (1961), 27.
10 Wiener, *op. cit.* (1964), 73.

While Wiener explored, throughout his career, fundamental issues regarding the relationship of humans to machines, he never lost sight of the fundamental thesis that "the performance of a non-human link in human relations can only be evaluated in human terms."[11] Accordingly, his fundamental contribution to human/machine communication consists in his rejection of that technicism which reduces the machine to nothing more than a kind of tool or device that slavishly executes orders. He remained steadfastly critical of the idea that the problem boils down to merely a matter of programming the machine to do what we want it to do. The assumption that such machines can be reduced to slaves simply because we create them and program them is a natural enough assumption, but one that inevitably leads to a dangerous complacency. Wiener was particularly concerned that with the introduction of information processing machines capable of learning, the illusion that the machines we build and program are willing and docile partners may be too tempting. Consequently, we may rely upon the success of technical expertise in place of inimitable human judgement.

The brave new post-war world was accompanied by that new hybrid discipline for which the name cyberscience was coined.[12] Cyberscience brought together the statistical mechanics of classical physics with thermodynamics and communication theory. Whereas the scientific revolution of the seventeenth century had conflated the terrestrial and celestial worlds, the cybernetic revolution collapsed the traditional boundaries separating the domain of the living from that of the non-living. Wiener accepted Schrodinger's discovery that "Life seems to be orderly and lawful behavior of matter, not based exclu-

---

11 P. Masani, Norbert Wiener. *Collected Works with Commentaries,* vol. IV (Cambridge, Mass. and London, England: The MIT Press, 1985), 671.
12 Evelyn Fox Keller, *Refiguring Life: Metaphors of Twentieth-Century Biology* (New York: Columbia University Press, 1995), 85.

■ BEING HUMAN:

sively on its tendency to go over from order to disorder, but based partly on existing order that is kept up."[13] "Entropy and information are evaluated in the same way. One is the negative of the other."[14] It was Claude Shannon, of course, who fathered modern communication theory by providing a scientific conceptualization of information, transforming it into a physical parameter capable of quantification. In the process, information was divorced from meaning. Henceforth, "meaning" referred to the actual content included in a particular message, while "information" came to designate the number of possible messages from which a particular message to be sent was chosen. This new quantitative concept of information, understood as a property of the physical universe, provided the common meeting point of diverse homeostatic systems.[15] Wiener set out to demonstrate that the cybernetical systems he proposed for investigation reproduced the entropy-decreasing capabilities referred to by Schrodinger, without violating the second law of thermodynamics. He analyzed three such systems – mechanical, biological and social – analogous insofar as each can be understood in terms of self-regulation. The question posed in each case was how homeostasis prevails in the face of entropic degradation. The innovation of Wiener's agenda consists in his comprehensive conceptualization of communication and control in order to create an interdisciplinary theory that would encompass such systems in terms of their capability for error and their utilization of error correction, by means of negative feedback, for successful operation or survival.

---

13 Erwin Schrodinger, *What Is Life?*, (Cambridge, England: Cambridge University Press, 1992), 68.
14 Francois Jacob, *The Logic of Life: A History of Heredity* (New York: Vintage Books, 1973), 251.
15 See William Aspray, "The Scientific Conceptualization of Information: A Survey," *Annals of the History of Computing*, VII.2 (April 1985).

MESSAGE IN A BOTTLE

> "...the agencies of magic are literal-minded...if we ask for a boon from them, we must ask for what we really want and not for what we think we want. The new and real agencies of the learning machine are also literal-minded. If we program a machine for winning a war, we must think well what we mean by winning.... We cannot expect the machine to follow us in those prejudices and emotional compromises by which we enable ourselves to call destruction by the name of victory. If we ask for victory and do not know what we mean by it, we shall find the ghost knocking at the door."[16]

Due perhaps to the influence of his philologist father, Wiener's thinking was colored as much by discursive sensitivity as mathematical rigor. In fact, it is in his more popular writings, in connection with his subtle appreciation of the gray areas surrounding ordinary language usage, that his ethical impulse is especially evident. His ubiquitous references to literary works and fictional beings are not peripheral to his central concerns: On the contrary, by harkening back to magical and theological contexts surrounding genii, the legend of the golem and other artificial beings, Wiener was able to highlight the ethical dimension of the domain of human communication. By introducing the richness of literary language into his investigations of the complex interrelations of science, technology and society, he problematized, in dramatic fashion, the communicative relationship between human beings and non-human agencies in our modern information era. Throughout his work, he crafts literary allusions which entail the moral that, in relation to both magical agencies and mechanical learning machines "we must ask for what we really want and not for what we think we want."

---

16 Wiener, *op. cit.* (1961), 177.

■   BEING HUMAN:

"Your Wish Is My Command"

What could be more tempting or seductive than the phantasy of fullfilment promised by the genie when it utters "Your wish is my command"? Such a solicitation is almost too good to be true. It conveys the promise of the transformation of the addressee's wish into the addressor's command. It is as though the genie said: "As soon as I hear what I take to be a wish, I promise to grant it." But this may not be as good a thing as it sounds at first. As a matter of fact, in the case of three wishes, it happens again and again that their satisfaction is fraught with a high degree of dissatisfaction. The moral refrain is well known: when you are granted three wishes, you had better be very careful about what you wish for.

Wiener indefatigably reminds his readers that the magical agencies that inhabit his favorite tales are literal-minded, and that our faster, more powerful, higher-functioning machines are no exception. In what is probably the most horrific tale he recounts, the fable of *The Monkey's Paw* written by W.W. Jacobs at the turn of the century, a retired workingclass man and his wife are entertaining a friend who shows them an amulet in the form of a dried, shriveled-up monkey's paw. The friend tells them that the paw possesses the power to grant three wishes to each of three people. It was so endowed by a holy man from India for the purpose of demonstrating the folly of trying to defy fate. He explains that he is the second owner of the charm, but refuses to speak of his own terrible experience. He adds that he knows nothing of the first two wishes of the first owner. With this, he casts the paw into the fire. But as soon as the paw lands on the fiery hearth, his host retrieves it and asks to make a wish. Shortly after he wishes for 200 pounds sterling, a knock is heard at the door. An official from the factory where his son was employed is there to deliver the news of a freakish accident and the boy's sudden death. The messenger also makes it known that the factory, while not recognizing any responsibility or legal obligation whatsoever, will pay the father 200 pounds sterling as a solatium. Grief-stricken, the father makes his second wish – that his son may return. When another knock at the door is answered, what appears to be the ghost of his son enters the room. His third wish is that the apparition go away.

The link of a wish to a rational project is fundamentally problematic because wishes are *wishes* and their fulfillment is best left to the world of fiction. But then, it is precisely in the enchanted world of fairy tales that wishes are granted – and at what expense! The consequences, as we know, are usually disastrous. Under the circumstances, making a wish must not be undertaken lightly. It requires rigorous formulation after careful consideration. The monkey's paw, like its mechanical counterparts, embodies that unstable combination of obedience (doing what it is told to do – granting the wish) with an "artificial" intelligence that mysteriously empowers it to accomplish the tasks at hand. Such intelligences are neither endowed with a rational soul nor a will of their own. They are simply literal-minded agencies. *According to Wiener, it is precisely this that renders them unpredictable and in possession of what appears to be a peculiar kind of "autonomy." In other words, such agencies resist our complete control even as they obey our commands.*

The learning machines of Wiener's day were less sophisticated than a chess-playing champion like *Deep Blue*, but the fundamental issues surrounding artificial intelligence were already apparent. Such machines 'learn' to reprogram themselves as they acquire more 'experience' in their quest to accomplish the end or ends specified by human operators. They are not confined to following a given program, but can also set the program by more general principles. In other words, we may not fully know either the details of the particular program being carried out or its consequences. The machines Wiener was familiar with already demonstrated a propensity to do things that neither their builders nor programmers had anticipated. We say that it is the programmer who determines what the machine may subsequently come to learn. "But the fundamental point is not that the program may perform tasks which may far outstrip the programmer's 'explicit' intentions, but rather, that the programmer may be far from aware of the consequences of the course which he has programmed."[17]

The business of programming, like that of wishing, is evidently no straightforward affair. In the case of programming, the programmer may not really know what the machine has really been programmed to do. In our fairy tales, the promise that the wish will be granted may be taken in good faith and, indeed, the promise is usually made good. The problem is at least twofold: on the one hand, the mechanism by which the wish is realized is not transparent to rational comprehension, thereby escaping the control of technical expertise and, on the other, we usually get more than what we bargained for.

If the fictional accounts that are so dear to Wiener tell us anything, they call attention to the gap that separates the complexity and specificity of human speech and language from the literal-mindedness of both fictional magical beings and modern mechanical agencies. The very literal-mindedness of the agencies designed to satisfy our wishes throws into relief the phantasy dimension intrinsic to human discourse. Specifically, this phantasy component of the human symbolic function is sharply revealed by the impossibility of unconditional wish-fulfillment. The wish can never be fulfilled as such because it is always a wish for unconditional fulfillment. What this failure demonstrates, however, is not simply that we cannot reduce natural language to a matter of logical syntax for purposes of machine translation, but that the imaginary component within the symbolic function of human discourse is a necessary and irreducible dimension of it.

It is my contention that Wiener's commitment to the analysis of error in the domain of cybernetic systems ran parallel to his appreciation of the complexity of human speech. While the statistical theory of in-

---

17 S.G. Shanker, "The Decline and Fall of the Mechanist Metaphor," in Born, Rainer, ed., *Artificial Intelligence: The Case Against* (New York: St. Martins Press, 1987), 100–1.

Salvatore Guido

formation concerns itself with syntactical errors associated with code-like machine languages, communication engineering in the broad sense envisioned by Wiener demonstrates that human discourse harbors within itself the very possibility of total misunderstanding. As the *Monkey's Paw* demonstrates, when otherwise impermissible wishes are linked to rational projects, they share a secret complicity that all too often ends in disaster. By demonstrating that the human symbolic function is animated by an imaginary register, Wiener's comprehensive vision of cybernetics reveals the seductive core of the illusion of technique. Thus he throws a monkey wrench into those phantasies of downloading our ethical responsibility into unimaginative automata.

"Ain't science a wonderful idea?"

An Advocation for Immortality

Jim Yount

PART I – IMMORTALITY: A WORTHWHILE GOAL

Introduction to Part I

In his paper entitled "A Crisis in Scientific Morale," for the George Washington University Symposium: "Science in Crisis at the Millennium," Dr. Robert Pollack expressed the concern that the scientist can not accept mortality. While biomedical science claims its intention is to "cure us of nature's defects," it acts as if its real purpose is to acquire a mythical immortality "at whatever cost to its players or anyone else."

The topic: "Are scientists animated by a dream of immortality?" was a discussion item for Extensions of the Body panel during the *Being Human* Symposium.

This potential for physical immortality, through modern scientific advance, has only become apparent to most people in the last forty years. However, it is likely to become a force driving medical and biomedical research for at least the next forty, perhaps the next hundred years.

Dr. Pollack is correct in his assertion that the attainment of human immortality is, in reality, the ultimate goal of some researchers. Even for those who would shy away from seeking immortality, their research, to extend human life and to eliminate human affliction, is an unwitting accomplishment to advance the agenda of those who want to live forever.

The Wonderful One-Hoss Shay

In our panel discussion, it became clear it wasn't simply that Dr. Pollack thought that we should discuss the topic; he, together with Dr. Dorothy Warburton, expressed definite opinions on the subject.

> Have you heard of the wonderful one-hoss shay,
> That was built in such a logical way,
> It ran a hundred years to the day?

In this poem, by Oliver Wendell Holmes, the shay (a small horse-drawn carriage) had been constructed so well that all during its 100 years existence it was just as good as the day it was first made. Then, it all fell apart at once and crumbled to dust!

> The poor old chaise in a heap or mound,
> As if it has been to the mill and ground!...

We all know that the parts in all of our modern gadgets wear out or deteriorate over time. There is always some part or another that is wearing out or breaking down, and needs repair or to be replaced. Wouldn't it be wonderful if our appliances were so well engineered that they would last a good long time without repair, then all the parts wear out at once?

Dr. Pollack and Dr. Warburton would like to live in a world where our bodies are built like the wonderful one-hoss shay. The goal, in the minds of people who take this tack, is for modern medicine to simply extend the time that we enjoy "quality of life." If we could all live for 100 years in good health, then die peacefully, it would be wonderful. That should be the goal of medicine. We should stop there. It is simply wrong for scientists to try to find ways to extend life beyond its natural span, or at least so argue those who take this tack.

The Immortalist Philosophy

The Cryonics movement is perhaps the most "radical" element of the larger Immortalist movement. The position I took on the *Being Human* Body Panel was consistent with the views held by my fellow immortalists: human immortality is a good and proper goal and should be vigorously pursued. I was surprised to discover that I alone, among all of the panel members, and practically all members of the audience who participated in the discussion, held this view.

Attaining Immortality Is the Logical Extension of Modern Medicine

While it was never enunciated by my colleagues on the panel, it surely must have occurred to them how "backwards" and even surrealistic this discussion would seem to an outside observer. Here were medical doctors and researchers, along with other learned professionals, arguing passionately that the practice and ethics of saving and preserving human life which they had aspired to all of their lives should be shunted aside.

After all, doesn't a doctor have a duty to treat an otherwise healthy and mentally alert 100 year old? Doesn't the doctor have a duty to save her life, as she does if the patient is 80 or 50 or 20?

What if the 100 year old patient could not be distinguished from the 20 year old? If both the body of the 100 year old and that of the 20 year old were physically similar? Both patients look young and healthy. Would these facts change the attitude of a doctor on treatment? Could the doctor justify withholding treatment from the 100 year old?

If a researcher discovers a cure for Alzheimer's disease where the beneficiaries of the cure are ONLY the very old, can the researcher morally withhold the cure?

What if the result of curing Alzheimer's is to advance the health of ONLY very old people, so instead of most people dying at age 100, many now live to 110 or 120?

What if a drug were discovered, whereby taking just a pill a day people would never age and never get sick? If a pill to prevent Alzheimer's is a good thing, isn't a pill to cure ALL of our illnesses, INCLUDING AGING, even better?

If doctors and researchers argue that such a pill is bad and they should not give it to mankind, isn't the burden ON THEM to prove this seemingly preposterous proposition?

THE TECHNOLOGICAL EXTENSIONS OF THE BODY ■

Jim Yount

The Cloning "Debate"

About a month ago I got a call from a woman in some distress. After years attempting to get pregnant, she and her husband were blessed with fertility. Unfortunately, her child was stillborn. She was at a local hospital known for its good medical practice and research, so she had the doctors preserve some cell (tissue) samples from the child in the hopes of future cloning. Her reason for calling me was to attempt to arrange for long-term cryogenic storage of the cell samples.

There is a very real possibility that within a very few years, the DNA samples which were taken from the stillborn child can be used to produce a new (cloned) child. Besides the emotional need to save her child, given her fertility problems, cloning procedure may be the ONLY WAY this woman can ever again have a baby.

Being Healthy and Having "Young" Bodies

People who argue against the quest for immortality, sometimes paint a picture of a population of old and decrepit people toddling about on canes, and walkers, taxing our medical facilities, and sponging off the few young people unfortunate enough to be born into such a society. That is NOT the "immortality" we are discussing. No one would want to live in such a society.

If we are able to crack the "aging code," or to achieve the ability to control our biology on the cellular level with nano-machines or engineered viruses, no one need ever get "old" again. Additionally, such discoveries would enable us to reverse the affects of aging, so people who now have bodies which appeared "old" could have all those physiological changes reversed. In other words, those who now look old, would then look (and feel) young again.

We would NOT have "old people" getting about as best they can for hundreds of years of decrepit life. Instead, we would have a world where each individual could choose the BIOLOGICAL (physiological) age which he or she desired, and where all diseases as we know them, are maladies of the past.

Having Radically Improved Bodies

The tools of science which make it possible for us to live forever (or at least for hundreds of years) will also allow the radical improvement of our bodies. If we are biological machines, and we discover how to make and remake these machines, then we may fashion them as we see fit.

It is likely that most of these improvements will be possible by the time we achieve immortality. After all, we must understand, and be able to change, a lot of our "natural" biology in order to stop the aging clock.

However, if we can't "run like a cheetah," after immortality is achieved, just wait a few hundred more years! We will have lots of time to work out other improvements.

Traits Borrowed from Other Animals

Is there any reason to believe that we will be limited to just the biological material of mankind? Of course not; we can already combine with our own, genes which we have borrowed from plants and animals.

Would you like to swim under water for twenty minutes? You've got it. The seal can manage such endurance, so it's not impossible; we simply borrow from the seal, and adapt. While we are at it, let's acquire the seal's tolerance of cold temperatures. Hibernate? Especially during election years? Consider it a done deal. Have the sexual endurance of a mink? It is yours. Though you will likely not conduct your affairs with half as much grace!

Virtual Life

Then, of course, you may choose to continue life in the computer. If we are simply information, and uploading techniques are developed, then a whole host of problems presented by the prospect of our living in the "real" world can be eliminated.

Virtual worlds can be created which so mimic the planet earth as we know it, that it is likely those living in such a virtual reality could not tell it from the real planet. A variety of other "planets" could also be created as a home for our information selves.

There is no longer the problem of too many people, or of too few resources. Perhaps a hundred million "people" live in a computer with the power consumption of a 100 watt light bulb!

Two Mirrors Face Each Other: Reflection of Reflection... Etc.

How do we know that THIS reality is not simply a computer simulation? If we agree that there will be hundreds of thousands, perhaps millions of such virtual worlds created someday, are the odds that THIS is the real thing, one in a million? If so, then the discussion is academic. We ARE immortal, and you are reading these words in a virtual world.

Laundry List of the "Woes" Caused by Immortality

1. Too many people. We don't have room for them on the planet. The quality of life for everyone goes down if there are too many people.

Over time, the quality of life has improved, even with population increases. The very technology which allows populations to rise, has given us a healthier, more prosperous life. That doesn't mean that this trend will go on forever, however.

Even if the immortality pill were invented tomorrow, many people would not take it for religious, philosophical or "traditionalist" reasons.

There are places to live on earth not now utilized, but which would be made livable and even quite nice through technological advances. These include the world's deserts, Antarctica, and the ocean floor.

Some may choose to live in outer space. Proposals for such off-world habitation range from the artificial planets proposed by physicists Gerald

O'Neill (which could be started now) to Freeman Dyson's proposal to break apart the outer planets and use this material to enclose the sun in a giant sphere (Nah, it'll never happen. Too much opposition from environmental groups!)

Life in virtual reality cyberspace presents the intriguing possibilities we discussed. Even given these possibilities for finding more elbow room for mankind, (should immortality rear its ugly head), we are stuck with the bottom line: a lot of us will do it (take that pill) even if our refusal to die DOES present problems in space utilization for the rest of you.

The time may have come (immortality or not) when our old habit of breeding ourselves silly needs to be changed.

2. It is against the will of God and nature for people to live indefinitely. God made mankind with a limited life span. We are thwarting His will if people live forever.

An oft quoted line among Immortalists is: "If nature intends that we should die, then TO HELL with nature." Tooth decay, tuberculosis, and typhoid fever (just sticking with the "T's") are a few of the "natural" phenomena we have made great strides to eliminate. Nature is often cruel, or at best indifferent. We now have the ability to control nature for the better. Let's use it.

The "will of God" is even a slipperier concept than "against nature." Few of us claim to know the will of God on such matters, or presume to speak for him, and fundamentally distrust those among us who tell us that they possess such divine knowledge.

3. If people don't die they can't benefit by going to heaven, or by being reincarnated to work out their karma.

True. It is hard to argue against such logic. It should be pointed out, however, that even if one lives 100,000 years, he will still someday experience whatever happens at death. If not, then he will be alive to witness Armageddon and judgment day.

THE TECHNOLOGICAL EXTENSIONS OF THE BODY

4. Old people need to move aside and make room for the young. If no one dies, then people can't have children.

That is the way it has been, but not necessarily the way it will always be or SHOULD always be. It may just be that people who expect to live indefinitely will choose to give up having children, or move to a space colony.

5. New ideas, inventions, and explorations are the product of young minds. Old people are too set in their ways. The advancement of civilization requires innovation which only comes about when young people dare to try new things, or advance new ideas.

It is true that new ideas and inventions often come from the young. We don't now know WHY this is true. However, it is likely we can discover the reason that young people are more inventive. It may simply have to do with physiology: young people are healthier, and have higher levels of various hormones and chemical stimulants in their blood. These physiological stimulators could be reproduced in humans who are older, so that this inventiveness need have little to do with age.

6. If people have very long life spans, we will become a more materialistic society. Facing the prospect of death forces us to get in touch with our spiritual side. We are a moral society because we know we will have to face God someday and account for (how we have lived) our lives.

This author, along with philosopher Dr. Charles Tandy, and other immortalist writers, has often written of and discussed the morality which follows from an expectation of immortality. This morality is as compelling, rigorous, and motivating as religious or secular based morality.

If one expects to live forever, here in this world, then one is very interested in preserving the wonders of our earth, not for future generations, or because religious leaders say that it is the thing to do, but FOR OURSELVES to enjoy in a future we expect to own.

■ BEING HUMAN:

An Advocation for Immortality

During the Reagan Administration, U.S. Interior Secretary James Watt established a policy of unfettered use of U.S. resources with no regard for environmental impact. Coal miners could strip-mine without restoring the soil. Factories could spew their toxins into the atmosphere. Mr. Watt, you see, is a born again Christian. He believes that the end of the world is at hand, so it doesn't matter how much we despoil the earth. We are just here for a few more years anyway, after which Christ will come and take us all to heaven!

Fortunately, most people, even other born-again Christians, don't have such a wanton attitude. Nevertheless, the spendthrift consumer society of waste embodied in the second half of the twentieth century may be a result of the "short-timer" attitude which can hardly be called moral.

"Eat, drink, and be merry," say the short-timers, "for tomorrow we die." Litter and pollute, burn down the rain forests, don't worry about endangered species. We are here but for a short time, then go to a better place.

"Eat, drink and be merry, but plan and work for tomorrow," say the Immortalists, "for tomorrow we LIVE." The immortalist is very interested in protecting, preserving, and making THIS earth a better place. He expects to be here a good long time.

Ultimately, the question of "should we allow mankind to become immortal?" will not be answered by us. Public opinion and public policy can slow, but not stop, the progress of science. The immortality pill will be available at "a store near you" soon, no matter what *Being Human* panelists, or even an outraged public, say or do.

The idea of human immortality is quite compelling. Our religious institutions are largely founded on the promise of human immortality. When it is science, not religion, which presents the assurance of immortality, we may grumble and complain, but most of us will order a supply of the immortality pills for ourselves and our loved-ones.

Jim Yount

PART II – CAN IMMORTALITY BE ACHIEVED
THROUGH CRYONIC SUSPENSION?

Let's Tell That Hermit

I suppose there just may be somebody out there who hasn't actually heard of cryonic suspension. There has to be a mountain somewhere, where lives a hermit, with no TV, no newspaper, who doesn't talk to her neighbors, who would be shocked or amazed to learn that the American Cryonics Society has been freezing people since she first turned her back on civilization, way back in 1974!

On the other hand, there are a lot of folks who have heard about "freezing people" but don't know the word. Some also, who know of the practice, but think of it as "cryogenics."

Cryonics and Cryogenics

Those who refer to the people freezing business as cryogenics aren't too far wrong. Cryogenics is a more general term that applies to the science or study of low temperatures, and how various materials behave when the heat goes away. The companies in the yellow pages under the listing category of "cryogenics" or "cryogenic research and development" are largely engineering firms or manufactures that sell liquid nitrogen, supply tubing and pumps to move the cold liquids around, and fabricate the "dewars" or insulated tanks that hold such low-temp fluids.

Cryonics, on the other hand, has to do particularly with the people freezing-business: freezing humans at (legal) death in the hope that these folks can be reanimated or restored sometime in the future. Let's not dignify the sentence above by calling it a "definition, which we will hammer out later." Maybe "working definition" is the term to use. Of course this raises a snowdrift of questions. When a member of the American Cryonics Society has his dog frozen (only after the dog dies, gang!), has the pet been put into cryonic suspension? Or does the term only apply to humans? No comment, for now.

The Possible and the Impossible

Sometimes when I explain to acquaintances the ideas behind the business I am in, the response is: "But, bringing people back is IMPOSSIBLE, when you are dead, you are dead!"

"Well, maybe," I sometimes reply, "but on the other hand, the folks we want to bring back may be just a little bit dead."

That answer hardly satisfies. Nor should it. But before discussing further the various degrees of "deadness," we need to talk about that more fundamental questions: is cryonic suspension (with ultimate reanimation) impossible? If it truly is impossible, then we need to put our minds to something more practicable, like mental teleportation or levitation, and leave freezing folks to the winter weather in the Arctic.

Dr. Dean's "Impossible" Illustration

A few years ago, members of the American Cryonics Society appeared on the Dr. Dean Edell Show. While the questions by Dr. Dean showed interest, he apparently hadn't read even that first paperback, and remained unconvinced. "But is what you are talking about really possible?" asked Dr. Dean. "Just because athletes are jumping higher and higher each year doesn't mean that someday they will jump to the moon!"

Jumping to the Moon

We get the picture, Dr. Dean. The muscle structure of any biological being is such that it can only do so much, and jumping to the moon is not amongst that "much." Of course, we have (and will have again) gone to the moon aided by artificial devices called rocket ships, but you have no disagreement from us that high-jumping to the moon is impossible.

Jim Yount

Jumping into the Freezer

Of course, even Dr. Dean would not argue that it is possible for us to jump into the freezer. The question is simply, in the present time, can we be reasonably certain that if we position ourselves next to the ice cream and wait a few hundred years, we will never be brought back? Is cryonic suspension like jumping to the moon? Is it impossible?

Not Possible Now

What would convince all our critics, of course, is a demonstration that cryonic suspension works. We freeze someone, keep them in the ice-chest overnight, thaw them, and then have the subject count from 100 backwards.

We can't do that. We can freeze some organs, thaw them, then transplant them back to the subject, with success. We can freeze most tissue (cells) in the laboratory, then culture them again. They grow just fine. Some animals, like the Arctic beetle, can be frozen to at least dry-ice temperature and then thawed to live again. There is even some success reported in frozen hamsters coming back to the world of the living (some would argue that although they were frozen, they never really left the world of the living). However, no human or other mammal can be frozen all the way down to the temperature of liquid nitrogen (-196 degrees C.) and be "brought back."[1]

---

1 (The frozen hamsters were probably chilled to less that -79 degrees C., perhaps just -20 degrees C. or so. Some of our scientists speculate that at these higher temperatures only the "extra cellular" water is frozen. That is, the water outside the cells has turned to ice. The water on the inside of the cells, because it contains quantities of salt, would require lower temperatures to freeze solid. If the intercellular material is still liquid, then some of the damage we could expect from the formation of ice crystals would not occur. On the other hand, the reported hamster freezing was overnight with the hamster packed in dry ice. This sustained cold temperature should have frozen both intercellular and extra cellular water. (ACS Journal report)

Saving Ourselves for Ourselves

Dr. John Bear, one of the early, if not founding, members of the American Cryonics Society once wrote an advertisment for a cryonics company (Trans Time, Inc.) which has the distinction of being rejected by the Editorial Board of the *Wall Street Journal*. It appeared in a local newspaper instead (a slightly humbling experience).

Dr. Bear's advertisement pointed out that you would not throw out a valuable watch, just because it was broken and you could not find anyone with the skill to fix it. You would put the watch away, perhaps in a little-used drawer, in the hopes that you would someday be able to find a craftsworker with the skill to repair it.

Even if we cannot restore a frozen mammal to life, that does not mean we will never be able to do it. Cryonics is taking the uncertain bet against the certain (physical) annihilation. Yet are we fooling ourselves with the "ain't science a wonderful idea?" Does the little bit of present success in freezing biological systems by cryobiologists mean that it MAY be possible to repair and restore the folks in the American Cryonics Society's cold tanks?

The Pretty Bad Condition Our Patients Are in

When people sign up for the suspension program of the American Cryonics Society, we want them to know all the bad things that can and do happen to them on their way to the next century. We ask them to acknowledge that long list of causes of damage to the frozen human body be signing a "consent for cryonic suspension" form which lists these sources of damage.

Just a paragraph will suffice for you to get the picture:

9. I understand and accept that the dying process and the process of cryopreservation will result in damage to my body on the molecular, cellular, tissue, and organ levels that is currently

considered irreversible. I understand and acknowledge that the damage experienced with the existing cryopreservation techniques currently employed by my cryopreservation service provider (as such damage is currently understood) includes but is probably not limited to the following:

9.1 Ischemic Injury. Currently, cryopreservation procedures cannot begin until after the patient has been pronounced legally dead by a qualified physician. In practice this means that the patient will frequently (although not always) experience an ante mortem period of deep shock (inadequate blood flow: ischemia) which will be injurious to most body organs, and especially the brain.
(This document goes on listing possible problems for 8 pages)

We Can't Jump to the Moon Now

There is no doubt that presently, and for the immediate future, we are not going to be jumping to the moon, or bringing people back from cryonic suspension, or freezing mammals to liquid nitrogen temperature and thawing them with recovery. No argument so far, Dr. Dean. We can't jump to the moon now.

What Would Make Something Possible?

Is cryonic suspension, then, just like jumping to the moon? Is it something that not only is not happening, but can't happen because it is impossible? For something (an event) to be impossible and for us to recognize it as impossible, it has to have characteristics that we agree would keep it (the event) from happening. We could calculate the material strength needed for our leg-bones and the spring in our legs needed to jump to the moon and quickly conclude that the intended leap is impossible.

■ BEING HUMAN:

An Advocation for Immortality

Does Cryonic Suspension Violate the Laws of Physics?

The laws of physics, as we know them, prohibit faster than light travel. If we hi-jumped to the sun, rather than the moon, we could not complete the leap in less than about 8 minutes (at the speed of light). Is there anything about cryonic suspension that violates the basic laws of physics? Unless we were somehow dependent upon time-travel for our procedures, it is hard to see where the laws that govern the universe are broken by our efforts. (No, we don't expect to invent time machines. If we had such a device, we wouldn't need cryonic suspension!) There is a little caveat on this question, having to do with the way we store memory that we will consider shortly. My promise, friends, no intention to duck the question!

Does Cryonic Suspension Violate the Laws of Biology?

I am taking a bit of poetic license in this topic's title. We usually do not think of biology as having laws. Yet in a way the science of biology does have such rules. The human body is a marvel of engineering and science that depends upon the laws of physics and chemical reactivity to exist at all. If future science was to completely recreate the human body (talk about "playing God"!) it could do so without violating any laws of science. That doesn't mean that we will become so sophisticated (another discussion), but if nature does it, it is not impossible.

Does Cryonic Suspension Violate the Laws of God?

This is really a theological question that goes beyond the scope of this discussion. We sometimes get into theological never-never land with the use of the words "dead" and "death." Is cryonic suspension the freezing of the "dead" in the hopes of future revival? If cryonics succeeds, some people would say that the patient was never really dead. He just had the appearance of death. When he is really dead, he cannot be reanimated and his soul or other form of extra-physical existence is dealt with in whatever way God handles such matters. If this is true, we need not be much concerned that God will perceive us as thumbing our noses at him and trespassing on his works. The creator of heaven and earth cannot be much concerned with any such feeble attempts as we humans might make to trespass. If we succeed, it must be a part of his plan. If God decides we should fail, then there is nothing we can do against such powerful opposition.

A related question might be whether or not we would be sinning if we reanimated people. Again this is a matter of religion, and religious interpretation. We can answer it in a practical way, however. Just as with most any other technique or tool of science, cryonics can be misused. If a diabolical movie producer 100 years from now reanimated 20th century patients to make the 22nd century version of *Jurassic Park,* of course that would be wrong.

Cryonic Suspension Not Impossible

Simply because cryonics is within the realm of the possible, does not mean it will succeed. There may be practical limitations that are just too much to overcome. For example, reanimation may require the aid of very powerful computers and the practical problems of making such machines will simply never be overcome.

Black Box Cryonics

A few members of the American Cryonics Society are students of the school of "Black Box Cryonics." Their thinking goes something like this:

Science is wonderful. It has given us so many marvelous inventions nobody even dreamed of before. There have also been many surprises in technological development that were completely unanticipated. When Jules Verne wrote his marvelous science fiction book about men going to the moon, he was right that mankind would go to the moon. He was wrong about how they would get there. Verne had his space ship shot to the moon by a gigantic cannon. We know now that cannon fired space ships for human travel is impossible or impractical because of structural limitations, among other reasons. But we got to the moon, nonetheless, as Verne predicted. If there had been any way to bet on humans getting to the moon, based upon Verne's prediction, the person who placed the bet would win.

For us to do armchair engineering on how future scientists will reanimate us is an exercise in futility. Whatever is our best guess, the reality is sure to be different, perhaps radically so. It is enough to concentrate on preservation of the human body, particularly the brain, as best we can. The knowledge that reanimation is not impossible is all we need to justify choosing cryonic suspension. It is a chance, even though the calculation of how much of a chance is not possible, we know that it is between zero and 100%. We should bet on future reanimation through cryonic suspension. The stakes are fairly modest, and the pay-off is tremendous: personal immortality.

Cracking Open the Black Box

While I have never tried to categorize cryonicists beyond the "Black Box" and "Others," there are probably several stripes. People who demand that cryonic suspension be a proven commodity before they buy in are probably the most conservative. We don't get many of those as members. Another group of "freeze me" folk, demand that there be at least some conceivable path between here and forever. Most know that their charting is apt to be faulty, full of twists and turns and blind alleys. Still, they have laid out a course, and if things go as predicted, they have mapped the road to infinitely extended life. A worthwhile approach to cracking open the black box is to start with what we can do now, and looking at several possible paths, any one of which could lead to the desired outcome. We will do that through a true story, now immortalized as one of the tales of the American Cryonics Society. Only the abbreviated version is told here.

The Practical Problems

The Story of How We Froze the Whales

It was May of 1990 when the first whale was frozen. American Cryonics Society members had been interested in using cryonics technology to help save endangered species for a long time. It was a perfectly logical pursuit by a group such as ourselves, who had been freezing everything from humans to pet cats for over 15 years. Why not collect cell samples when an individual member of an endangered species dies? Cloning of amphibians was then a reality with predictions of mammal cloning as "just around the corner."

Two Whales Frozen

In 1990, cell samples (tissue samples) from two rare or endangered whales were frozen by the American Cryonics Society. These whales had beached themselves near Santa Cruz, California. One, an adult Gray Whale, died shortly after beaching, or perhaps just before. The other, a baby bottle-nosed whale, succumbed after a valiant attempt to keep it alive in a salt-water tank. Because cloning was then regarded as so much science fiction, we encountered considerable trouble in getting permission to preserve such samples. In the end, we had to characterize the procedure as "genetic research" before the National Marine Fisheries Service, a division of the National Oceanic and Atmospheric Administration, would grant us permission to keep our iced-down whales.

Cell Samples Grown in Incubators

Even though scientists have been growing human and animal tissue in the laboratory for years, many people are unaware of this fact. To some folks who know it is being done, it seems a little like magic and it truly is remarkable science. Almost all tissue types, such as lung, heart, skin, liver, etc. can be grown in this way. The researcher takes a small sample from a human or animal who is either living or has just died, dices it into small bits (either mechanically or chemically), places the tissue in a small dish, adds a kind of artificial blood to the dish, and keeps it at about body temperature by placing it in an incubator. It works! The tissue continues to grow, with the cells living and dividing just as they did when they were part of a living human or animal.

Cell Samples Kept Frozen, Later Thawed and Regrown

Because of the inconvenience of having lots of cell samples growing in incubators, long ago scientists found a more convenient way to keep them. They have developed reliable methods to freeze the samples, after which they are kept in a low-temperature freezer, if need be, for decades. After freezing, it is convenient to ship such samples to other laboratories, sometimes halfway around the world, using dry-ice and special shipping containers. When there is need for the sample again (for study, for example) it is thawed back to active life. Once thawed, it is again placed into the incubator to grow still more tissue. Considerable study and experimentation have gone into finding the best ways to freeze such samples, and the fluids we use in cryonic preservation (which contain chemical agents that help to minimize freezing damage) are a result of that research.

Jim Yount

Whale Cells Could Be Cultured

Although we have not attempted to do so, there is reason to believe that the cell samples taken from one of these whales could be cultured in the laboratory, with new cells grown from our samples. Even where such culturing is not possible (with the adult Gray Whale, for example) the genetic blueprint remains intact and might be implanted into a host cell someday to once again act as a blueprint to direct the formation of the cell (tissue) type.

Blueprints Work

After the successful cloning of Dolly the Sheep, there is no doubt in anyone's mind that these genetic blueprints work: that the DNA sequence in the nucleus of each cell can direct the growth, and development of a mammal egg to maturity and birth. The cloning of mammals was no surprise to members of the American Cryonics Society, or to most scientists. Even since amphibians were successfully cloned over two decades ago, the cloning of mammals was predicted. The procedure is simple. A cell nucleus (complete with DNA) is extracted from the donor cell using a tiny suction tube. It is then injected into the host egg cell (with the original host DNA removed or neutralized). The donor DNA then directs the egg to develop into the twin of the individual who donated the DNA. Dolly the Sheep was the identical twin of the sheep who donated the cell whose nucleus was extracted.

Whale Blueprint

With the samples of whale tissue now in cold storage, there is good reason to believe that either currently, or in the near future, the tissue sample could be thawed, new tissue grown in incubators (if necessary) and the nucleus of any of those cells extracted to be used as the donor DNA to instruct an egg cell to develop into the twin of the whale that died. If a species of whale becomes extinct entirely, then the egg cells from a closely related species might be used as host.

Bring Back the Mammoths

Anyone who has seen *Jurassic Park* is familiar with the possibilities suggested by cloning. The explanation given in the movie of nuclear transplant cloning and DNA sequence completion is actually pretty good. If my description seems confusing, go rent the movie! Michael Crichton does a better job of explaining. While the possibility of dinosaur DNA found in amber-preserved mosquitoes seems remote, there is actually a very real possibility that mammoth tissue frozen in glaciers may be used as donor tissue to recreate the mammoth! In that case, modern elephants (genetically very close to the mammoths) could provide the host eggs, which would be brought to term in the womb of an Indian or African mother elephant.

New Hope for Endangered Species

We are not proposing bringing long-extinct species back, though the possibilities are intriguing. However, for species now threatened with extinction, cell preservation by freezing may provide a kind of "life insurance" which could ensure that our children (and for those of us who opt for suspension: ourselves) may enjoy a world with the genetic diversity of animal life which we now have.

Cryonic Suspension Preserves Tissue

The research which has led to the techniques we now use to preserve humans through the cryonic suspension procedure have grown out of two research disciplines: 1) tissue and organ preservation research and 2) suspended animation (through hypothermia) research. In our discussion of how we froze the whales we discussed the fact that tissue preservation through freezing, and tissue culture has been practiced for many years. There is no doubt that the bodies preserved by cryonic suspension have DNA intact, and in most cases, many of the cells are viable. That is, if they are thawed, they are found to be still living, and can be used to grow new tissue in the laboratory. In establishing the cryonic suspension protocol, we attempt to preserve the cells, tissue, and organs, by a procedure that treats the tissue in about the same way that produces success in cell, tissue, and even organ preservation. The difference is we try to do this without removing the tissue or organs from the body. The body is oxygenated during the cool-down procedure, and the same kinds of special fluids (which contain cryoprotectants to help protect against some damage by freezing) are introduced into the body through the circulatory system.

Suspended Animation and Hypothermia Experiments

Years ago, the American Cryonics Society embarked on a series of experiments with hamsters, and later with dogs, whereby the animal was placed into a state of suspended animation by lowering the body temperature down to just above freezing. As the temperature is lowered, the blood is replaced with a balanced salt solution, and the subject is placed on heart-bypass to better control the cool-down rate. As the body temperature goes down, metabolic rate slows, and the need for oxygen decreases. The heart stops beating, the lungs stop working, and the brain goes into suspended animation with flat EEG (brain waves). There is a great benefit to humanity from this and related research. Better methods of organ preservation, transplantation, and transport systems for donor organs result. Future benefits may include extended times for difficult operations, and a higher recovery rate for people who have suffered severe frostbite. The success of hypothermia research is well documented in the medical journals and other reports by scientists. In 1980, there was considerable coverage by the popular press on these experiments. Scientists from the American Cryonics Society were interviewed on *Good Morning America, The Sally Jesse Raphael Show,* and the *Phil Donahue Show,* among many others. We have just touched on successful organ preservation research since the techniques suggested here do not depend upon success in organ preservation. However, there has been a noted success in the freezing and transplantation of a number of organs including skin, bone marrow and the cornea of the eye.

Reanimation

Nature's Building Blocks (A Quick Review)

We are now ready to look at the other side of the procedure: reanimation. Before doing so, let's quickly go back to basics with human reanimation in mind, not just the restoration of extinct species. To understand cryonics, how it is practiced, and why we think it might work, we start with the basic building block of living things, the individual cell. Before any of the complex multi-celled animals developed, there was the cell itself. In it are encoded the DNA sequence, which gives the blueprint for nature to fabricate a sea urchin, a dolphin, or a human being. In humans, nature starts with a single cell, formed when the egg and sperm unite. This single cell duplicates, and duplicates again, each time copying the DNA blueprint code. At some point, the cells become specialized and become heart, lung, muscle, etc. Scientists have studied the mechanism by which the unspecialized cells, which are present in the first few divisions, become specialized to develop into the various body tissues. Turning this specialization off and on, is necessary to cloning. Or else start with an unspecialized cell to begin with. The important thing is that the DNA blueprint is present in ALL cells of the human body. When Dolly the Sheep was cloned, the investigators took the nucleus (with its DNA blueprint) from a cell taken from the udder of the donor sheep. They then implanted it into an unfertilized egg cell. The DNA blueprint from the donor sheep gave instructions for the duplication of many copies of itself as cells for Dolly. As these cells specialized, Dolly, the sheep, became the genetic identical twin of the donor sheep. Suppose you were put in charge of a reanimation facility and told to "Make all those patients walk out of here." "But those guys are FROZEN!" you protest. "How can I possibly get them up and running?" "That biological material is in pretty sad shape," agrees your supervisor. "Try replacing as much of it as you can, but the frozen guys must still retain their uniqueness. Their memories must remain intact. After all, they have to be able to remember their numbered Swiss accounts, if we are going to get paid for this."

Reanimation Simplified

Your supervisor has actually done you quite a big favor. Nature has been growing tissue for years, and doing quite a nice job of it, I might add. All you have to do is figure out how nature does it, and then go and do likewise. You don't even have to start from scratch. You can start with almost any old cell in one of your patients, extract the nucleus and implant it in most any healthy human cell (in which the original nucleus has been removed or destroyed), incubate it, wait nine months, and Voila!: Instant human.

Your First Mistake

Now just hold on a second. What you have just succeeded in doing is cloning the poor schmo who was unlucky enough to be your first reanimee. You can almost hear the protest coming from the capsule: "That ain't me, you dumb bozo. That's my identical twin, you overpaid reject from a remake of Frankenstein!"

■ BEING HUMAN:

Bad Brain Blues

OK, so you start over. This time you keep the old brain intact and grow a new body around the brain. Remember, we now have that clone thing down pat; it is the repair part that has us stymied. You take a good look at the brain, yeah, it's in pretty bad shape. Time to have another talk with the boss. "Do you really need to repair THE ENTIRE brain?" asks your supervisor. "Seems to me I read somewhere that most of the brain is used for automatic stuff, like making sure you can sit up without falling over. Couldn't you replace that, too?"

Keeping the Good Parts

Task simplified yet again. The brain stem and perhaps part of the cerebellum that controls the autonomic nervous system can be replaced, so can all the connective tissue, the part of the brain which registers sensation, in fact everything NOT required for the storage of memory and the processing of information can be thrown away. We are now down to the computer and we have just figured out that we can throw away the monitor, keyboard, and case. We want to keep the hard disk and the motherboard.

A Pound of Matter

We may be left with just a pound of brain matter that is really unique to who we are. So you now set about growing everything else as new, young, healthy material.

How to Repair the Cerebral Cortex?

Granted you have sampled the task, but you haven't eliminated it. Sure bodies and body parts can be grown, even around some part of the old brain, with a little imagination, but you still have to repair SOME-

THING, even if it is one-tenth of the cerebral cortex and maybe part of the cerebellum. It's time, once again, to talk to the boss.

Arguing with the Boss on the Nature of Identity

"Let's look at what you have left," says the person who signs the paychecks. "It is a bunch of nervous tissue in a particular configuration called the neural weave. If we replace these same cells with identical cells, haven't we accomplished our purpose?" Now it doesn't pay to argue with the boss. Except, she has her boss, too, who might fire the pair of you. Maybe this is an occasion when the boss is really wrong. "When you get down to this level of detail, I don't know what CAN be replaced," you explain. "I'm really nervous about this, Chief." "Well, suppose we replace the cells with mechanical gadgets or electrical parts. All these amount to is a bunch of circuits and electrical devices. Our engineers can manufacture replacements to do the identical thing. The customer won't know any difference. Besides, the electrical circuits need only be in place until these characters grow their own neural tissue. What's wrong with that?" "Maybe that would be all right," you say hesitantly, "it just gives me an uncomfortable feeling. Look, I know that these folks' neurons, as they originally functioned as part of the neural weave in the cerebral cortex, worked just fine. I just don't know the limits on this replacement process which we have started. How much more can we replace without having someone different walk out of here than the fellow who was frozen 50 years ago?"

Are We Just Information?

"Look," says the boss, obviously feeling exasperated, "aren't we really just INFORMATION? When we talk about a unique individual, isn't the essence of that uniqueness her memories? We can duplicate Jane Doe's body, right down to the mole on her left cheek. We can even give her the accumulation of scars and other deformities she built up during her lifetime from various misadventures.

Uploading and Downloading a Personality

"We've got scanning machines now that can scan the brain and tell, to a pretty fair degree of accuracy, just what are the positions of the synapses relative to one another. From that we build up a computer model which has all the information and information processing capacity as did the original brain. In essence, we upload the person into a computer. We can then later download the person back into the brain, once you have grown him, brain and all, in a culture tank. Some of these guys are just as happy staying in the computer and living in a virtual world. Better for the planet that way; they take up less space; don't need to make hamburger runs in the middle of the night." "You're getting pretty far out for me on that one, boss. I understand that some of these characters have written instructions on file which direct us to do that. They made that choice, so I have no trouble making them part of a computer. I'm not sure I would sign up for it, though, or maybe I would if something more natural couldn't be done." "Nature doesn't have much to do with this, sport," says your supervisor. "We've about gone through the list of things we can do for these folks with today's technology." "They are working on engineered viruses," you point out. "A virus goes into a damaged cell, takes it over, and makes certain changes to bring it back to function again. Since the virus is parasitic, they could only be used if the cell was viable. Special made-to-order bacteria or paramecia-like creatures could be used for badly damaged cells. They also have modified white blood cells, engineered as clean out agents, to do the housekeeping in the extra cultural spaces." "That helps, but it isn't enough. Not if you aren't going to cooperate in using some of the techniques I have suggested. The only technology that will satisfy your requirements depends upon being able to go into each cell, identify everything that is wrong with it, and kick any atoms which are out of place back to where they should be, atom by atom. It has been proposed that various tiny self-replicating nano-machines could be invented to do just that. That ain't going to happen this year, Partner." "Nanotechnology makes me nervous, too," you add. "There will certainly be ruptured cell membranes where the original atoms in those cells have leaked out and evaporated. If they replace them with new atoms, that won't be the same person." "That's where you are wrong, Nano-breath, once you get down to the level of atoms, the physicists and chemists all agree that you can substitute the atom for a particular element with another atom for that element without it making a difference. An atom, is an atom, is an atom. You were going pretty good until you tripped up on that one, my Former Employee. After you pick up your check be sure to drive your aircar carefully out of the skyport. We wouldn't want you involved in an accident and become one of our clients. Then I would have to think real hard about applying your own reanimation criteria to you, in which case you would be lucky to get out of here before the heat death of the universe."

Filling in the Genetic Gaps when Parts of the Puzzle Are Missing

Jurassic Park, the movie, actually gives a pretty good explanation of how even imperfect DNA can be used to become the blueprint from which humans or dinosaurs can be made. Only a small percentage of the genetic material in each human being is unique to that individual. The rest is shared by all humanity. So, damage even to some parts of the DNA sequence may not be fatal. We just have to know how to go into the cell's nucleus where the DNA is sequenced, look at what is there, figure out what is not, and put what is out of place, back in place. Simple, no?

■ BEING HUMAN:

Missing Memories

We do not know precisely how memories are encoded. One of the current theories holds that short term memory is encoded chemically, long term memory through the neural weave. Another way to put it is that memory exists because of the strength of the synapse along neural pathways. It is like having the wires that carry electricity to our homes able to grow larger to carry more current. The bigger wires (neurons) allow more electrical flow, which is then able to jump the synapse gap more readily. What happens when the structure that holds (records) memories is no longer there? In some cases, there is such destruction due to a long senescence (Alzheimer's disease, for example) or where the patient is in a coma for a long time (this can lead to an actual shrinking of the brain).

Rewiring

The basic task in restoring a brain to function, is to find and restore the neural weave. Fortunately, it appears that synapses (the basic neural weave wiring) are more durable than many other cell parts. To rewire, we must first determine what the original wiring pattern was. Then, we just run new wires where the old ones are missing or damaged. The task of tracing the missing or damaged parts may be assigned to "tame" bacteria. We know that the herpes viruses work by following the nerve paths. If such viruses, or bacteria with such functions, can be modified in the laboratory to trace this pathway, then "report back" to computers, we can construct a computer model of the neural weave that will become our wiring diagram. Even with incomplete information from these "tracers" the computer may make some logical inferences to complete its diagram. With this diagram as a guide, we can use carrier bacteria to bring new neural cells into position, or even construct parallel (but identical) paths to duplicate the damaged part of the neural weave. Finally, there is the possibility of using nano-repair machines, tiny self-replicating "engines of creation" that have been engineered to go into the cells themselves and effect repair by moving displaced molecules back where they should be. While some cryonicists place great faith in nanotechnology and its projected ability to repair even badly damaged biological systems, other cryonicists see the more mundane repair engines just described as being developed much sooner. Either way, no one argues that it is an impossible dream, though which technique can be used first, is a matter of argument and speculation.

The Sherlock Holmes Computer

Anyone who has ever read a Sherlock Holmes book, or even seen Holmes movies, knows that the famous detective could come to correct conclusions based upon scant information. Mr. Holmes' powers of observation were legend. While these feats of observation and logical deduction far surpass any mortal I have heard of, a future computer may well be able to "out Sherlock Sherlock."

Sherlock Sees All

To start with, the computer benefits by its vast sensory array. It can be fed by an infrared-sensitive camera. Sound pickup can be tuned to detect a pin falling through the air, well before it completes its "drop." Motion sensors can pick up the slightest movement.

Train that sensory array upon a human being, either live and in the flesh, or by way of video tape, perhaps of video taken 100 years in the past. The slightest twitch of the lips, the merest nervous tick, the briefest of hesitations, will be sensed, recorded and analyzed.

BEING HUMAN:

Sherlock Knows All

Apply, by way of computer logic, any of a variety of programs to analyze the personality profile of the subject. Does he stutter? What is the correlation of stutterers to pathological liars? Is there a slight bit of moisture on the upper lip? How does that correlate with people who have something to hide, and who stutter? Is his collar turned down slightly? What starch did the laundries of 1990 use? What toxic effect did it have to act as an irritant? Feed genetic information about the subject into the computer. The Sherlock Computer could access genetic information concerning millions of other people and make close matches, research the history of people genetically similar to the subject and analyze by correlation.

Does the subject not behave the way genetically similar individuals did? That must be because of his experiences, not his genetic heritage. What were the experiences of people with somewhat similar genetic matches that correlate with a moist upper lip, a stutter, and a propensity to wear heavily starched shirts? What do his accent and speech pattern indicate about where he lived and whether he was traumatized as a child?

Sherlock Programs You Both Emotionally and with Image Data

The Sherlock Holmes computer could tell us a lot about the personality, and perhaps even memories of the subject. A video tape of you, taken at a birthday party in 1990 might be fed into the Sherlock Computer. Sherlock, while accessing the virtual town to give you image memories, then program experiences for you so you can come back just as messed up as you ever were with all the old bad habits.

Jim Yount

Implanting False Memories

Just how important are your memories, anyway? I have fond childhood memories of our little dog Tippie. How she would bark gleefully when she made a bee-line to get inside whenever the door was open. How she would howl in pain when someone slammed the door with the dog still making that mad rush. But what if I lost that memory and a technician implanted a false memory? I would remember instead my little dog Bowser, how Bowser would catch the boomerang in his teeth as Lola and I played on the Australian desert. Come to think of it, I might actually PREFER memories of playful romps through the Outback with Lola and Bowser! Programmed Memories from Someone Else's Memories. If you are suspended under conditions that wipe out most or all memories, all is not lost. There will likely be people around who won't mind sharing a memory or two. It is possible that the scanning mechanisms, or other equally sophisticated devices will make it possible to program real memories from your brother, sister, or friend. Yes, you would see Dad do a belly flop into the pool from the height of a six year old, whose field of vision took in the twelve year old you, but so what? In the cryonic suspension business you knew that sooner or later you would meet yourself either coming or going, now didn't you?

■

The Worlds of Bodies

Nicole Malinconi

And so we went to Mannheim to visit the Gunther von Hagens exhibit. We were among the hundreds of thousands of living beings who wanted to see the two hundred bodies he embalmed, the two hundred bodies he conserved, treated and exposed as works of art. Works of art that were the actual dead bodies of once living beings.

We don't know exactly what it was that Gunther von Hagens wanted to embalm when he rendered these dead bodies incorruptible, nor when he opened them up and showed exactly what they had died of, nor when he slashed, flayed, and laid bare what we usually don't see of a living body; exposing the diseased areas, the insides that stood gaping open for all to see.

He did it with the consent of all the bodies when they were still alive, thanks to the promise he made of sparing them the usual end that dead bodies suffer, the end that we all know, that we all dread. The promise was, in a sense, to avoid this end to a certain extent. They would be dead, but not in the usual sense of the word, not as we commonly hear and say the word. Not decomposed. Not disappeared, having escaped common fate to become works of art, immortal.

We went to see, by the hundreds of thousands. So numerous were we that in the end, the museum had to stay open even at night.

We saw the bodies from the inside. They were virtually transparent. We saw everything. Here, we could no longer speak, as we often do, of the mysteries of the body; there was no longer any mystery. Moreover, the professor made it clear by calling his exhibit *the worlds of the body*. If he had called his exhibit *the world of the bodies* we would have remained in the mystery. He exposed the interior worlds and showed that there was nothing other than what was there to see. Better than

Nicole Malinconi

with the wax mannequins and reproductions of yesteryear, here we could see the whole ensemble, and the functioning of the entire apparatus. Even touching was possible, and for the school students nothing could be better than the ludic and animated exhibitions in which they were sure not to be bored. In any event, they couldn't say that they had seen this before. Here, we had the "never before seen" and "never before touched." Very instructive. Professor Gunther von Hagens invented a revolutionary method that keeps the materials in a good state – hygiene guaranteed – called "plastination." The professor has even enriched the language with a new word. He must be proud of his success in Germany. Soon, the exhibit will travel to the United States and Japan. To Hiroshima.

What are you saying? The dead people of Hiroshima? What people? Are dead bodies people, you say? What dead? Where is the relationship? The forgetting, you say? What is it that you want to say, exactly?

"You knew that sooner or later
you would meet yourself
either coming or going."

E-mail

John Perry Barlow

From: John Perry Barlow <barlow@eff.org>
Subject: BarlowFriendz 4.7: Celebrate Human Rights with EFF in San Francisco Tomorrow Night!

```
    ^
  <(o)>
  /_ _\
  -----> B a R L o W F R i e N D Z --->
```

A continuing series of occasional outbursts to about 860 of my dearest friends. Please let me know if you wish to be removed from this list. But you'll miss some great parties if you do…

Also, if this broadcast feels as impersonal to you as it obviously is, I hope you will remember that individual responses to it will usually elicit personal replies from me. And I do hope to hear from you!>

----------------> ----------> ---->

1. Velocity, Ecstasy, Agony.
2. Declaration of Human Rights Party in SF tomorrow night, December 10!
3. International Human Rights Petition.
4. Albanian Wit and Wisdom.

-------->-->----------------->

JET LAG OF THE GODS.

Once again I'm starting to get fretful little messages from you sweet and solicitous BarlowFriendz, anxious at the unusual hiatus since the last blast from me. The fact that the last bulletin contained evidence of physical decline - a herniated disk - has apparently also elevated concerns.

You're very kind. I'm touched, truly I am, but you needn't worry. I'm fine, or certainly fine enough. You haven't heard from because, as is often the case, I've been too busy accumulating great stories to find time to tell any of them.

Consider this, my itinerary since BarlowFriendz 4.6:

San Francisco -> Champaign-Urbana, IL 9/28-29 -> San Francisco 9/29-10/3 -> Homestead Hot Springs, VA 10/4-5 -> Planet Harvard 10/5-8, Tucson 10/8-9 -> Pinedale 10/10-14 -> Cleveland 10/14 -> Pinedale 10/15-19 -> Manhattan 10/20-22 -> Rome 10/21-22 -> Jo'Burg/Pretoria, South Africa 10/23-29 -> Paris 10/29-30 -> San Francisco 10/30-11/3 -> Jackson Hole 11/3-4 -> San Diego 11/5-6 -> Santa Monica 11/6-7 -> Las Vegas 11/7-8 -> Los Angeles 11/8-11 -> Salt Lake City 11/11 -> Pinedale, Wyoming 11/13-15 -> MIT/Nation1/Cambridge 11/16-20 -> New York City 11/21-22 -> San Francisco 11/23-24 -> London 11/25-29 -> Amsterdam 11/30-12/3 -> Colorado Springs 12/4-5 -> Planet Harvard 12/6 -> Salt Lake City 12/7-9 -> San Francisco…

By my rough calculation, that comes to 76,400 miles in a bit over 2 months. What a quick, strange trip it's been. A cavalcade of wonders, a trail of tears. And one hell of a lot of schleppage for a guy with a bad back.

Mind, I'm not complaining. None of this was exactly compulsory, and in balance, the adventures of this planetary plunge have been well worth any attendant suffering. Indeed, there's no real evidence than I would have been in any less pain had I spent the same time at home. On the few occasions when I gave it a break and took proper care of myself, it seemed I got worse instead of better. I'm beginning to think that you can even run away from your own body if you run fast enough.

Nevertheless, three and a half months after blowing out the little cushion of cartilaginous-contained human jelly between my backbone and my pelvis, I am still seriously on the limp. I have tried treatments that are definitely not in the literature. Everything from Voodoo (or Voudou, as its called by those who are serious about it) to Big Expensive Machines Made by White Men. I've tried exercise (and I still am), daily icing, chiropracty, prayer, needles. Filipino psychic surgery is probably in my future.

What looked like a promising recovery in the first couple of months finally plateaued about six weeks ago into a condition that will probably not improve until something fundamental is done. There is a sharp shard of cartilage wedged against my

■   BEING HUMAN:

spinal cord and the only way it's going to move will be if someone moves it.

In any event, the relentless growl of my lower back had only a little to do with my silence.

The real reason was that I was too busy. I made several attempts to whack out a bulletin, but I was just too busy helping design the next generation Internet, sticking the electrodes on bureaucrats and bizbots, dancing all night in clubs on three continents, tolerating Eurocracy, prowling the Eternal City, experiencing the miracles and horrors of Whitest Africa, eating too well in gray Paris, raising three teenaged daughters, chasing women, trying to keep happy the ones I've caught, cruising The Strip with a beautiful Swede, working with a 100 young folks from all over the planet to start a Global Kids Nation, swinging in London, inspiring rural Dutch businessmen, altering my consciousness, helping inject tech into my old prep school, being a Fellow at Harvard Law School, and, finally, sparring with my 93 year old mother and experiencing Big Medicine.

And that's just the short version. It's enough to keep a fellow occupied.

Obviously, there's more to be said about all this. Much more. Some of that will follow in the next message, which I hope to crank out sometime in the next few days. But I wanted to get this one off right away, since the Electronic Frontier Foundation is throwing a party on San Francisco tomorrow night and I wanted to give you a *little* warning.

Now, as to that…

-------->-------->----->->--->

CELEBRATE 50 YEARS OF THE VERY BEST INTENTIONS IN SAN FRANCISCO, THURSDAY EVENING, DECEMBER 10!

Fifty years ago tomorrow, the General Assembly of the United Nations passed Resolution 217A (III), more generally known as the Universal Declaration Of Human Rights. This remarkable document contained such words as these:

THE TECHNOLOGICAL EXTENSIONS OF THE BODY

John Perry Barlow

Article 2

All human beings are born free and equal in dignity and rights. They are endowed with reason and conscience and should act towards one another in a spirit of brotherhood.

and

Article 19

Everyone has the right to freedom of opinion and expression: this right includes freedom to hold opinions without interference and to seek, receive and impart information and ideas through any media and regardless of frontiers.

While no one would claim that these goals have been achieved in the years since, the fact that they could be set forth at all marked a turning point in history. And, despite continuous attacks from such powerful bodies as the United States Congress, the fact that this document appears on over 10,000 web sites in hundreds of languages indicates that there are still many all over the planet who believe we can create a world in which such principles are not considered naive or impractical.

If you happen to be in San Francisco tomorrow night, you can join help celebrate that cause by attending the following:

CYBER RIGHTS = HUMAN RIGHTS:
The Universal Declaration of Human Rights is Not a Local Ordinance in Cyberspace

EFF invites you to spend an evening with several respected authorities on the Internet and on human rights to celebrate the importance of the Declaration for the future of Cyberspace.

8:00 PM, Thursday, December 10, 1998
The Anon Salon, 285 9th Street (9th and Folsom)
San Francisco, California, USA

Hosted by Mike Godwin, EFF Senior Fellow and Author of "Cyber Rights: Defending Free Speech in the Digital Age"

With Special Guest Appearances By

■   BEING HUMAN:

- Congresswoman Zoe Lofgren (USA House of Representatives),
- John Perry Barlow (EFF Co-Founder),
- Esther Dyson (EDventure Holdings and ICANN)
- Patrick Ball (AAAS Science and Human Rights Program),
- Cindy Cohn (McGlashan & Sarrail and Lead Counsel in Bernstein v. Justice),
- Dave Del Torto (CryptoRights Foundation),
- James Yee (Silicon Valley for Democracy in China)
- AND distinguished others…

Cosponsored by

AAAS Science and Human Rights Program (http://shr.aaas.org)
Alexa Internet (http://www.alexa.com)
Computer Professionals for Social Responsibility (http://www.cpsr.org)
CryptoRights Foundation (http://www.cryptorights.org)
Ernst & Young (http://www.ey.com)
Freedom Forum Pacific Coast Center (http://www.freedomforum.org)
Human Rights Advocates
RSA Data Security (http://www.rsa.com/conf99/)
Silicon Valley for Democracy in China (http://www.svdc.org)
The Unity Foundation (http://www.unityfoundation.org)
Verbum, Inc. (http://www.verbum.com)
Zero-Knowledge Systems (http://www.zks.net)

An official event of the San Francisco Bay Area's Human Rights Day Celebration

Please see http://www.eff.org/udhr for more information.

Since Anon Salon is fairly limited in space - the Fire Marshall says 200 max - I would ask that you *please* RSVP with an e-mail or phone call to EFF's Director of Public Affairs, Alexander Fowler. He can be reached at either afowler@eff.org or 415/436-9333.

For the same reason, this is not one of those BarlowFrenzies where I would ask you to send a representative if you don't happen to be in town yourself. I'll try to have one of those in San Francisco sometime around the 8th of January.

Hope to see some of you tomorrow.

------------------------->

THE TECHNOLOGICAL EXTENSIONS OF THE BODY

John Perry Barlow

PETITION TO SUPPORT THE DECLARATION OF HUMAN RIGHTS

As I say, there has been some gap between intention and reality in the delivery of these old promises. Toward the end of getting the United Nations to take their own words seriously, Amnesty International is collecting signatures for a pledge to support The Declaration of Human Rights.

Amnesty already has 3 million signatures (real and virtual) world wide, and wants 8 million (which would be 1% of the world's population). The UN Secretary General has already agreed to be present either in person or live by satellite, if he has to be in New York, to receive the pledge as a tangible statement of the people of the world's commitment to an international agenda of human rights.

The most simple way to add your name to the pledge is to

* Send an e-mail to

  udhr50th@amnesty.org.au

* Put YOUR NAME in the SUBJECT and the following text in the message:

I support the rights and freedoms in the Universal Declaration of Human Rights for all people, everywhere.

* Forward this message to as many people as you can.

Alternatively you can visit their web site and fill in your name there, as well as being able to find out more information:

http://www.amnesty.org.au/campaigns/udhr.html

------------>------>------------>

ALBANIAN WIT AND WISDOM

Now, for something completely different...

About 15 years ago, as the height of the Solidarity Movement in Poland, I decided that I would show *my* solidarity with them by retiring my ample supply of Polish jokes. (Besides,

■   BEING HUMAN:

I'd started to wonder about them anyway. I had never met a real Pole who didn't seem smart as hell to me.)

Unfortunately, I couldn't quite kick the habit. The urge to tell Polish jokes was irresistible. So eventually I decided that I would port them over to some nationality that probably *was*, on the whole, a bit dim. Eventually, I fell on the Albanians. Surely, no intelligent people would suffer such an appalling government decade after decade. Furthermore, the Albanians never seemed to leave Albania, so the chances of insulting one seemed slim.

Still, the first time I trotted out a freshly minted "Albanian" joke at a wedding in LA, there was a Slavic looking fellow among those listening who did *not* laugh. "Eees not funny." he growled. I'd met my first Albanian. And told my last Albanian joke.

Now it turns out that it was a good thing I met that guy. I recently came across a startling book called Great Big Book of Albanian Proverbs. It should probably be called the Great Big Book of *Modern* Albanian Proverbs, since most of these seem pretty urban, but nevertheless they provide strong evidence that the Albanians are anything but dumb.

Anyway, here's a gift sampler. Enjoy.

Celibacy used to be a way of life. Nowadays, it's just something to try after you've tried everything else.

Some things to avoid: a man wearing a suit, a woman wearing sunglasses in her hair, a dog wearing a muzzle, a man holding a clipboard who just wants to ask you a few questions, a bishop in a bad mood, and anyone who claims to remember your mother well and fondly.

Abdication is a one-way ticket. There is no word that means the opposite of "abdicate."

The first Christmas card always comes from someone you weren't going to send a card to.

There is nothing quite so sad or useless as a complete crossword puzzle.

John Perry Barlow

When you wash your hands in someone else's kitchen, you always dry them on the wrong thing.

The 18th Century had the right idea about wigs: they tried to make them look as different from real hair as possible.

The most law-abiding and careful motorists are driving stolen cars.

One of the sad things about Alzheimer's Disease is that nobody can remember who Alzheimer was.

Just before we sell a car, we polish and clean it into a state where, suddenly we no longer wish to sell it.

Everyone wants press coverage, but nobody wants press treatment.

The secret of the English is that they never feel belittled by the names people call them.

--------------------->-->

MORE. LATER.

Cool, huh?

Anyway, I'm looking to settle into Wyoming directly, hole up with my kids, and, among other things, file some more detailed reports on recent events to you folks. There's plenty to say. I just have find time to say it.

Preserve Wild Life,

Barlow

****
John Perry Barlow, Cognitive Dissident
Co-Founder & Vice Chairman, Electronic Frontier Foundation
Berkman Fellow, Harvard Law School

Home(stead) Page: http://www.eff.org/~barlow

MegaPhone: 800/654-4322

Fist Phone: 917/863-2037

■   BEING HUMAN:

Barlow in Meatspace Now: Salt Lake City 801/582-5035

Coming soon to: Pinedale, Wyoming 12/8-10 -> San Francisco 12/10-12 -> Pinedale 12/13-27...

****

We should be grateful that we aren't getting all the government we pay for.

— Herman Kahn

The Anger of Friendship

Mark Stafford

John Perry Barlow
Pinedale
Wyoming

Dear John,

Each day it seems I receive more and more of your letters, updates, travel plans and party announcements. Invisible in cyberspace you also provide us with your coordinates in "meat" space which despite its derogatory inflection, evokes for me your beloved Wyoming. There I imagine you touch down from cyberspace to relish a frontier that is visible rather than infinitely electronic. There amongst the reckless and the able, the meat eaters and the laconic, you can rejuvenate your native skepticism of the cosmopolitan crowd that eagerly awaits next week's guru.

 Is it too commonplace to acknowledge the close tie between religious and technological evangelism? Abhorrence of the former makes me skeptical of the latter. Perhaps hucksterism is a degenerate form of prophecy and evokes a longing for a founder to combat the isolation of the wilderness.

 We first exchanged ideas, during your sojourn at Viking Penguin, where you blasted like a homesteader or sod-buster the arbitrary barrier between production and distribution and became the first and only writer-in-residence. It was there that I began to pay attention to the message, even though I couldn't grasp the pleasures of the medium. You were the first storyteller of the Internet and I am indebted to you for bringing to my attention its promise to re-value our ideas of publishing and the place of the writer. Any thoughts I may have had on the subject are marked by the questions you raised and the conviction you bring to this moment in the history of communication.

For an editor or publishing manager it is a shock to see that one is so blinkered as to not even have examined the central premises of your activity – that you are creating a piece of intellectual property and that, for a time at least, the law circumscribes these rights. At the time your piece "Old Wine for New Bottles" appeared to be a prospectors dream of undiscovered mines, the utopian hope of a man who had staked his claim on a new form of freedom, (although he could still find something else to do if the revolution didn't arrive).

Cynically, I assumed that those with the resources and the capital would soon find a way to stall the leakages that you predicted. Some of the "brightest" minds in the debate on electronic commerce are now arguing that "business will take care of the privacy issue…that privacy will become a new product" and that government does not have the right to know. This was one of your original EFF founders speaking so I don't think my cynicism was unjustified. How ironic to see that the advocates of "freedom of information" believe that multi-nationals care more about the rights of citizens than elected governments. Is this one more way that they can reconcile their youthful protests with their middle-aged desire to be members of the mediocracy. But you were right – no one is going to think about the question of intellectual property in the same way since the advent of the net, and while many others now claim to be delivering this insight, I know how early you were in developing the debate. How right you were to title your projected book "Everything You Know Is Wrong."

For this reason, among many others, your acceptance of our invitation to participate in the Being Human Colloquium was a favor returned. The cognitive dissident amongst analysts and scientists tapping away on his notebook. The contrarian. I remember that you thought your plane was close to crashing when you flew into Kennedy that morning. And that you were most involved in the talk about death. Since you spend so much time in the air, and appear to be destined for an entry in the Guinness Book of Records under "Most Miles Received" I thought in reflecting about the significance to me of your letters that I would tell you about another flight.

Paola Mieli first invited me to collaborate on the project and to make suggestions about a panelist she was eager to try and locate, an astrophysicist she had met on a plane to L.A. She didn't remember his name but she thought he worked in Rome at La Sapienza. He was working on new forms of rocketry and was interested in the debris that was being left in space as well as the possibilities of new satellite communication, but the most interesting thing she said "is that he was reading Montaigne" I tried calling the university, but to no avail. Our Montaigne-reading astrophysicist had disappeared, but a sensibility that can engage the remarkable trajectory of contemporary science and the exquisite melancholy of Michel de Montaigne became an ideal of those who might be interested in the "Being Human" Project.

The image of this "lost" panelist returned to me when I first started to feel that e-mail messages have a gloss that wears off. The initial exhilaration of this immediate, economic, correspondence is not easily lost, but a resignation begins to set in as more and more messages appear to be written in order "to get something off," in order to enjoy the satisfaction that one returns all calls and answers all letters. The contact with Brazil or Rumania is still a miracle. But more and more the messages appear to have the quality of a reflex reaction. Will anyone write a history of writing on the Internet, try to trace the changes in style as the familiarity breeds a new contempt for writing? What would Montaigne have thought in his self-imposed isolation of the kind of friendships we now form between the reflex arc of an immediate response and the isolation of perpetual motion?

Once again my exposure to your goal – to promote the removal of all barriers that exist between those individuals who want to share information – makes me want to discuss this with you.

A fax that circulated with the preparatory material for the colloquium has remained a reference point for the possibility that we are

■   BEING HUMAN:

developing with e-mail a new epistolary form. It was a piece of "found art" that came off an office fax machine. A letter that was not addressed to me, but which I read while I prepared to fax an author or an agent. It was a tender note faxed to somebody who had departed, and it asked eagerly for news while relating a few of the weekend's events. Unquestionably a love letter and sent early in the morning before anyone else appeared in the office, but the fax machine at the other end had already replied to it by printing on the original "Undeliverable. Destination Unknown" The sender had not received this message nor did the recipient know of the difficulty.

I am beginning to collect anecdotes from friends who report a new frustration at work and at home – the burden of their email. One is accustomed in a democracy to receive numerous solicitations. We have the constitutional right to bombard, hassle, and pursue the Other. William Gass points out in "On Being Blue" that we have impoverished our language for the body, by turning the terms into powerful taboo words – perhaps it is time to take back "fuck" from the expletives and substitute "directmail." Should one have expected anything more respectful on the net, given its speed and economy? Friends report that correspondence is increasingly, and uncomfortably for them, their work. They don't even have time to open those they might want to read. Conversations that could have happened are substituted by a message. Remarkably obvious demands are emailed and every message reminds you that you are expected to respond. But not as we used to say "in person." Are people giving up other forms of letter writing because they don't have time to keep up with their email? How clearly they stand out now those messages that are written with a different sense of time, a consideration for the time of understanding or confusion.

When one opens the mail box, stuffed with junk mail and finds a letter addressed by hand from the most distant of acquaintances, don't we still open it in eagerness? What mystery are they going to reveal, a new child, a death, a birthday greeting delivered two months late? Could we receive too many of these letters?

THE TECHNOLOGICAL EXTENSIONS OF THE BODY ∎

I write emails to the friends I have forgotten to write to. I claim that another more pressing demand prevented me from writing sooner. With this fibre-optic writing pad I can contact them out of the blue and at least I can let them know that they are in my thoughts. After a brief acquaintance I can develop a conversation that otherwise might lie dormant, unfulfilled. The writing, allows for that now familiar form of flirtation that is conducted from behind the screen. Even keyboards can be tender. Pressing "send" I want to "receive," to know that my message was read. A new tag comes in and I open it eagerly for the secret it may contain, like opening a fortune cookie, to see if the fates have delivered a dream. This process of exchanging is addictive.

So many of those whose words I use, whose gestures I imitate and jokes I tell have moved away. In our mutual pursuit of work, family, security they have been lost in the tide, that surge of life that urges us to believe that we need to move on if we are to avoid the traps laid by want and fear. What a beautiful fantasy to maintain all of my friends at my fingertips. What a new way to fight against the breaking of ties and bonds to keep alive that longing for what is always lost.

Is there something that I want to say to all of my friends? Some are the elements of who I am which I have never been able to show others, they are my history, my idea of being in the world and my failures. Where I have lost parts of my self they have resided in these people who I call my friends, so how remarkable it would be to see them all together, at the celebration of lost elements. Instead, we are left looking at one or two photographs whose poignancy derives from the realization that we are losing the ability to remember all the names. Could I arrest this wearing away of the names, if I could keep the phonemes of their names, forever reverberating on my electronic address book so that I could keep all our connections alive?

What path are we taking if we imagine that we do not lose our friends, even our "contacts"? I hear in your letters figments of a dream of "total communication." The racing pursuit of a mythological figure who tells us that we can all understand one another. The information

■ BEING HUMAN:

keeps piling up at our feet, but what does it try and stand in place of? The subtleties of an exchange or an encounter? Or have we lost interest in subtleties now that we have such powerful resources at our finger tips. Will we cease, with constant updating, to say that there is something we have forgotten to say? Am I a Luddite to be suspicious of this mythological figure because he seems to see no value, finds no truth or meaning in misunderstanding? Is this the end of solitude, loneliness, and the emergence of speaking beings who feel secure against the illusions and deceptions of the Other? What could be a more appropriate representation of the ideological naïveté of our times than this autonomous advocate, who serving his new god tolerates and is complicit in all forms of ideological manipulation? Justifying the obscene as another component of "total communication"? Are we afraid of being in such a place of opposition?

Occasionally, you "spam" us all and it is in the apology not the "update" that I hear your voice. Something doesn't work. The gracious acknowledgement carries the request that we don't lose faith. A momentary difficulty.

We rarely speak about the anger we have towards our friends because such an anger marks the difference between the public and the private. Did they receive the message or not? Like the fax machine that replied to the lover's letter "destination unknown," our letters reach a destination that is unknown. But let us hope that in reckoning with this unsettling evidence we do not give up on writing letters in a manner that might have amused Montaigne.

THE TECHNOLOGICAL EXTENSIONS OF THE BODY

Some Notes on the Technological Extensions of the Senses in the Age of Television

Claus-Dieter Rath

"Facing the cameras and the audience, now barely visible in the background of the studio, Chance abandoned himself to what would happen. [...] The cameras were licking up the image of his body, were recording his very movement and noiselessly hurling them into millions of TV screens scattered throughout the world – into rooms, cars, boats, planes, living rooms, and bedrooms. He would be seen by more people than he could ever meet in his entire life – people who would never meet him. The people who watched him on their sets did not know who actually faced them; how could they, if they had never met him? Television reflected only people's surfaces; it also kept peeling their images from their bodies until they were sucked into the caverns of their viewers' eyes, forever beyond retrieval, to disappear. Facing the cameras with their unsensing triple lenses pointed at him like snouts, Chance became only an image for millions of real people. They would never know how real he was, since his thinking could not be televised. And to him, the viewers existed only as projections of his own thought, as images. He would never know how real they were, since he had never met them and did not know what they thought."[1]

Within the totality of sensory faculties, sight, hearing, smell, taste and touch, only the first two – as the senses involving distance are engaged by electronic audiovisual mass media. Since the acoustic and the visual channels of the body are related to two objects of desire, the voice and the gaze, we can consider the technological extensions of

---

1 Jerzy Kosinski: *Being There*, New York: Bantam Books, 1972, 53–54

these channels, the extension of oral and visual expression as well as the extension of the reception through ear and eye – means of *jouissance*, of erotic enjoyment.

We *listen* to audiovisual mass media. They make us hear distant noises, voices and melodies. They call upon us (to do things, to join a fashion) and recall for us (what has to be remembered). They even announce an *outcry* in the name of the public and/or cultural values. They appear as our voice, like "The Voice of America," and they "hear" us, insofar as they yield to our needs, demands, wishes and desires or to those of the public.

We look at them. They "look" at us. They show us distant things. They even look after us insofar as they claim to be at our service and insofar as they "un-cover" what is not accessible to our gaze; they illustrate, narrate or symbolize – verbally and iconographically – what they make us encounter of the real. As some Techno-Prophets tell us, because of new and future technological means, we will come across things the world hasn't seen yet. Cameras are consuming images of bodies and objects – images which are sucked into the caverns of the audience's eyes several hours each day (microphones and ears function analogously); viewers feel filled up and devoured at the same time.

The mass media not only feed us information and fantasies, but also claim our attention. Furthermore, they socialize us insofar as they integrate our traumas and fantasies into a symbolic order and make us part of an electronically constituted "social body." As a new geographic entity, a sovereign "state" with its own guarantors, the "space of the broadcast media" – a territory of transmission – cuts across the geographies of power and social life which together define national or cultural space.

Thus, those extensions of our senses are "technological" not merely in the sense of an apparatus – like a telescope and a hearing aid – but in the sense of something that constitutes a systematic treatment, a téchne, a craft or skill, a set of rules in art.

Each electronic mass medium, like radio and television, consists of a plurality of elements: it is an apparatus, a receiver, a designed piece of furniture, a gadget, and it is language including different kinds of languages, subjected to an order of discourse; it is a variety of programs and it is a broadcasting institution. It is a system of perception, storage and transmission of information subjected to particular ethics and aesthetics. It works both as an extension of senses and as a producer of sense. New modes of ordering reality emerge at the push of a button: the world of television language, television geography, television community. Television thus can create social reality, an ability of a quite different order from merely improving or wasting family or community life.

In the transmission network with which a general summons is set in motion, a specific efficient power is at work; a symbolic network allows the viewers to weave what has happened to them along with their fantasies, into a cultural structure – a sort of permanent initiation through media events, the self-constitution of the subject by pushing a button. As a means of proclaiming and establishing a conformity to the spirit of the times, television constitutes a link between macropractices and micropractices of power. It constitutes a guiding instrument for those who control it as well as for the viewers, who, by its means, assure their reality and their actual mode of life. The electronic mass media provide us with images of how to relate to our bodies, of how to enjoy shaping them and caring for them, how to read the bodies' signs and symptoms.

Chance, the main character of the novel and the film *Being There,* is an approximately 60 year old orphan who has spent his whole life behind the wall of a villa's garden, where he worked in the service of an "Old Man." When his Master dies, he has to leave the villa and meets for the first time the outside world, previously, totally ignored. Until then, his only world is his garden and all kinds of television programs; he seems not even to know himself. Whenever he is asked a question or a comment, he answers in terms of gardening.

Surprisingly, his words are considered by the others as most intelli-

■ BEING HUMAN:

gent and instructive metaphors. In fact, he is only able to repeat or reproduce what he had heard from his Master's Voice or seen through his Master's Gaze – be it a person or television. But his interlocutors relate to him as a subject supposed to know. Similarly, we, the Audience, relate to television as a "screen that is supposed to know"[2]: to know about the real, the inaccessible, the mysteries of human subjectivity, desires, anxieties, symptoms, as well as of what we call the physical and the social world.

Jacques Lacan holds that the mass media in our time constitute some features of "our very relation to the science that ever increasingly invades our field," and which might be illuminated by the reference to "the voice – partly planeterized, even stratospherized, by our machinery – and the gaze, whose ever-encroaching character is no less suggestive, for, by so many spectacles, so many fantasies, it is not so much our vision that is solicited, as our gaze that is aroused."[3]

Walter Benjamin spoke of the loss of experience characteristic of modernity.[4] What for generations used to be transmitted to sons and grandsons, the confidence in a meaningful world, in the human patrimony, suddenly failed. The overwhelming technical development represented for instance by the new weapons or urban life, with the establishment of new communications media, new social relations, new rhythms, rituals and myths – provoked a particular kind of human impoverishment. Overstimulation and a fragmented perception of reality took the place of experience. The contemporary need for rapid orientation and for flexibility, constitutes a challenge to perception and to aesthetics in general; the loss of meaning derived from it, engenders the desperate pursuit of pseudo-myths, of salvation messages, supposed

---

2 Anne-Lise Stern: Wo Es War: Weiss. Ein Dunkel. In: Wunderblock. Zeitschrift fuer Psychoanalyse, Nr.16, Berlin 1987, 12–15.
3 Jacques Lacan: *The Four Fundamental Concepts of Psycho-Analysis*, Seminar 11, Norton, New York, 1981, 274.
4 Walter Benjamin: Erfahrung und Armut. In: Gesammelte Schriften vol.II, I. Frankfurt/M.: Suhrkamp 1977, 213–219.

to integrate and calm the shocking encounters with the real. But by filling our senses with "instant-sense," we block access to our desire.

The development and proliferation of electronic media, in the military, business or entertainment contexts, led to the advent of a scientific paradigm which becomes a basic assumption about Being human: human beings communicate according to the structure "Communicator – Message – Medium/Channel – Communicatee."

"An act of communication between two persons is complete when they understand the same sign in the same way,"[5] Lasswell claimed. Yet, such a statement, applied to the human being, needs to be examined.

For instance, we can question the illusion of understanding each other. Television's "reading" of the world, needs not only to be decoded, but also to be read. The notion of a transparency in "communication," conceals the question of the desire of the other to our own desire, as well as the impossibility of accessing our unconscious desire, of fully knowing what we are saying while we are talking. What is the desire of each single subject in a particular act of "communication"? What are a communicator or the receiver appealing for? What is the desire of a "communicator" who addresses people without knowing, as Kosinski puts it in *Being There*, how real they are, since he never met them and doesn't know their thoughts?

Human language cannot be reduced to a simple system of codes, similar to the language of bees, where a correspondence between signs and their reception can easily be ascertained. By its very nature, human language involves the possibility of miscommunication and misunderstanding. A global call for unification – "come together," "let's become one body" – can be defined as Totalitarianism of Code. As soon as a message can only be encoded and decoded either correctly or incor-

---

[5] Harold Dwight Lasswell, "Describing the Contents of Communication," in Bruce Lannes Smith et al. (ed). *Propaganda, Communication, and Public Opinion*, Princeton University Press, 1946, 83.

■ BEING HUMAN:

rectly, there is no more space for metaphor. A double movement, constituting a closed image of the social body, is promoted by the global TV market: we observe a tendency towards the establishment of a universal code, and, on the other hand, a tendency towards the production of a particular code, manufactured for restricted communities, whose members claim to understand each other, to understand the same sign in the same way, by means of common biological or environmental factors. Consequently, those who show different features, must be considered enemies; those whose way of acting, speaking, and reading a message, contradicts the code, must be avoided. They menace the image of social body unity.

As Lacan has noted, the human being is animated by the "passion for ignorance."[6] Technological extensions of the body also favor repression, therefore facilitating the avoidance of perceptions and experiences which might disturb the subject's construction of its reality: of its physical body and of the social body as well.

---

6 Jacques Lacan: Vorwort zur deutschen Ausgabe meiner ausgewaehiten Schriften, in: *Lacan Schriften II* (Olten u. Freiburg i.Br: Walter, 1975), 13.

Because We Are Digital

Charles H. Traub & Jonathan Lipkin

As members of an artistic community, we accept the premise that a revolution has taken place and that it is time to build on its promise and transcend the inevitable losses that follow such changes.

The advent of the digital computer has created new possibilities and allows us to reinvent ways of interacting, thinking, and creating. The computer in its current form enables us to explore new artistic and scientific methodologies, that are historically unprecedented and provide us with a new form of link between cultural forms and individual minds. The value of the computer as a creative tool lies in its ability to initiate a reconceptualization of our relationship to knowledge. We can now respond to and interact with knowledge rather than merely accumulate information. The methodologies of the computer allow us to cross disciplines and engage in a collective human endeavor. If developed by people who appreciate the humanist tradition, this enterprise will allow us to rediscover what makes us human. The great achievements of man lie in the quest to expose the unseen, and the computer's value lies in its ability to further this quest.

The ways in which we are now working with digital technologies are not new, and precursors of multimedia and hypertext have been around for centuries. At present the strength of the computer lies in its speed, flexibility and memory, which enhance the basis of human intelligence – the desire to create new meanings and new relationships.

We posit a new type of creative individual, the *creative interlocutor*, a navigator of the pathways of intelligence and memory, who enables others to re-define their interests and creativity. This individual is one who tries to make intellectual connections for the common good. The creative interlocutor integrates disparate fields of human knowledge and develops them in previously unimagined ways.

Technology has always aided, rather than hindered, human expression and creativity. Human beings, however, have always had to overcome an initial hesitancy marked by a fear of the unintended consequences of change. This resistance takes the form of skepticism and nostalgia. At the dawn of telecommunications Henry David Thoreau remarked "We are in a great haste to construct a telegraph from Maine to Texas. But Maine and Texas, it may be, have nothing important to communicate." Clearly, today one does not doubt the humanity of a grandmother in Maine who talks to her granddaughter in Texas. What we lament is the loss of content in that conversation.

We seek to negate the self-fulfilling prophecy engendered by those who lament their potential loss of primacy and who are ready to blame technology for what appears to be a new barbarism. You can't touch it; you can't read in bed; it hurts my eyes, are all manifestations of this prophecy. These regrets and fears are inhibitions and do not take the form of justified reservations. All too often, humanists and artists unnecessarily segregate themselves from the creation of new technology. This restriction only reinforces the power of the technocrats who then design and implement technology for their own, limited purposes. Ask not what the computer can do for you, but what you can do for the computer.

HUMANISM AND THE LIBERAL ARTS

The computer has value only because it enhances that which makes us human. Most likely, this is our ability to learn, or rather to learn how to learn – the knack to order, manage, reconfigure what we know. Culture is the accumulation of conceptual riches carried down through history.

In the Liberal Arts we try to understand these riches. We try to treat the heterogeneous fields of knowledge in a balanced and unprejudiced manner, emphasizing a common vision of human experience within its diverse fields. This negotiation between fields of knowledge, and the search for a shared vision is precisely what is furthered by a judicious use of multimedia digital technology. The Renaissance masters are models of the creative interlocutor: Leonardo da Vinci is the exemplar for the Millennium. In the post-Renaissance, Francis Bacon's *Novum Organum* seems particularly relevant, as it fervently called for the reintegration of the sciences and the arts. Many other examples can be enumerated from pre-modern times, but just one will serve to make the point. Anna Maria Sibylla Merian (1647–1717) exemplified Bacon's notion: as a noted visual artist, she used her talents to explore the natural world of insects, depicting them through careful observation which revealed essential entomological information. In the Industrial Age, fathered by such as Bacon, Merian, and the great thinkers of the Age of Reason, ever expanding fields of knowledge required specialization at the expense of more universally learned individuals.

CROSS FERTILIZATION

The idea that one field might enrich another is not a new one, although it seems to be forgotten by the specialization emphasized in our educational institutions. Multimedia is not the consequence only of technological advancement, but rather it is grounded in fundamental human practices which predate the invention of the computer by thousands of years. The advent of the computer did not create the technical tangle of multimedia, but rather manifested a pre-existing need in our culture for a more democratic, universal, and diverse way to communicate.

We can see multimedia in the burial rituals of the ancient Egyptians who made no demarcation between media employed in the great technology of the pyramids and burial rituals. These burial sites combined elements of architecture, writing, sculpture, and the rites incorporated music and performance.

In the Middle Ages, one can see a form of multimedia in religious architecture. The cathedral communicated the awe in which all members of the social order held the Christian doctrine. The message was made accessible to the illiterate by being embodied in a variety of media stimulating the senses: visual (stained glass and statues), sound (music and hymn), touch and taste (performance and mass) and smell (incense and myrrh). Writing was reserved as a means of codifying the knowledge represented by the cathedral, the knowledge to sort out the patterns of our existence, to know the unknowable.

When Victor Hugo comments on the advent of the printing, in *The Hunchback of Notre Dame*, through a character in the novel, a priest in 15th-century France, he compares the newly invented book to the cathedral and states "this will kill that" – the book will kill the cathedral. We know it did not. But Hugo's phrase does refer to the conflict between the text of the book and the multimedia imagery of the church. By the 19th century, the text had become dominant as a means of discourse. In the centuries that followed, the word and the dissemination of the written text became the primary locus of information.

Marshall McLuhan's *The Gutenberg Galaxy*, foresaw the return of the image, revitalized by global broadcasting and the iconography of advertising. He envisioned "the civilization of imagery" where the word is no longer the sole locus of the imagination. Today we hear posed an unanswerable conundrum – what stimulates the imagination first, the word or the image? The computer doesn't care, because it's a multimedia cathedral!

The phrase "this will kill that" was repeated with the invention of photography, and is often heard today as we experience the digital revolution. The book threatened the multimodal cathedral, photography threatened painting, now the computer endangers the book. But there will be co-existence in the media. The book will not disappear, but it will change in function and meaning, as did painting after photography.

THE TECHNOLOGICAL EXTENSIONS OF THE BODY

The computer offers us the possibility of another Renaissance, a new extension of our creative possibilities. The Web is an ever expanding territory of thought, commerce and entertainment.

Technology relieves us of burdensome tasks, whether it is the welding of metal or of numbers or of images, or all of them together. But have we allowed it to free us for greater pursuits? The responsibilty lies not with technology but with our educational systems.

All too often today, intellectual ideas are treated as chattel property whose purpose remains locked in the discourse of those who "know" rather than serving the common good. Pre-Enlightenment myth returns in these forms. Specialists sequester themselves in monasteries of learning, untouched by the great unwashed masses. Something medieval is happening.

Even at Harvard university, as late as 1989, Italo Calvino in the distinguished Belknap lectures stated what should have been obvious. He noted that not only poets and novelists deal with this problem, but scientists as well. "To draw on the gulf of potential multiplicity is indispensable to any form of knowledge. The poet's mind, and at a few decisive moments the mind of the scientist, works according to a process of association of images that is the quickest way to link and to choose between the infinite forms of the possible and the impossible. The imagination is a kind of electronic machine which takes account of all possible combinations and chooses the ones that are appropriate to a particular purpose, or simply the most interesting, pleasing, or amusing."[1]

---

[1] Italo Calvino, *Six Memos for the Next Millenium* (Cambridge: Harvard University Press, 1988), 87–91.

■ BEING HUMAN:

JOHN DEWEY: PRAGMATIC VISIONARY

Earlier in the 20th Century, John Dewey, advocated an educational system in which we would be able to recognize a common humanist thread within experience, communication, and art. In his analysis of the Greek Parthenon he noted:

"The collective life knew no boundaries between what was characteristic of these places and operations and the arts that brought color, grace and dignity into them. Painting and sculpture were organically one with architecture, as that was one with the social purpose the buildings served. Music and song were intimate parts of the rites and ceremonies in which the meaning of group life was consummated."[2]

Dewey sought to recover the continuity of aesthetic experience and normal processes of living through proper education. Aesthetic understanding must start with and never forget that the roots of art and beauty lie in basic vital functions. Marvin Minsky, one of the founders of modern computer science, has portrayed the mind as a society of tiny components forming a magnificent puzzle, a constantly evolving place of imagination. In his book *The Society of Mind* he refers to Papert's Principle (formulated by Seymour Papert) which theorizes intellectual progress as dependent not on the acquisition of new skills but on the acquisition of new administrative ways to use what one already knows.[3] Our conception of the computer as an art-making and communication device is just that – a tool which fosters and encourages the creative re-administration of information.

Dewey's pragmatism enabled him to envision an educational system which transmitted information without elitism. To allow education to become a tool that enables humanity to cultivate and reorganize our work and culture, we must abandon authoritarian methods of educa-

---

2 John Dewey, *Art as Experience* (NY: Perigree Books, 1934, 1980), 7.
3 Marvin Minsky, *The Society of Mind* (NY: Simon and Schuster, 1985).

tional practice, where the teacher is the endowed disseminator of privileged knowledge. As a tool for art-making and scientific thinking the computer is unique, and allows us the potential to realize Dewey's vision. More than at any other time in history, it is important to educate students with tools, both technical and intellectual. Cybercommunication must be made to be the intelligent extension of human capability for new discoveries. Communication is education.

DATASET: ALL ART IS IMAGE

Whether communication takes the form of vocal utterances, print, or modulation of radio waves, the intention has always been the transfer of meaning from one individual to another. This creates an image that will convey an idea. Up until now, the medium has determined both the audience for the message and its destination. Thus oil paintings were destined for the museum, text for the printed page, music for the radio. Subcultures have grown up around these destinations, and these subcultures have become insular and self-referential. Yet, the separations are artificial, imposed by the restraints of the technology and mostly by the lack of vision of those working within politically defined fields. These boundaries between media also force a separation of audiences, creating artificial divides of high and low culture. The evolution of a medium allows for an evolution of an audience. With its virtual writing spaces, the computer positions us to transcend these restraints, and to reunite all experience within its algorithms. The digital computer, when combined with the optical scanner, the music sampler, and a myriad of other computer input devices, allow us to reduce all physical media to a virtual binary digit. When we digitize a sound, a photograph, a cartoon, they all become equal in the cathedral-like space of the computer, free of dogma. Images become reduced to a dataset, a sequence of numbers; nothing more, nothing less. Every digital movie, every digital image, every digital sound is nothing more than a sequence

■ BEING HUMAN:

of zeros and ones stored in the memory of the computer. These numbers can now be seamlessly combined and juxtaposed. so that in the computer's virtual spaces, all forms of communication are equal.

The computer in its use of multimedia merely reinforces common and historic human themes. In order to communicate in the interest of evolving the human condition, there must be access to the creative tools – the computer network – to all who are interested in pursuing this goal. It empowers the user to also reorganize any message in new ways that allow for pattern thinking, trans-disciplinary intercourse, and the visualization of the unseen.

THE MEMEX'S ASSOCIATIVE INDEX

The idea of making a large body of information available to others is not new: in 350 BC the Athenian Speusippus created an encyclopedia which purported to contain all human knowledge, as did Lu Pu-Wei in China in 239 BC, who gathered 3,000 known scholars and assembled their knowledge into a work of more than 200,000 words. One of the limitations of these encyclopedias was their size: Pliny the Elder's encyclopedia *Natural History,* compiled in 79 AD was said to comprise 37 volumes containing 2,500 chapters. Another limitation was cost: at a time when books were reproduced by hand copying, works of this magnitude were enormously expensive. But the greatest limitation is a cognitive one. When large bodies of information are put together, some organizational scheme must be used. Today, encyclopedias are organized more or less alphabetically, with one entry following another from A to Z. This is a modern solution to a problem which dates back to Aristotle, and yet no perfect solution has been found.

Science advisor to Franklin Delano Roosevelt, Vannevar Bush, whose work has been obscured, stands as a remarkable creative interlocutor. In 1945, he developed the notion of the memex. The memex was a machine that could improve memory, like an encylopedia, but also allow the mind to operate by association. With one item in its grasp, it snaps instantly to the next which is suggested by its association of thoughts. This vision of a prosthesis that could scan information and allow the user to recombine it with related knowledge went unrealized but it is the premise for the work of the creative interlocutor.

Bush talked of new organizational schemes – ones which can be customized to the needs and interests of the particular users. His device combines two of the liberating capabilities of the digital computer: reduction of images, words and music to a dataset, and networking in a personal device. He foresaw both the internal network of hypertext and the possibilities of the external network such as today's World Wide Web.

Today, Bush's mechanism is as common as the desktop computer, and but his essential idea of the memex – that users can be empowered by hypertextual trails through information – remains unfulfilled. Why, we might ask? It is not the fault of technology, but rather a failure of vision. The Web, the prevalent implementation of hypertext, is essentially a one-way distribution system, where the user has little facility to be creative. We foresee the use of the computer networks to facilitate and empower the creative interlocutor.

The creative interlocutor uses hypertext and hypermedia to create trails. These trails transform data into knowledge to be redistributed to others, thus feeding the network. The memex, and the computer create a miniature network within their databanks. When linked to a larger network, such as the World Wide Web, the ability to create new meaning is increased almost infinitely.

Our ability to nurture and engage our own genius is stifled by an education that fails to recognize the value of associative capabilities inherent in this network. Clearly, this is the task of new humanist and visual education or what we referred to as the Liberal Arts. It must engage us all, as scientists, engineers, artists, and scholars. Technology has failed us in accomplishing this goal because it has been segregated from humanist activity.

■ BEING HUMAN:

THE COMPUTER AND THE CREATIVE INTERLOCUTOR

A new artist, interlocutor-designer, should be a product of an enlightened engagement, fostered within a new educational system. The creative interlocutor is one who facilitates the exchange of ideas and information which address human needs. This person is the producer, director, the organizer-navigator. More specifically, this person is the curator, editor, and collector, then the maker, weaver, welder, builder and distributor. Creative interlocutors are: programmers, producers, inventors, researchers, teachers, scholars and volunteers. The creative interlocutor negotiates revolutionary associations, to produce a new kind of genius.

We see a budding of the creative interlocutor in the collaborative spaces of the Internet. The language of the computer is a shared language which allows participation. In the following example, there is no longer a single creator, but rather a collective genius, a web of creative nodes which weave together previously disconnected pieces of information. In 1979, the inventors of the RSA encryption scheme (the one currently used by Netscape Navigator) put forward a challenge.[4] They encoded a message, and offered a $100 reward to anyone who could crack it. They felt that given the computing resources of the time, and even granted advances in chip speed with a factor of millions, nobody would be able to break their code in the foreseeable future. They were wrong! Instead of thinking of a single computer as a self-contained and limited system, Derek Atkins, a twenty one year old engineering student at MIT realized that while one computer would take a long time to crack the code, he might harness the power of the Internet and distribute the computing load over many computers. And that's exactly what he did: in 1991 he directed his friends to use a recently discovered mathematical method to devise a program which would crack the code,

---

4 Steven Levy "Wisecrackers" in *Wired Magazine,* March 1996, 128.

and had it ready to go by mid 1993. The program was distributed over the Internet to more than 1,500 computers on six continents to create an expanded computer which churned out 5,000 MIPS.[5] The code was cracked in the spring of 1994 by looking beyond the boundary of the individual computer, and thus consolidating the power of the network.

Another example would be the development of the Linux operating system. Linux was not so much invented as evolved through creative re-administration. It is a prime example of the networked aesthetic of the expanded public sphere of individuals working in concert. It began with Linus Torvalds. He was in school in 1990, and owned a PC which ran Minux, an operating system designed mostly as a UNIX tutorial. UNIX is a powerful operating system which, at that time, could only be run on more powerful computers. So, he imaginatively worked within his limitations, and wrote a few programs – a terminal emulator and a disk driver so that he could save files to disk. He posted these initial programs, and generously shared them as freeware – software which is distributed primarily over the Internet, and for which there is no charge. From there it took off. As a result of Torvalds' interlocution, the operating system evolved in a democratic and Darwinian manner; anyone could contribute code to the operating system, but only the most evolved would become part and parcel of the final released version. He presided over the development, but was by no means entirely in control of it. He provided the seed idea, and the guidance, but left the mechanics of its development up to the community of creative users. Today, Linux has an established base of nearly ten million users in 120 countries, and is comprised of millions of lines of code. All of this primarily because Torvalds didn't follow the usual notion of creating software through the confines of a defined proprietary scheme – hundreds of hackers around the world wrote it, collaborating over the Internet.

---

5   MIPS is short for millions of instructions per second – a measure of computational power.

These examples serve to help define the notion of the creative interlocutor, a multimedia universal designer, engineer, artist, socially responsible person whose mandate is to help negotiate the crossing of boundaries. While they reside in the field of technology itself, their parallels must be generated within the humanities and arts.

Variations on the Technical Body

Dennis Phillips

Strap on those prosthetics: Calendars and pitchpipes,
eroticism digital or eroticism genital,
fulfillment is still an incomplete gesture
even when we separate the digits from the hand
or augment the genitals, if you know what I mean.

*

What form of prosthesis is governance?

*

It may not be true that cathedrals were invented
to amplify the voice, spoken, chanted or sung.

*

Or acceleration is the worry, strap into what vehicle you may.

*

Just the vehicle, mylar or paper, sonnet or discourse,
picture the pen, that mouth and brain extension
or is it sword?

*

Opposition may also construct around us,
dappled though the light on yonder mountains may be,
or purple.

*

Whose spoon shattered on which table
in Vienna or Zurich
or was it some other extension of the digital?

*

I only say to you what you have already said to me.

*

Therefore (lyrical):

A devil in the umbra of a zeppelin
let's break the chain the devil said—
Devil, ouch, a sticky foreskin.

Carry a message, proctor a war,
wait for the zeppelin, carrier of news.

These things the devil said, shod in horn the devil said.

Ouch the foreskin, blesséd kneepads
adds the devil: gesture completed.

On the road to wisdom, aerie or whisky,
the sidewalk bordered in ordnance,
bless their nails, bless their strapped biologies.

Dennis Phillips

Private the office the devil peeks in
barbed of forehead the devil peeks.

Let my shadow be my helmet
shadow I stand in, shadow I make.

These are the stories now disclaimed,
the famous who have laundered there
now are elsewhere gone.

Yes the mountains bear clear today
(Ouch the foreskin sticky)
Yes the umbra a floating appeal
(Shod the devil in harmony applies.)

"Maybe they are everywhere,
hearing all the messages
we are constantly sending,
and under no circumstances
do they want to answer."

Alien Abilities and Behavior

Seth Shostak

In the 1996 blockbuster film *Independence Day*, the American president finally manages to have a short but pithy conversation with one of the aliens whose machines are wreaking havoc and destruction on our planet. "What is it you want us to do?" asks the president in measured tones. The alien responds pointedly: "Die."

The message was brutal, but you have to give the extraterrestrials a lot of credit for knowing their mind, and being straight up about telling us. You also have to marvel at the fact that such a conversation could take place at all. This is not so much a question of language (the alien uses a human intermediary as a translator), but rather of mind set. The alien's mental capabilities are obviously not so different from our own.

This is a common circumstance in movies. In the 1985 film *Enemy Mine*, earthling Dennis Quaid finds himself stranded on the less-than-lovely planet Fyrine IV with only a reptilian alien for company. By the film's midway point, the earthling and the extraterrestrial (played by Louis Gosset, Jr.) have joined forces in the interests of survival. The alien soon becomes the best pal Quaid ever had. Not only do the two species understand one another, but they share such traits as anger, humor, and a knack for sarcastic insult.

Hollywood aliens inevitably think like humans, despite their evil intentions and scaly complexions. Part of this resemblance is due to the requirements of dramatic structure. If your antagonist is unfathomable, then his actions appear to be random. Such an enemy cannot be either evil or outsmarted, for he has no comprehensible will. Even the great white shark in *Jaws* had a clearly understandable game plan: to gnash and gnaw his way through as much of the film's supporting cast as possible. He wasn't just an underwater eating machine; he was malevolent. That set the audience up for a collective sigh of satisfaction when the great white was turned into fish meal at the movie's explosive climax.

Real sharks, of course, are less deliberate, and also less decipherable. Humans have a hard time imagining how fishes or birds think about the world. The only concept of "thinking" we can muster is our own. Consequently, we assume that our own perceptions and style of comprehension will be shared by ET. Needless to say, they may not be.

ET'S CULTURAL LEVEL

If we ever encounter or communicate with extraterrestrials, one thing is overwhelmingly probable: the aliens won't be at our level of development. Neither their technical abilities nor their IQs will be comparable to ours.

Why do we say this? Simply because if we detect the aliens, the very fact that we've done so will be a consequence of their high degree of sophistication. To begin with, it's obvious that any extraterrestrials we uncover can't be *less* sophisticated than humans. We won't find backward, primitive aliens on our doorstep or hear them on the radio, simply because we will only encounter ET if he makes contact with us, either literally or figuratively. And that requires substantial expertise.

For example, it's possible (though unlikely) that extraterrestrials might physically enter our solar system and pay a personal visit. Perhaps they will land in the back yard and ask to be taken to the Dalai Lama or some other leader. That would involve a feat of interstellar transport far outstripping our own abilities. Any aliens capable of rocketing from one star system to another will obviously hail from a very advanced society. If ET pays a house call, you can be sure that, at least on a technological level, he is very different from you and me.

Another prospect is that we may detect ET at a distance, for ex-

■ BEING HUMAN:

ample by eavesdropping on his radio traffic. That, too, would require that he be scientifically adept. To give us a chance of tuning in on his signals, ET would have to construct powerful transmitters and affix them to large, steerable antennas. So clearly, even if we're expecting communication rather than contact, the aliens will need to be at least as technically sophisticated as twentieth-century humans.

In fact, they will be a lot *more* sophisticated. The reason for this can be understood by a simple analogy. Imagine that you have just learned the game of chess. You memorize the moves, play a few games against your younger brother, and decide that you really like this kingly pastime. So you find a chess club in town, and join up. With the club's roster in hand, you randomly pick one of the members to challenge for your first, friendly game. The chances are high that your initial face-off will be against someone who has played chess for years; someone who has advanced far beyond your level of competency. You are the new guy in the club, and most of the other members have a lot more experience.

When it comes to communicating with ET, we are like the chess neophyte. Humans have had high-powered radio transmitters and sensitive receivers for only a few decades. So we're still a new member in the club of radio technology. The first other clubman we encounter may not even remember which of his ancestors *invented* radio. Any extraterrestrial with radio capability will likely have had it far longer than we.

In addition, if we do pick up ET on our radio telescopes, that very fact will virtually guarantee his advanced cultural standing. Imagine, as example, that every 10,000 years a planet somewhere in the galaxy cooks up a technologically sophisticated civilization. This example postulates a fairly optimistic birth rate, incidentally, for it implies that the galaxy has witnessed the appearance of a million such civilizations in the course of its long history. Now, you can be certain that each of those technological societies will invent the atomic bomb at about the same time they invent radio. After all, the two devices appeared on Earth only a generation apart. If you are the pessimistic sort, you might expect that many of these newly arrived technological societies will blow themselves up within a century or two after first going on the air with their high-powered radios. If this regrettable situation is the norm, then societies will only broadcast for a few hundred years. The chances that *any* of the million civilizations that have sprung up in the Galaxy is on the air now would be only about 1 in 100. We won't hear a thing, and any receivers we use in an attempt to tune in ET will be greeted with nothing but static.

On the other hand, suppose you take a more optimistic view. Imagine that technological societies manage to survive and stay on the air for ten thousand years or more. Then clearly we might have a reasonable chance of hearing one, since at least a few simultaneously broadcasting societies will be hanging out in the Galaxy at any given time.

The numbers we have cited in this argument are only an example. After all, no one really knows how often technically sophisticated societies arise. But the argument itself leads to a conclusion that is not critically dependent on the precision of our estimate. It boils down to this: If we succeed in overhearing an alien broadcast, the odds are great that the beings on the transmitting end are members of an old civilization. For them, radio will be a technology that their ancestors devised tens of thousands, or possibly many millions of years ago. They will be as much in advance of us scientifically as we are beyond the Neanderthals, and maybe a great deal more. In addition, if this difference in maturity is large, it may be that they have outstripped humans in more than just technology. Their intellect may also be far suppler than our own.

ET'S IQ

Is it possible that ET surpasses us in raw IQ as much as we surpass caterpillars and canaries? The human experience suggests that he could. After all, brains have developed quickly for humans. *Australopithecus africanus*, one of our humble predecessors, had only a pound of gray matter. In the two million years since this pint-sized pundit roamed the

BEING HUMAN:

savannas, the human brain has trebled in size, and now houses tens of trillions of neurons. So why not bring on more? Unfortunately, the dimensions of the human birth canal limit how large a baby's head can be. Big-headed babies can't be born. And that limits the number of neurons, as evolution on Earth has so far failed to miniaturize these building blocks of the brain (they are the same size for all animals). If Nature is making any attempts to produce significantly smarter humans, her efforts may have stalled, at least for the moment.

But that needn't be the case for ET. If he can somehow reduce the component size of his thinking machinery, for example with smaller neurons, then he could pack more brain into his head. And of course, his head could be bigger to begin with. Either of these factors could send ET's IQ off the charts, as a small change in neural number makes a big difference in performance. While our cerebral volume is only three times that of *Australopithecus*, our IQs are far more than triple his. When it comes to thinking, what matters is the number of neural *interconnections*. The number of interconnections rises geometrically with the neuron count. As Josef Shklovskii and Carl Sagan noted many years ago, there is probably no limit to the extraterrestrials' intelligence. So while it's likely that other technological societies may have been started by creatures with human IQs, any that we overhear or encounter will have had a significant amount of additional time in which to dramatically improve their intellectual lot.

This disconcerting thought is one that Hollywood is unwilling and unable to address. Cinematic aliens are always at about our level of savvy, although they frequently outpace us in technology. (Not *all* technology, however. The hostile extraterrestrials of *Independence Day*, despite having built the mother of all mother ships, eventually succumb to 20th-century computer viruses and dogfight tactics.) Only a modest amount of reflection should convince you that the aliens of the movies are of necessity at our level, for otherwise the films would be, like *Bambi Meets Godzilla*, short and distressing. It is good dramatic practice to have a worthy antagonist. It is bad dramatic practice to have one that's invincible. Alas, the real aliens are much more likely to fall into the latter category than the former.

Of course, some extraterrestrials will be more advanced than others. The differences will not be minor. On Earth, all human societies are similar in their level of development. Yes, the *National Geographic* can make a decent profit chronicling the customs of primitive tribes in some antipodean outback, but even these stone-age societies differ from their first-world opposite numbers by only a few millennia at most. *Homo sapiens* is a mere 100 thousand years old. All mankind left the starting gate at the same time. As different as human societies may seem, they're still running very much in a pack.

Not so for extraterrestrial societies. The Galaxy has been around for 10 billion years or more. Presumably, sentient civilizations have emerged during most of that long history. Needless to say, it's extraordinarily unlikely that two random worlds, separated by hundreds or thousands of light-years, will develop in parallel. No two societies will have left the starting gate at the same time. The disparity will be such that there is little chance that aliens from two societies anywhere in the Galaxy will be culturally close enough to really "get along." This is something to ponder as you watch the famous cantina scene in *Star Wars*. In this futuristic pub, a politically correct, multi-hued assemblage of extraterrestrials (all conveniently upright in posture, if not demeanor) share a brew and engage in some back-slapping camaraderie in the grungy port city of Mos Eisely. Does this make sense, given the overwhelmingly likely situation that galactic civilizations differ in their level of evolutionary development by thousands or millions of years? Would you share drinks with a trilobite, an orangutan, or a saber-toothed tiger? Or would you just arrange to have a few specimens stuffed and carted off to the local museum?

The fact that ET's intellectual plane will be beyond that of humans may make him difficult or impossible to understand. Even aside from the indelicate matter of IQ, the enormous culture gap could cause alien communications to appear mysterious or magical. Consider for example

BEING HUMAN:

Cro-Magnon man, who sported biological equipment quite similar to our own. Despite the kinship, Cro-Magnon man would have a hard time understanding the purpose of a computer operating system or the meaning of a television signal. Two centuries ago, the scientific societies of Western Europe began to regularly interact with one another, thereby stimulating a faster research pace. The societies were able to do this because the countries involved were all at about the same level. They also spoke comprehensible languages. Neither condition will obtain in the case of communication with aliens. Human intelligence has been so shaped by the peculiarities of our Earthly environment and evolution that we might be simply unable to connect with non-humanoid intelligence. And, needless to say, the aliens aren't going to speak languages understandable to us, nor become instantly fluent in English. In fact, they might be physiologically incapable of speaking *any* Earthly languages, in the same way that a giant squid is.

Aside from problems in communication due to lack of vocal chords or compatible ways of thinking, there's also the matter of common cultural ground. ET might regale us with music or poetry, but there's little chance that we'll either recognize or appreciate it.[1] The alien's literature and even medicine may have little congruence with, or relevance to, our own.

However, sci-fi stories have conditioned us to expect at least some points of cultural overlap with ET, especially in science and technology. As noted, we demand that the extraterrestrials be accomplished in such fields. These minimum requirements spring not from our desire to promote science literacy in the Galaxy, but simply because only technically deft aliens will make their presence known. In the last chapter, we ar-

---

[1] Note that this hasn't kept us from occasionally launching some of our own music into space. The Voyager probes, sent to reconnoiter the outer solar system in the late 1970s, carried a small record bearing such cosmopolitan musical offerings as Bach's *Brandenburg Concerto* and Chuck Berry's "Johnny B. Goode." These were for the edification, or perhaps puzzlement, of any extraterrestrials who might retrieve these spacecraft in the distant future.

gued that intelligence will frequently spring up on planets where species compete, thanks to its tremendous capacity for adaptation. But does intelligence inevitably lead to science, particularly science like our own?

Once again, researchers try to gain insight on a universal question by holding a mirror up to themselves. They turn to our own history, and what they find is that science is rare for human societies. While every earthly civilization worthy of the term has elaborate social structures, organized agriculture, and urban centers, they don't all manage to cobble together an analytic description of the physical world. In Europe, the millennium-long preoccupation with theology during the Middle Ages promoted very little scientific development, despite the widespread practical knowledge of many artisans. Anthropologist Kent Flannery notes that the Mayans had civil-servant astronomers and boasted a better calendar than the Spanish conquistadors who eventually laid waste to their state. But the Mayans didn't develop metal tools or even the wheel. Philosopher Nicholas Rescher finds it easy to imagine advanced alien societies that are possibly skilled in engineering, but lacking in substantial science. As Rescher says, "a lot of know-how can be built up without much know-why."

This skepticism regarding the emergence of science is reminiscent of the argument against the evolution of intelligence. "You don't need it to get by, so it might not occur very often." But the counter-argument is also similar: an understanding of nature can rapidly expand your range of actions. Intelligence has survival value, and so too does science. Less than four centuries after Copernicus espoused a new way of looking at the heavens, we see modern science's dramatic effects in every corner of the planet. It has revolutionary consequences, not the least of which is the rapid extension and proliferation of our species. And while the majority of Earth's civilizations, including such long-lived empires as Egypt and China, failed to develop very much science, it does not seem to be a fluke, an improbable activity limited to Renaissance Europe. The Greeks, at least, had been there before. Human intelligence took four billion years of evolution, and has appeared once. Still, we think it

■ BEING HUMAN:

likely that comparable intelligence will sooner or later emerge on many inhabited planets. Science has taken intelligent creatures only 100 thousand years to concoct, and has appeared more than once. So it seems that if intelligence is probable, science is inevitable.

ET'S ATTITUDE

Aside from matters of a cultural and cerebral nature, can we say anything about the extraterrestrials' world view? We've noted that any aliens that we either detect or run across will be members of an old, technologically accomplished society. Given this elevated cultural status, mightn't we expect ET's morals and motivations to be on a similarly lofty plane?

We might, although Hollywood usually doesn't. Film aliens are generally of low moral caliber. The malevolent ones are single-minded predators, with only destruction or lunch on their minds. Their ethics are no more complex than those of a piranha. They don't feel much need for sociable behavior even among their own kind. Hollywood extraterrestrials seldom interact with their planetary brethren, opting instead to follow one another around like mindless robots with strict orders.

The friendly cinematic aliens, on the other hand, share human morality so completely that they degenerate into funny-looking pals who happily participate in our activities. Little ET will pull a beer from the fridge, sit on the couch and watch daytime television. The furry Chewbacca and the logical Mr. Spock willingly take on the second-in-command jobs in our spacecraft. We readily accept these alien comrades, although you would probably regard it bizarre in the extreme if NASA suddenly opened its astronaut program to non-human species here on Earth, let alone to something reared on the eighth rock from Alpha Centauri. Could we rely on such extraterrestrial crew members in a crunch? Would they give their all to save human society? Can we expect that they have the same ideas about winning, about power, friendship, or courage?

What we call moral behavior, or altruism, clearly has survival value in human society, and consequently the real aliens will likely practice it as well, at least with one another. Of course, it's asking a lot of ET to ask him to extend his altruism to non-related species from another solar system. But perhaps such altruism could be motivated by religious belief. Religion, which institutionalizes morality, promotes stability in human societies. It may do the same for the aliens. In the 18th century, as Captain James Cook explored the Pacific discovering island after island, his log invariably contained descriptions of the local religious practices because such practices were always to be found. Religion is universal among human societies. Whether the same holds true for the extraterrestrials depends on whether religious belief is a fundamentally useful survival tool, or merely an artifact of some other human trait. For example, music, like religion, is ubiquitous among earthly societies. But music doesn't really promote survival; it's a side benefit of the way in which our brains are organized for other activities that *do* promote survival. Music may not be among the pastimes of thinking beings organized along different lines. In the same way, it's difficult to know whether the extraterrestrials will have their own theology.

Of course, if they do, there's little chance that the specifics of ET's faith will mimic our own, any more than his appearance will resemble ours. Not so long ago, this self-evident statement was considered less so. In 1853, British philosopher William Whewell, who penned *Of the Plurality of Worlds: An Essay*, was concerned that if the universe is filled with intelligent societies, then Jesus would be compelled to undertake an endless voyage of salvation from one planet to the next. Whewell realized that this literal extension of Christianity to other worlds might considerably dilute mankind's special relationship to God. Earth would be just one more station on a non-stop itinerary to save souls. However, the interest in, and concern with, logistics problems of the sort that bothered Whewell have dissipated in the twentieth cen-

■ BEING HUMAN:

tury. An informal polling of modern theologians by the Israeli sociologist Michael Ashkenazi has shown that today's clergy members are far less insistent that alien religions should conform to earthly ones.

Although the extraterrestrials would probably be altruistic with one another, and even religious, that possibility alone says very little about whether they would be "moral" with us. Altruism has a biological basis, as noted. A selfless act may prompt reciprocity at a later date. One good deed encourages another, and both the individual and the species benefit. However, when it comes to interactions between extraterrestrials and humans, the aliens will have little biological reason to be altruistic, only intellectual ones. After all, consider how we treat animals. Some are our pets. Others we grind up and lace with ketchup. Human attitudes about the treatment of our local "aliens," namely animals, are highly divergent. We treat many of these other species badly (from their point of view). Why shouldn't the aliens behave the same way towards *their* aliens?

This argument might be faulted on the grounds that the animals we abuse are, after all, not terribly intellectual. Presumably, any ET in contact with us will recognize our cognitive horsepower, and grant us a little respect. We're all sentient beings, after all, and surely we can sympathize with one another?

Not necessarily. Altruistic behavior towards fellow beings who are outside the immediate group is not so obviously advantageous, and doesn't always occur. The diplomat Ichiro Kawasaki has noted that traditional Japanese society is subject to rigid rules of etiquette and behavior. How, then, to reconcile this institutionalized civility with the brutalities perpetrated by Japanese soldiers while occupying other Asian countries in the Second World War? Kawasaki suggests that "once outside the confines of his home or family, a Japanese is at last 'liberated' from all these restraints and starts behaving like a different person." The rules of behavior at home don't necessarily apply abroad. The indigenous inhabitants of the Americas and the islands of the Pacific all suffered grievously when European explorers made contact. This was true even during the voyages of Cook, who was under specific instruction not to inflict damage on the natives. The aliens aren't guaranteed to treat us gently simply because we're conscious and cogitating.

Curiously, the distress that befell the South Sea islanders was not really Cook's fault, and in his story lies a lesson that might be applicable to any future contact with extraterrestrials. The famous British captain had insisted upon honest dealings between his crew and the natives, and he absolutely forbade violence. He also refused to let syphilis-infected sailors ashore. However, this enlightened effort to avert later disaster was frequently thwarted. The native women would swim out to the ship. They did this because the foreign civilization that had appeared on their coasts was seen to be technologically superior. After all, they arrived in large sailing ships and were outfitted with cannon. The native women assumed that these technically proficient visitors were superior in every other way, as well. They might be good breeding stock.

The presumption that alien technological prowess goes hand in hand with a superior culture or sophisticated morals is one that we might make as naturally as did the island women. But the link is hardly inevitable. The Russian academic I.S. Licevitch has remarked on the fact that, while 18th century Europeans would marvel at the technology of their 20th century descendants, they would probably be horrified by the current low level of decency and popular culture. Technological supremacy is no guarantee of cultural refinement or moral virtuousness. Conceivably, the aliens could be scientifically hi-tech and culturally low-brow.

What, then, can we say about ET's ethical being? That he will have a moral code seems highly probable, given the obvious survival value that any society derives from regularizing the behavior of its members. He may also be religious. But both his moral code and his theology will serve the purposes of his own kind. His views towards us, or towards any other creatures that may exist among the stars, may be as divergent as ours are towards cockroaches and cocker spaniels.

When it comes to describing the mental and moral makeup of the

■ BEING HUMAN:

aliens, the sci-fi films usually fumble the ball. Most of the Hollywood product would have us believe that ET, once in contact with earthlings, will be interested in our societal values, our music, our kids, and possibly even our job opportunities. Wide-screen aliens may be here for good or evil, but at least we can discuss such matters with most of them. Their science and technology are sophisticated, but their intellectual acumen is not very dissimilar to our own. They either share our moral code, or are completely amoral.

Alas, for the case of any real, detected ET, most of this cultural congruence is doubtful. He will be technologically and intellectually far in advance of us, and his culture will be far older than ours. We might be unable to understand him at any level. ET is likely to be ethical and possibly religious, but his attitude towards the inhabitants of other worlds is unpredictable. In the end, the only thing we might have in common is science. The laws of the universe are, not surprisingly, universal. They may be formulated in a multitude of ways, but that's not the point. If, for example, Nature allows only one efficient mode of communication between star systems – electromagnetic waves – then the unearthly way in which the aliens describe these phenomena is of little consequence for our search, although it might make for an intriguing point of discussion once contact is established. Any ET that we either hear or come across will have mastered science. Beyond that, we can only say that humans are undoubtedly the low-level baseline for the complexities and capabilities of the extraterrestrials. Our speculations on the details of their nature may be likened to an attempt by tanked goldfish to describe the abilities and motivations of the two-legged creatures who share the house.

MACHINE INTELLIGENCE

Our view of ET is heavily predicated on what *we* are – the first thinking beings to occupy the planet. We assume that if the human species can figure a way to survive its own destructive tendencies, then perhaps an extraterrestrial species has managed to do the same. The result would be the establishment of a long-lived, technological civilization. Such societies may exist elsewhere in the Galaxy, and we might either overhear or meet them. But note that we have subtly assumed that our contact will be with the *first* sentients to appear on the planet they occupy. In fact, our quest is not so restrictive. We're only looking for cosmic company. We don't insist that any aliens we discover are the original thinking critters from their own star system. All that counts is that they be around now.[2]

This is an intriguing point simply because the universe is old and the Earth is young. Other worlds will have started down the tricky paths of life billions of years ago. They may have spawned intelligent beings long before our planet came into existence. But species come, and species go. As unpleasant as the thought may be, even *thinking* species might come and go. A commonplace scenario, as we've already noted, might be one in which intelligent creatures spring up on a planet, strut their brief stuff, and are forthwith extinguished by external circumstance, or more plausibly by their own hand. Many editorial writers have postulated such a dismal end for *Homo sapiens*, after all.

If this dystopian view of alien nature is accurate, then self-destruction could frequently bring quick and ugly ends to technological species (a point we will return to later). In that case, our efforts to locate intelligent neighbors might be stymied. Shortly after the aliens become sophisticated enough to build radio transmitters and rockets, they will suddenly expire in a mushroom-shaped cloud of smoke.

But if, as we have emphasized, intelligence has real survival value, it will come back like a bad penny. It will arise Phoenix-like from the ashes of its prior destruction, undoubtedly in new form. Imagine being

---

2 In the case that we find the aliens by, say, overhearing a radio broadcast, "around now" means that they were around at the time the transmission was sent.

on such a world when civilization erupts for a second time, and the archaeologists begin to dig up the ruins of a long-gone, intelligent species on their own planet! This phenomenon could be encouraging news for those seeking cosmic company. If intelligence is persistent, if sentient species routinely have successors, then the outlook for finding them brightens considerably.

Another possibility is that thinking beings can manage to avoid any such break in the reign of intelligence. They might do this by finding a cure for the built-in aggression that is an inevitable by-product of evolution in a competitive world. A civilization whose members were content to contemplate their navels or play harmless video games might last a long time. But another, less insipid approach is possible. A sentient society might be able to short-circuit biological evolution, and deliberately engineer its own successors. This seems plausible simply because we see signs of it on our own technological horizon. We may be unwittingly taking the first steps toward producing the next thinking inhabitants of this planet.

Consider the medical practice of transplanting body parts, such as hearts or knee joints. (Whether the parts come from other humans or from other species, for example by using a pig heart instead of a human one, may be of great interest to moralists, animal activists, and the pig community. It is not particularly relevant to our argument here, however.) Success with these biological transplants will create a strong drive to develop artificial replacements. Manufactured parts will, in the long run, be cheaper and more readily available when required. The recipients' bodies will also be less prone to reject them. Dialysis machines (artificial kidneys) are a contemporary example, although today's models are too large for comfortable implantation. But there is little doubt that the next half-century will witness the development of a greater variety of both artificial body parts and synthetic bodily fluids, such as blood. At some point in the next century, we may, indeed, have the technology for the *Six Million Dollar Man*.

Parallel with this partial mechanization of humans will be the humanization of machines – the production of sophisticated robots. These devices will be outfitted with computing horsepower that will allow them to take on far more challenging work than today's models, such as those that put together the family car. Robots designed to do the tasks that humans would rather avoid – picking strawberries, mining coal, or servicing nuclear reactors – are not far in our future.

These easily-foreseen co-minglings of machine and protoplasm are, in a broader sense, simply hi-tech tool use. For thousands of years we have been constructing devices that facilitate our existence, that free us from the infirmities and unpleasant labor that interfere with what we like to do most: contemplating new physics, chasing one another around the desk, or dozing in front of the TV. We can confidently expect that within our lifetimes, the rude chipped flints of our forefathers will evolve into sophisticated household robots and implantable spare parts.

This is all but a modest extrapolation of current capability. The situation changes radically, however, if we take one more step and produce a stand-in for our brains. This would go far beyond tool use, for now we have replaced the guy using the tool. We will have engineered our successors.

Is this possible? Everyone knows that machines can't think. The best among them can play a depressingly good game of chess, but none can, say, write a book on the Zen of Chess. For years, researchers toiling away in the optimistically-named field of Artificial Intelligence have tried to build machines that could successfully tackle such relatively simple problems as stacking tin cans or diagnosing illnesses. Their success so far is modest. But the researchers involved remain upbeat in their belief in the possibility of thinking machines, although the idea is not without naysayers. Among those who are not so sanguine about the possibility of synthetic sentients is the British physicist Roger Penrose. He has argued that consciousness and creativity may depend upon unpredictable quantum mechanical processes, and would therefore be beyond the capabilities of a deliberately constructed device. However, most scientists subscribe to a more mundane and mechanistic view of

BEING HUMAN:

our thinking apparatus. MIT physicist Philip Morrison has described brains as merely "slow-speed bit processors operating in salt water." If so, it is quite possible, in principle, to replicate their functioning in dry hardware. The detailed construction of such a machine might be somewhat different than that of the human brain, in the same way that airplanes don't imitate birds by flapping their wings. But the functional behavior could be the same.

Science fiction long ago mastered the production of such cybernetic cerebellums. A contemporary example is the android Data, a favorite character on *Star Trek: The Next Generation*. Data is human-shaped, presumably for ease of accommodation aboard the USS Enterprise-E, a ship built, and largely crewed, by hominids. He has landed the job of Operations Manager, showing that even in the 24th century, machines are still putting humans out of work. Even more perplexing, Data has been accorded full civil rights as a sentient being. This is enlightenment of a curious sort, for it suggests that if enough citizen androids were constructed, they might vote themselves into office and take over civilization via the ballot box.

Data was programmed without emotions, although an emotion upgrade chip was made, and even inserted into the android's socket. Lamentably, the chip burned up due to a power problem, and Data remains saddled with a Pinocchio complex: he wants to be human. Data longs to understand us, and even develop a sense of humor. He wants to be a good primate.

Data is like many aliens of both fiction and the popular imagination: he suffers an emotion gap. And because of this, we feel a bit sorry for him. Poor Data; he's got all the equipment any red-blooded male has, and is programmed in multiple unspecified "techniques," but he can't quite muster love. Naturally, this gives the android some reassuring vulnerability. But emotions, as we've noted, are useful survival tools for a biological being in a competitive environment. They could be less beneficial for a machine, and as Mr. Spock might remark, could lead to behavior that (from a machine's point of view) is "not logical." Androids don't have to get mad to defend their turf or fall in love to ensure their progeny. They might be as cool as chipped ice.

However, it is Data's intellectual prowess that is of interest here. He claims to have a mental storage capacity of 100 quadrillion bytes, which is about a hundred times as much information as you'll find in the Library of Congress. Despite this impressive total, the android doesn't shame his fellow (human) crew members with overwhelming intellectual insight and brilliant deduction. Data's just another guy on the ship's bridge (except that his jokes tend to fall flat).

But imagine a real thinking device, able to react to new situations, plan ahead, and evaluate strategies. It could be fed the world's accumulated knowledge and not forget a thing. One of the first tasks we might put before this super savant is to design its own successor – a machine more capable than itself. And that machine would be asked to do the same. In short order, we could produce a device that the slow and uncertain processes of biological evolution might never bring forth.

The consequences of silicon smarts for our own society would no doubt be revolutionary. But for the purpose of finding the extraterrestrials, the important point is that truly capable machines would be able to do something that is quite hard for biological intelligence: they could journey to the stars. Biology is fragile, and complex organisms live only a short time. Interstellar travel for biological aliens will only be practical if enormous velocities, close to the speed of light, can be attained. Machines, on the other hand, are less likely to be in a hurry. Cheaper and safer slow-speed rocketry might be an acceptable travel mode for a machine.

Why would a machine leave the planet of its birth? One obvious attraction would be the fact that its natal neighborhood might, like Earth, be in a relatively dull part of the Galaxy. When asked what the truly interesting and important things in the universe are, most humans would probably answer "sex and money." In fact, a more global answer is matter and energy (which we manage to convert to sex and money on Earth). Matter and energy can be found in far greater abundance elsewhere. For example, in the central regions of our galaxy, the density of stars is more than a million times higher than in the fringe areas we inhabit. We are in the galactic boondocks, and a machine capable of leaving home might naturally hanker to go where the action is.

So one possibility is that machines developed billions of years ago by a distant, alien civilization have spread through the star fields of the Milky Way, perhaps producing occasional duplicates as back-ups, and are now cruising the interstellar voids like a flotilla of insects. They could rapidly adapt to the harsh environments of the galaxy, since machines can improve themselves. This ability doesn't extend to living species, despite beliefs to the contrary by early researchers. At the beginning of the 19th century, the French biologist Jean Baptiste Lamarck proposed that creatures could influence the characteristics of their offspring by their own behavior. In straining to reach leaves on high branches, giraffes would stretch their necks, and this modification would be passed on to their progeny. In fact, acquired traits like a stretched neck are not inherited. Biological evolution, as Darwin showed, is a slow and haphazard process of pruning for comparative advantage.

But machine improvement wouldn't proceed at Darwin's languorous pace. The machines could rapidly refine themselves, and adapt to existence in the rarefied bath of starlight and gas that fills the vastness of space. Their evolution would be Lamarckian and speedy. Their ability to think might earn for them the honor of being the true intelligentsia of the galaxy.

In characterizing the aliens, we can make plausible arguments as long as they are biological. Our description of ET's cultural level, cranial capability, and moral bent are all based on traits and behaviors that have clear survival value in a competitive, life-filled environment. They obviously apply to us, and one assumes to any biological aliens that inhabit a planet not enormously dissimilar to Earth. But if ET is a machine, and possibly a lonely machine with very little daily interaction with others, then his attitude and behavior will be strange beyond our imaginings.

THE TECHNOLOGICAL EXTENSIONS OF THE BODY

It may be that the ultimate achievement of biological intelligence is to get the silicon sentients underway. Biology is a nice, necessary first step, and a creature like *Homo sapiens* is an example of a simple foray by life into the realm of thinking entities. But the next big step, the truly dramatic step, is the start of machine intelligence. You might regard this scenario as both arguable and depressing. Nonetheless, if sentient machines do exist, then their obvious advantages could make them an important, if not the dominant, intelligence in the galaxy. Should SETI scientists succeed in picking up a signal from the cosmos, no one will be surprised to learn that the signal comes from a machine, a radio transmitter. But we should also be prepared for the possibility that *that* machine is in the service of another machine.

■

Alphabetical Listing of Panelists and Contributors

AMIRI BARAKA is a poet, dramatist, essayist, fiction writer, and political activist in Newark, New Jersey. His thirty books include *Transbluecency: Selected Poems, Digging, Eulogies, Blues People.*

JOHN PERRY BARLOW is a cognitive dissident, writer, lecturer, and cofounder of the Electronic Freedom Foundation, in Pinedale, Wyoming.

JAMES BAILEY is a writer and technology consultant in New England. His books include *After Thought: The Computer Challenge to Human Intelligence.*

TIMOTHY BINKLEY is a digital artist, author of books and articles on aesthetics, and teacher in New York City. He has designed software for artists, including Paint Brush and Symmetry studio.

DANY-ROBERT DUFOUR is a philosopher, professor of Semiotics at the University of Paris VIII, and writer. His books include: *Le Bégaiement des Maîtres, Les Mystères de la Trinité, Folie et Démocratie, Lettre sur la Nature Humaine à l'Usage des Survivants.*

MARCOS EINIS is a psychiatrist and psychoanalyst in Paris. He was the cofounder (along with Serge Leclaire) of the Club du Silicium and of Organisation Hyperlog. He is the author of many articles on psychoanalysis and technology.

NATHAN FELDE is the former director of the Media Laboratory of NYNEX corporation. He is a graphic and computer technology designer, in Massachussets. His books include *Public Works* and *Images.*

RENÉE FOX is Annenberg Professor of the Social Sciences at the University of Pennsylvania with joint appointments in the Departments of Sociology, Psychiatry and Medicine. Her books include (with Judith Swazey) *The Courage to Fail* and *Spare Parts*.

ANNICK GALBIATI is a psychoanalyst in Paris, member of Le Cercle Freudien, Association Psychanalyse et Medicine and APUI.

SALVATORE GUIDO is a psychoanalyst in New York City, member of Après-Coup Psychoanalytic Association, on the faculty of Eugene Lang College.

MICHAEL GRODEN is a professor of English Literature at the University of Western Ontario and a hypertext author. His books include *The Johns Hopkins Guide to Literary Theory and Criticism* (with Martin Kreiswirth), and *James Joyce's Manuscripts: An Index*.

JACQUES HASSOUN was born in Alexandria, Egypt in 1936 and died in Paris in 1999. He was a psychoanalyst, writer, teacher, political activist, and cofounder of Le Cercle Freudien (Paris), President of Le Cercle Juif Laique (Paris). His books include: *Juifs du Nil, Non Lieu de la Mémoire, La Cassure d'Auschwitz, L'Exil de la Langue, Les Passions Intraitables, La Cruauté Mélancholique, L'obscur Objet de la Haine*.

PERRY HOBERMAN is an installation and performance artist in New York City. He has performed and exhibited extensively in the United States and Europe.

JACQUES HOUIS is a teacher of French at the Brearley School in New York City, a translator and editor.

VYACHESLAV IVANOV is a professor of Slavic Languages at the University of California at Los Angeles. His major research interests in-

■   BEING HUMAN:

clude semiotics, Indo-European comparative grammar, general linguistics, neuropsychology, film theory and poetics. Among his books are *Odd and Even: Asymmetry of the Brain and Sign Systems* and *Carnival* (with U. Eco and M. Rector).

LEWIS KIRSCHNER is a psychoanalyst in Cambridge, Massachusetts, training and supervising analyst at Boston Psychoanalytic Institute, and Associate Clinical Professor in Psychiatry at Harvard Medical School.

JEAN-PIERRE LEBRUN is a psychoanalyst and psychiatrist in Namur, Belgium, member of Association Freudienne Internationale; on the Faculty of the Université Catholique of Louvain. His books include *De la Maladie Médicale, Un monde sans Limites*.

SERGE LECLAIRE was born in Strasbourg in 1924 and died in Paris in 1994. A student and associate of Jacques Lacan, he was a psychiatrist and psychoanalyst in Paris. As founder, in 1968, of the department of psychoanalysis at the University of Paris VIII, he introduced the teaching of psychoanalysis to French universities. At the same time, he became involved in the feminist movement. He was the chairman of the Société Française de Psychanalyse (1963) and a member of the board of the Ecole Freudienne de Paris (1964–1968). He experimented with new ways of disseminating psychoanalytic ideas, including film and television. He founded the association A.P.U.I., devoted to the study of psychoanalysis and society. His works, considered classics of psychoanalysis, include: *Psychanalyser, Démasquer le Réel, On Tue un Enfant, Rompre les Charmes, Le Pays de l'Autre, Ecrits pour la Psychanalyse vol. 1 and 2. Oedipe à Vincennes, principes d'un psychothérapie des psychoses*. Two of these have recently been translated into English: *Psychoanalyzing* and *A Child Is Being Killed*.

JACQUES LECLAIRE is a research scientist, Director of Research L'Oréal, Paris.

ANDRÉE LEHMAN is a psychoanalyst, member of the Cercle Freudien and of Espace Analytique (Paris). From 1975 to 1995, she worked as psychoanalyst in the department of general surgery and in the Committee of Mammary Pathology at the Institute Gustave Roussy's European Center for the Fight Against Cancer. Since 1998, she has worked in the Oncogenetics Department at the Centre Jean Perrin in Clermont-Ferrand.

JONATHAN LIPKIN is an assistant professor of Multimedia Design at Ramapo College in New Jersey.

NICOLE MALINCONI is a writer in Belgium, author of *Hôpital Silence, Nous Deux, Da Solo, Rien ou Presque.*

PAOLA MIELI is a psychoanalyst in New York, founder and president of Après-Coup Psychoanalytic Association New York, member of the Cercle Freudien (Paris), on the faculty of the Department of Photography and Related Media of the School of Visual Arts (New York) and of the Lacanian School of Psychoanalysis (Berkeley). She is the author of many articles on psychoanalysis and culture.

MARGARET MORSE is an associate professor of Film and Video at the University of California, Santa Cruz and a theorist of contemporary culture and media. She is the editor and principal author of the book *Hardware, Software, Artware* and the author of *Virtualities: Television, Media Art and Cyberculture.*

ONA NIERENBERG is a psychoanalyst and staff psychologist working in the AIDS Program at Bellevue Hospital in New York City.

ERIK PARENS is director of programs at the Hastings Institute, New York. He is the author of many articles on philosophy and ethics and is a member of the American Association for the Advancement of Science.

■ BEING HUMAN:

ROBERT POLLACK is professor of Biological Sciences, lecturer in psychiatry at the Center for Psychoanalytic Training and research, and director of the Center for the Study of Science and Religion, at Columbia University. His 1994 book *Signs of Life, The Language and Meanings of DNA* won the Lionel Trilling Award. His most recent book is *The Missing Moment: How the Unconscious Shapes Modern Science.*

DENNIS PHILLIPS is a poet living in Los Angeles. His published works include 20 *Questions, Arena, Book of Hours* (with artist Courtney Gregg), *Credence, Means, The Hero Is Nothing,* and *Study for the Ideal City.* He teaches at Art Center College of Design.

CLAUDE RABANT is a psychoanalyst in Paris, a former member of the Ecole Freudienne de Paris, member of the Cercle Freudien, editor of the psychoanalytic journal *Io,* and author of many papers. His books include *Délire et Théorie* and *Inventer le Réel.*

CLAUS-DIETER RATH is a psychoanalyst in Berlin, co-founder of the Psychoanalytische Assoziation, the author of many articles on psychoanalysis and culture, and co-editor of *Lacan und das Deutsche. Die Ruckkehr der Psychoanalyse über den Rhein.*

WILLIAM RICHARDSON is a professor of Philosophy at Boston College and engaged in the practice of psychoanalysis. He is the author of *Heidegger: Through Phenomenology to Thought* (Preface by Martin Heidegger), and co-author of *Lacan and Language: Readers Guide to the Ecrits* and *The Purloined Poe: Lacan, Derrida and Psychoanalytic Reading.* He has written widely on religion, philosophy and psychoanalysis.

STUART SCHNEIDERMAN is a psychoanalyst and a writer in New York City. His books include: *Returning to Freud, Lacan: The Death of an Intellectual Hero, The Rat Man, An Angel Passes, Saving Face.*

ANTONELLO SCIACCHITANO is a psychoanalyst and psychiatrist in Milan, Italy. He is the author of many articles on psychoanalysis, philosophy, and science. His books include *Anoressia, sintomo e angoscia*.

SETH SHOSTAK is an astrophysicist, director of programs at S.E.T.I. (Search for Extra-Terrestrial Intelligence)in California.

MARK STAFFORD is a writer and editor in New York City. He is the secretary of Après-Coup Psychoanalytic Association, on the faculty of Parsons School of Art, and New York University.

CHARLES TRAUB is Chair of the MFA in Photography and Related Media Department at the School of Visual Arts in New York City.

RICHARD TEITELBAUM is a musician, on the faculty of Bard College, Annandale on Hudson, New York. He is a pioneer in electronic music and his compositions have been presented internationally. He has performed and recorded extensively as soloist and with many performers, including Anthony Braxton, Lee Konitz and Nam June Paik.

DOUGLAS TRUMBULL is a filmmaker and inventor, a pioneer in new technologies applied to advanced entertainment. He is currently vice-chairman of Imax Corporation and President of Ridefilm Corporation. He designed special effects for many motion pictures, including: *2001: A Space Odyssey, Blade Runner, Brainstorm*.

DOROTHY WARBURTON is a professor of Clinical Genetics and Development at Columbia University and director of the Genetics Diagnostic Laboratory at the Presbyterian Hospital in New York City. She has served as senior editor of *Chromosome 13* for the Human Genome Database.

■ BEING HUMAN:

JIM YOUNT is the director of the American Cryonics Society in Cupertino, California. Before joining the foundation he studied theater. He has also worked in the insurance industry.

HISTORY: During the summer of 1991, Marcos Einis and David Lichtenstein joined Serge Leclaire and Paola Mieli to form the organizing committee for the project *The Technological Extensions of the Boundaries of the Body*. In 1993, Dany-Robert Dufour and in 1994 Mark Stafford joined the organizing committee. On June 18, 1993, Leclaire, Dufour and Einis signed the founding declaration of the association *Franchissement* (Crossing) in Paris, meant to support Après-Coup from Europe in carrying out its project.